3D Printing of MEMS Technology

3D Printing of MEMS Technology

Editor

Andrea Ehrmann

MDPI

Basel • Beijing • Wuhan • Barcelona • Belgrade • Novi Sad • Cluj • Manchester

Editor
Andrea Ehrmann
Faculty of Engineering and Mathematics
Bielefeld University of
Applied Sciences and Arts
Bielefeld
Germany

Editorial Office
MDPI
St. Alban-Anlage 66
4052 Basel, Switzerland

This is a reprint of articles from the Special Issue published online in the open access journal *Micromachines* (ISSN 2072-666X) (available at: https://www.mdpi.com/journal/micromachines/special_issues/3D_Printing_MEMS_Technology).

For citation purposes, cite each article independently as indicated on the article page online and as indicated below:

Lastname, A.A.; Lastname, B.B. Article Title. *Journal Name* **Year**, *Volume Number*, Page Range.

ISBN 978-3-0365-9770-6 (Hbk)
ISBN 978-3-0365-9771-3 (PDF)
doi.org/10.3390/books978-3-0365-9771-3

Contents

Andrea Ehrmann
Editorial for the Special Issue on 3D Printing of MEMS Technology
Reprinted from: *Micromachines* **2023**, *14*, 2195, doi:10.3390/mi14122195 . . . . **1**

Sindhu Vijayan, Pravien Parthiban and Michinao Hashimoto
Evaluation of Lateral and Vertical Dimensions of Micromolds Fabricated by a PolyJet™ Printer
Reprinted from: *Micromachines* **2021**, *12*, 302, doi:10.3390/mi12030302 . . . . **3**

Jiaqing Xie, Haoran Pang, Ruqian Sun, Tao Wang, Xiaoyu Meng and Zhikang Zhou
Development of Rapid and High-Precision Colorimetric Device for Organophosphorus Pesticide Detection Based on Microfluidic Mixer Chip
Reprinted from: *Micromachines* **2021**, *12*, 290, doi:10.3390/mi12030290 . . . . **17**

Jianhao Zhang, Rongbo Wu, Min Wang, Youting Liang, Junxia Zhou, Miao Wu and et al.
An Ultra-High-Q Lithium Niobate Microresonator Integrated with a Silicon Nitride Waveguide in the Vertical Configuration for Evanescent Light Coupling
Reprinted from: *Micromachines* **2021**, *12*, 235, doi:10.3390/mi12030235 . . . . **31**

Mahmuda Akter Monne, Chandan Qumar Howlader, Bhagyashree Mishra and Maggie Yihong Chen
Synthesis of Printable Polyvinyl Alcohol for Aerosol Jet and Inkjet Printing Technology
Reprinted from: *Micromachines* **2021**, *12*, 220, doi:10.3390/mi12020220 . . . . **39**

Seongbeom Ahn, Woojun Jung, Kyungho Ko, Yeongchan Lee, Chanju Lee and Yongha Hwang
Thermopneumatic Soft Micro Bellows Actuator for Standalone Operation
Reprinted from: *Micromachines* **2021**, *12*, 46, doi:10.3390/mi12010046 . . . . **55**

Abid Ahmad, Bhagyashree Mishra, Andrew Foley, Leslie Wood and Maggie Yihong Chen
High Permeability Photosintered Strontium Ferrite Flexible Thin Films
Reprinted from: *Micromachines* **2021**, *12*, 42, doi:10.3390/mi12010042 . . . . **65**

Jian-Chiun Liou, Chih-Wei Peng, Philippe Basset and Zhen-Xi Chen
DNA Printing Integrated Multiplexer Driver Microelectronic Mechanical System Head (IDMH) and Microfluidic Flow Estimation
Reprinted from: *Micromachines* **2021**, *12*, 25, doi:10.3390/mi12010025 . . . . **77**

Deepak Kumar, Saikat Mondal, Yiming Deng and Premjeet Chahal
Wireless Battery-Free Harmonic Communication System for Pressure Sensing
Reprinted from: *Micromachines* **2020**, *11*, 1043, doi:10.3390/mi11121043 . . . . **93**

Frederik Kotz, Markus Mader, Nils Dellen, Patrick Risch, Andrea Kick, Dorothea Helmer and et al.
Fused Deposition Modeling of Microfluidic Chips in Polymethylmethacrylate
Reprinted from: *Micromachines* **2020**, *11*, 873, doi:10.3390/mi11090873 . . . . **105**

Tomasz Kozior and Jerzy Bochnia
The Influence of Printing Orientation on Surface Texture Parameters in Powder Bed Fusion Technology with 316L Steel
Reprinted from: *Micromachines* **2020**, *11*, 639, doi:10.3390/mi11070639 . . . . **117**

Ishak Ertugrul
The Fabrication of Micro Beam from Photopolymer by Digital Light Processing 3D Printing Technology
Reprinted from: *Micromachines* **2020**, *11*, 518, doi:10.3390/mi11050518 . . . . **135**

Elkana Bar-Levav, Moshe Witman and Moshe Einat
Thin-Film MEMS Resistors with Enhanced Lifetime for Thermal Inkjet
Reprinted from: *Micromachines* **2020**, *11*, 499, doi:10.3390/mi11050499 . . . . **149**

Van-Thai Tran, Yuefan Wei and Hejun Du
On-Substrate Joule Effect Heating by Printed Micro-Heater for the Preparation of ZnO Semiconductor Thin Film
Reprinted from: *Micromachines* **2020**, *11*, 490, doi:10.3390/mi11050490 . . . . **161**

Tomasz Blachowicz and Andrea Ehrmann
3D Printed MEMS Technology—Recent Developments and Applications
Reprinted from: *Micromachines* **2020**, *11*, 434, doi:10.3390/mi11040434 . . . . **173**

micromachines

MDPI

Editorial

Editorial for the Special Issue on 3D Printing of MEMS Technology

Andrea Ehrmann

Faculty of Engineering Sciences and Mathematics, Bielefeld University of Applied Sciences and Arts, 33619 Bielefeld, Germany; andrea.ehrmann@hsbi.de

Microelectromechanical systems (MEMS) combine electrical and mechanical functions and are nowadays broadly applied in many technology fields, often as sensors or actors [1]. Opposite to microelectronics and micromechanics, standardization of the production processes is still relatively low. While this poses challenges for the development of new MEMS for specific applications, it also calls for the use of additive manufacturing with its high degree of freedom in design and the large chance of individualization to produce new MEMS [2].

Additive manufacturing, or 3D printing, belongs to the emerging technologies of our time. While previously they were mostly used for rapid prototyping, the technology has been in rapid production for a long time, especially for complicated objects or those with small lot sizes [3]. Most recently, new 3D printing technologies enable the printing of the smallest features on micro- or even nano-scales [4]. At the same time, well-known problems like the waviness of fused deposition modeling (FDM) printed parts, the missing long-term stability of some typical printing materials, or the reduced mechanical properties of 3D-printed objects still exist and have to be investigated in detail to enable the optimization of these parameters [5].

This Special Issue focusses on all topics dealing with the 3D printing of microelectromechanical systems (MEMS), such as new or advanced features enabled by 3D printing compared to conventional technologies, but also the challenges which still exist of using 3D printing technologies for MEMS and new approaches for how to overcome them.

One of the challenges addressed by the papers in this Special Issue is dimensional accuracy and surface roughness, e.g., in PolyJet printing (Contribution 1) or powder bed fusion (Contribution 2). A high mixing efficiency (Contribution 3) was aimed to be created, as well as the creation of specific optical properties (Contribution 4). Amongst the investigated materials, polyvinyl alcohol (PVA) was investigated as a possible sacrificial material for MEMS devices (Contribution 5), as well as photo-sintered magnetic strontium ferrite samples (Contribution 6), polymethylmethacrylate (PMMA) microfluidic chips (Contribution 7), as well as the in situ annealing of semiconducting ZnO thin films (Contribution 8). Specific structures (Contribution 9) are investigated as well as sensors (Contribution 10), actuators (Contribution 11), and electronic devices (Contribution 12), partly combining an experiment and simulation (Contribution 13). Additionally, a review of the recent developments and applications of 3D-printed MEMSs technology is given (Contribution 14).

To conclude, the papers collected in this Special Issue provide an overview of different additive manufacturing methods, such as fused deposition modeling, photolithography, PolyJet, aerosol jet or inkjet, and materials including diverse polymers and metals. They describe the development of microfluidic and colorimetric devices, micro-resonators and micro-beams, sensors and actuators, and resistors and micro-heaters, as well as the corresponding challenges and proposed solutions. We hope that these papers will inspire more research in the highly topical research area of 3D printing MEMS.

Conflicts of Interest: The author declares no conflict of interest.

Citation: Ehrmann, A. Editorial for the Special Issue on 3D Printing of MEMS Technology. *Micromachines* **2023**, *14*, 2195. https://doi.org/10.3390/mi14122195

Received: 27 November 2023
Accepted: 29 November 2023
Published: 30 November 2023

List of Contributions:

1. Vijayan, S.; Parthiban, P.; Hashimoti, M. Evaluation of Lateral and Vertical Dimensions of Micromolds Fabricated by a PolyJet™ Printer. *Micromachines* **2021**, *12*, 302.
2. Kozior, T.; Bochnia, J. The influence of printing orientation on surface texture parameters in powder bed fusion technology with 316L steel. *Micromachines* **2020**, *11*, 639.
3. Xie, J.Q.; Pang, H.R.; Sun, R.Q.; Wang, T.; Meng, X.Y.; Zhou, Z.K. Development of Rapid and High-Precision Colorimetric Device for Organophosphorus Pesticide Detection Based on Microfluidic Mixer Chip. *Micromachines* **2021**, *12*, 290.
4. Zhang, J.H.; Wu, R.B.; Wang, M.; Liang, Y.T.; Zhou, J.X.; Wu, M.; Fang, Z.W.; Chu, W.; Cheng, Y. An Ultra-High-Q Lithium Niobate Microresonator Integrated with a Silicon Nitride Waveguide in the Vertical Configuration for Evanescent Light Coupling. *Micromachines* **2021**, *12*, 235.
5. Monne, M.A.; Howlader, C.Q.; Mishra, B.; Chen, M.Y.H. Synthesis of Printable Polyvinyl Alcohol for Aerosol Jet and Inkjet Printing Technology. *Micromachines* **2021**, *12*, 220.
6. Ahmad, A.; Mishra, B.; Foley, A.; Wood, L.; Chen, M.Y.H. High Permeability Photosintered Strontium Ferrite Flexible Thin Films. *Micromachines* **2021**, *12*, 42.
7. Kotz, F.; Mader, M.; Dellen, N.; Risch, P.; Kick, A.; Helmer, D.; Rapp, B.E. Fused Deposition Modeling of Microfluidic Chips in Polymethylmethacrylate. *Micromachines* **2020**, *11*, 873.
8. Tran, V.-T.; Wei, Y. F.; Du, H. J. On-Substrate Joule Effect Heating by Printed Micro-Heater for the Preparation of ZnO Semiconductor Thin Film. *Micromachines* **2020**, *11*, 490.
9. Ertugrul, I. The fabrication of micro beam from photopolymer by digital light processing 3D printing technology. *Micromachines* **2020**, *11*, 518.
10. Kumar, D.; Mondal, S.; Deng, Y.M.; Chahal, P. Wireless Battery-Free Harmonic Communication System for Pressure Sensing. *Micromachines* **2020**, *11*, 1043.
11. Ahn, S.B.; Jung, W.J.; Ko, H.H.; Lee, Y.C.; Lee, C.J.; Hwang, Y.H. Thermopneumatic Soft Micro Bellows Actuator for Standalone Operation. *Micromachines* **2021**, *12*, 46.
12. Bar-Levav, E.; Witman, M.; Elnat, M. Thin-Film MEMS Resistors with Enhanced Lifetime for Thermal Inkjet. *Micromachines* **2020**, *11*, 499.
13. Liou, J.-C.; Peng, C.-W.; Basset, P.; Chen, Z.-X. DNA Printing Integrated Multiplexer Driver Microelectronic Mechanical System Head (IDMH) and Microfluidic Flow Estimation. *Micromachines* **2021**, *12*, 25.
14. Blachowicz, T.; Ehrmann, A. 3D printed MEMS Technology—Recent Developments and Applications. *Micromachines* **2020**, *11*, 434.

References

1. Tanaka, M. An industrial and applied review of new MEMS devices features. *Microelectron. Eng.* **2007**, *84*, 1341–1344. [CrossRef]
2. Blachowicz, T.; Ehrmann, A. 3D printed MEMS Technology—Recent Developments and Applications. *Micromachines* **2020**, *11*, 434. [CrossRef] [PubMed]
3. Ben-Ner, A.; Siemsen, E. Decentralization and localization of production: The organizational and economic consequences of additive manufacturing (3D printing). *Calif. Manag. Rev.* **2017**, *59*, 5–23. [CrossRef]
4. Ertugrul, I. The fabrication of micro beam from photopolymer by digital light processing 3D printing technology. *Micromachines* **2020**, *11*, 518. [CrossRef] [PubMed]
5. Kozior, T.; Bochnia, J. The influence of printing orientation on surface texture parameters in powder bed fusion technology with 316L steel. *Micromachines* **2020**, *11*, 639. [CrossRef] [PubMed]

micromachines

MDPI

Article

Evaluation of Lateral and Vertical Dimensions of Micromolds Fabricated by a PolyJet™ Printer

Sindhu Vijayan [1,2], Pravien Parthiban [1] and Michinao Hashimoto [1,2,*]

1 Pillar of Engineering Product Development, Singapore University of Technology and Design, 8 Somapah Road, Singapore 487372, Singapore; sindhu_vijayan@mymail.sutd.edu.sg (S.V.); pravien_parthiban@outlook.com (P.P.)

2 Digital Manufacturing and Design Centre, Singapore University of Technology and Design, 8 Somapah Road, Singapore 487372, Singapore

* Correspondence: hashimoto@sutd.edu.sg

Abstract: PolyJet™ 3D printers have been widely used for the fabrication of microfluidic molds to replicate castable resins due to the ease to create microstructures with smooth surfaces. However, the microstructures fabricated by PolyJet printers do not accurately match with those defined by the computer-aided design (CAD) drawing. While the reflow and spreading of the resin before photopolymerization are known to increase the lateral dimension (width) of the printed structures, the influence of resin spreading on the vertical dimension (height) has not been fully investigated. In this work, we characterized the deviations in both lateral and vertical dimensions of the microstructures printed by PolyJet printers. The width of the printed structures was always larger than the designed width due to the spreading of resin. Importantly, the microstructures designed with narrow widths failed to reproduce the intended heights of the structures. Our study revealed that there existed a threshold width ($w_d{}'$) required to achieve the designed height, and the layer thickness (a parameter set by the printer) influenced the threshold width. The thresholds width to achieve the designed height was found to be 300, 300, and 500 μm for the print layer thicknesses of 16, 28, and 36 μm, respectively. We further developed two general mathematical models for the regions above and below this threshold width. Our models represented the experimental data with an accuracy of more than 96% for the two different regions. We validated our models against the experimental data and the maximum deviation was found to be <4.5%. Our experimental findings and model framework should be useful for the design and fabrication of microstructures using PolyJet printers, which can be replicated to form microfluidic devices.

Keywords: microfluidics; PolyJet 3D printing; fidelity of 3D printing

Citation: Vijayan, S.; Parthiban, P.; Hashimoto, M. Evaluation of Lateral and Vertical Dimensions of Micromolds Fabricated by a PolyJet™ Printer. *Micromachines* **2021**, *12*, 302. https://doi.org/10.3390/mi12030302

Academic Editors: Andrea Ehrmann and Nam-Trung Nguyen

Received: 29 December 2020
Accepted: 9 March 2021
Published: 13 March 2021

1. Introduction

This paper describes the characterization of polymer-jet (PolyJet™) 3D printers for their ability to produce microscale structures. Previous studies reported that the CAD drawing and the microstructures printed by PolyJet printers differ in the widths due to the spreading of the photoresin [1–3]. However, the effect of the spreading on the height of the printed structures has not been investigated. In this research, we studied the capability of PolyJet printers to print microstructures in terms of width (w_p) and height (h_p). We have experimentally studied the deviation in dimensions and developed mathematical models to explore the dynamics of printing. Models were developed for two different regions considering the dynamics of printing in the low and high widths. The experimental findings suggested that the spreading of the resin contributed to not only increasing the width of the features but also decreasing the height of the same features when the designed width of the features was below particular thresholds. These findings explain the inherent limitation of the printer to produce microstructures with accurate dimensions. The models can serve as a tool to predict the outcomes of PolyJet printing. The knowledge obtained in our study set out guidelines to obtain the designed microscale features by PolyJet printing, which shall be of interest to the microfluidic community for the fabrication of the micromolds.

1.1. *Microfluidics*

Microfluidics is the science and technology of fluids in small scales, with dimensions in the range of tens to hundreds of micrometers. Microfluidic systems offer advantages such as the requirement of low volumes, high sensitivity, fast response, small footprints, and precise control over the experimental parameters [4]. Such advantages can be exploited in a wide range of applications including chemical synthesis, biochemical analysis, drug delivery, detection, and sensing [5–9]. Microfluidic devices to perform such operations are initially fabricated by photolithography [10] and machining [11]. Those devices are fragile and found to be unsuitable for handling fluids. The advent of novel polymeric materials in the latter half of the 20th century resulted in the production of materials suitable for the microfabrication of devices. Microfabrication demanded polymers with desired properties such as low glass transition temperature, non-hygroscopic nature, chemical and mechanical stability, optical transparency, and flexibility. A method of fabrication collectively termed soft lithography was adopted to use such polymers for microfabrication [12–14].

1.2. *Soft Lithography and 3D Printing*

Soft lithography has been widely adopted for the fabrication of microfluidic devices. Soft lithography involves replica molding to transfer the features of the micromold to castable elastomeric resins and other resins [9,15] Polydimethylsiloxane (PDMS) is commonly used for replica molding in soft lithography due to its desirable properties such as flexibility, optical clarity, chemical inertness, and gas permeability [13]. Microfluidic devices fabricated by soft lithography are primarily planar with uniform height. Multiple steps are involved to fabricate non-planar 3D microchannels with varying heights [16]; non-planar microchannels are a requisite for applications such as droplet generation and manipulation [17–21], biological flow units [22,23], and soft robots [24]. For example, an additional dimension of non-planar channels permits the complex actuation of robotic structures; non-planar channels also allow handling fluids with different wettability without any modification to the surface. 3D printing has evolved as a rapid prototyping tool for 3D microchannels [25–31]. 3D printing can be employed to fabricate micromolds [28–35], sacrificial molds [36], and fluidic devices [37–39]. 3D printing of micromolds replaces the need for the photolithography used to fabricate a silicon master mold, which is adopted in the research community of microfluidics. In particular, stereolithography (SLA) printers and PolyJet 3D printers are suitable for creating molds for replica molding of PDMS and subsequent bonding with flat substrates [23,28–34]. Most of the commercially available 3D printers produce dimensions different from the manufacturer's specification. There is a systematic deviation between the designed and printed dimensions, and the deviation varies with the types of printers. For instance, the minimum attainable dimension of the structure is ~200 μm for PolyJet printer, while the specified resolution is 42 μm (deviation is ~376%) [3,40]. In this work, we particularly focused on understanding this deviation observed for PolyJet printers.

1.3. *PolyJet 3D Printing to Fabricate Micromolds*

PolyJet printing has exhibited the potential to fabricate microscale structures including microfluidic molds [31–33]. PolyJet printing is a process to build 3D structures in layer-by-layer manners using photocurable polymeric ink. Owing to the smooth surface finish of the printed molds, PolyJet printing has been increasingly used to fabricate 3D molds for replica molding where the relief structures are printed on the mold [31–33]. The previous characterization of PolyJet printers revealed that the microstructures were fabricated with dimensions different from the design [1–3,22,23,33–35]. The printed structures were also found to exhibit leaning sides and rounded corners instead of straight lines and sharp corners [2,3,23,33]. For large structures (in centimeters), the discrepancy between the designed and printed dimensions may not be pronounced well relative to its original dimensions. However, change in dimensions and shapes of the microstructures (in submillimeter) result in large deviations affecting their usability. The discrepancy between the designed and the printed microstructures printed by PolyJet printers was attributed to uncontrolled reflow and the spreading of the resin [1–3]. A previous study on the characterization of PolyJet printers reported that the optimal width and height that can be produced was ~300 μm with a deviation >40% from designed geometry [23]. A comparison of four inkjet-based printers from Stratasys and Projet showed that all the printers produced structures with dimensions larger than the dimensions reported by the manufacturers [3]. This comparative work studied the variation in lateral dimension and devised a formula with correction factors to predict the printed width. These reports highlighted the deviation of the printed features in the lateral dimensions and predicted the expected width using linear relationships. However, practically, we should consider the contribution of both the designed dimensions (width and height) in achieving the printed dimensions. Indeed, the effect of the spreading of the resin on the vertical dimensions has not been examined. Understanding the fidelity in the printed height is particularly important when we fabricate molds for narrow microchannels with low aspect ratios. It is required to account for all the designed dimensions to develop a general predictive model for PolyJet printing.

To address this gap, this study revisited the characterization of the microstructures fabricated by PolyJet printer. A model PolyJet printer (Objet30 Prime™) was tested for its ability to fabricate cuboid-shaped microstructures with varying designed height (h_d) and width (w_d) (100–3000 μm). Crucially, we studied the effect of spreading on both the lateral and vertical dimensions. Our study revealed that the fabricated height of the microstructures was influenced by the designed width of the features. We found the threshold widths (w_d') below which the heights of the features were not printed as designed. In addition, we studied the effect of the layer thickness on the printing accuracy; we found that the layer thickness was also an important parameter to determine the degree of deviations in the width and height of the printed features. Further, we identified that the printer deposited excess resin in a particular mode of printing (High Speed (HS) mode; with layer thickness = 28 μm). In this mode, the height of the structures was not compromised as much as in other modes of printing. From the experimental observations, a mathematical model was developed to predict both the lateral and vertical dimensions of printed features of HS mode. The model was bifurcated to represent variations in printing in the regions of low and high widths. The accuracy of the model was tested by validating it with the experimental data. This study contributes to establishing a practical understanding of the inherent limitation of the printing mechanisms of PolyJet printing. Such an understanding shall be useful for the fabrication of microstructures by PolyJet 3D printers by identifying the design that could potentially offset the deformation of the microstructures.

2. Materials and Methods

2.1. Research Aims and Approach

This research aimed to characterize the dimensions of microscale features printed using a PolyJet 3D printer. The experiments included printing of cuboid-shaped microstructures designed with a wide range of w_d and h_d (100, 200, 300, 400, 500, 1500, and 3000 μm) and quantification of w_p and h_p; the subscript d denotes the designed dimension, and the subscript p denotes the printed dimension. The experiments were performed with a model PolyJet (Objet30 Prime) printer using Veroclear (an acrylate-based photoresin). Veroclear™ was suitable for fabricating microfluidic molds due to its rigidity and non-sticky surfaces after simple post-processing. Veroclear was used as a representative material of Vero™-based materials. Other Vero-based materials should exhibit a similar degree of deviation through the same mechanism as observed for Veroclear. PolyJet printers are a closed system that uses proprietary resins and settings for printing. We designed our study to characterize the printer in its original setting without any modification. Studying the commercial printer with its original settings and representative materials would be beneficial to gain an in-depth understanding of the printer's performance. In this work, we studied the deviation between the designed and printed dimensions (width and height) in terms of (1) designed dimensions, and (2) the layer thickness of printing.

Firstly, the deviation between the designed dimensions (w_d and h_d) and the printed dimensions (w_p and h_p) was studied. The microscale structures were designed with rectangular cross-sections to observe the formation of curvatures after the spreading of the resin. While the effect of spreading on w_p was discussed in detail in previous studies [1–3], detailed characterization on h_p was not reported. Secondly, the influence of the layer thickness on the spreading of resin was tested. The layer thickness of the printing (preset by the printer in three different modes) determines the amount of resin deposited in each printing cycle. We hypothesized that the layer thickness would also influence the profile of spreading that affected the fidelity of printing. Importantly, the relationship between the designed and printed dimensions was studied to verify the hypothesis that the features need to be sufficiently wide to generate the designed height of the structures. Further, based on the experimental data, two mathematical models were developed to predict the printed dimensions for given designed dimensions. The models were developed to empirically represent the dynamics of PolyJet printing in the regions of low and high widths. For the low widths (below w_d'), the height of the features failed to reach the intended height, and we observed a steady increase in h_p with an increase in w_d. Above w_d', h_p attained h_d and remained stagnant. To represent these two different behaviors, we used two separate quadratic models to fit the experimental data. Our models can serve as a practical guideline to predict the outcomes before printing.

2.2. Design of Microscale Features and PolyJet Printing

A 7 × 7 array of cuboid-shaped microstructures with w_d and h_d of 100, 200, 300, 400, 500, 1500 and 3000 μm were designed over a flat base using AutoCAD® 2016 (Autodesk, CA, USA). The length of the cuboids was 1 cm and a gap of 0.5 cm was provided between each of the structures. The designed structures were 3D printed in Objet30 Prime™ (Stratasys, MN, USA) with Veroclear and support SUP705 as a model material and a support material, respectively. The structures were printed in three different layer thicknesses of 16, 28, and 36 μm, each of which is termed as high quality (HQ), high speed (HS), and Draft modes of Objet30 Prime printer, respectively.

2.3. Post-Processing and Replica Molding

3D printed master molds were subjected to post-processing to remove the uncured resin and support material. Post-processing involved the removal of the support materials in a water jet followed by the soaking of the printed models in deionized water for 2 h. Afterward, the molds were baked in an oven at 60 °C for 24 h. The replica of the 3D printed mold was obtained by casting PDMS (Sylgard 184 Silicone Elastomer kit, Dow Corning, USA) mixed in a 10:1 ratio (by weight) of the prepolymer to the curing agent. Molds cast with PDMS were cured at 60 °C for 3 h. The cured PDMS replica was peeled off with the relief structures replicated from the master mold.

2.4. Characterization of Printed Features

To investigate the deviation between the designed and printed dimensions, cross-sections of the PDMS replica at three different places along the length of cuboid-shaped microchannels were taken. Precisely, the cross-sections were obtained from the center region ignoring the ends. The cross-sections were flipped to show the open rectangle and imaged using the MU500 AmScope (Irvine, CA, USA). The physical parameters (w_p, h_p, and cross-sectional area) were measured from the cross-sectional image using ImageJ (ImageJ, National Institutes of Health (NIH), Bethesda, MD, USA). The deviation in the dimensions between the designed and printed structures was examined with respect to the printing conditions (e.g., the designed dimensions and the layer thickness set by the printer).

2.5. Mathematical Modelling

Surface fitting of our experimental data was performed to predict the printed dimensions for HS mode using SciPy module in Python 3.8. We used a second-order polynomial to fit our data. The data represented two different regimes: (1) low widths (100, 200 μm) with h_p less than the intended height and (2) high widths (300–3000 μm) where the intended height was printed. Owing to the two different behaviors observed, we used two different models to fit the data. The mathematical models predicted the dimensions of printed structures based on the input of designed dimensions. We validated the models against the experimental data and reported the accuracy of the models.

3. Results

3.1. Fabrication of Microstructures by PolyJet Printing

We first fabricated samples with microstructures by Polyjet printing. PolyJet printing is derived from inkjet printing. Similar to printing inks, PolyJet uses photocurable polymer resins to form 3D models. Briefly, PolyJet printing is based on the following process: (1) deposition of an array of photopolymer droplets, (2) leveling of the deposited droplets by a heated roller, and (3) UV curing resulting in solidification of a single layer (Figure 1a). The process continued sequentially and repeatedly, and the 3D structures were built in a layer-by-layer manner. Figure 1b shows the overlaying sketch of the cross-section of designed and printed structures obtained from Objet30 Prime printer, and Figure 1c shows the microscopic image of the same feature (with $w_d = h_d = 500$ μm). We defined the maximum width (at the horizontal baseline) and the maximum height (at the vertical centerline) of the printed structures as the measurement for w_p and h_p. The printed width (w_p) deviated from the designed width (w_d) by the spreading of the base that created curved corners. In contrast, in this example, the printed height (h_p) was comparable to the designed height (h_d) at the center of the feature. The printed cuboid attained a plateau at the top of the feature with a maximum height (that was comparable to h_d). This observation led to the hypothesis that the feature must be sufficiently wide to attain the plateau with the designed height. We, therefore, studied the fidelity of printing for varying widths including narrow features.

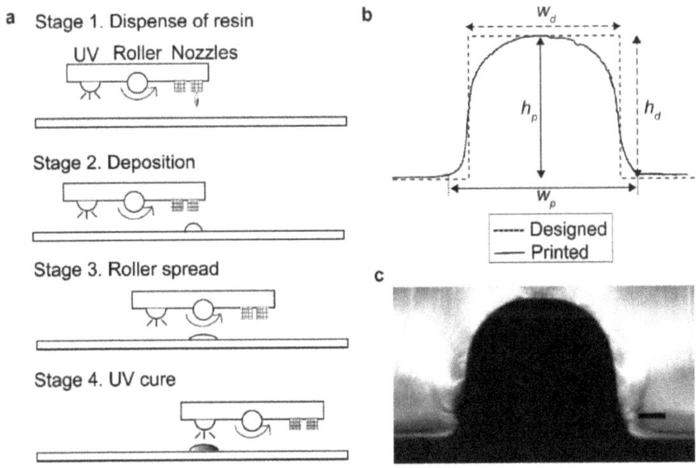

Figure 1. (**a**) Schematic illustration of the process of Polyjet printing. (**b**) Illustration showing the cross-sectional view of printed and designed structures. (**c**) Corresponding microscopic image obtained for the microscale structure with $w_d = h_d = 500$ μm. Scale bar = 100 μm.

3.2. Deviation between the Designed and Printed Dimensions for HS Printing Mode

Objet30 Prime offers three preset modes of printing based on layer thickness; namely, HQ (16 μm), HS (28 μm), and Draft (36 μm). Initially, we used HS mode to evaluate the fidelity of the printing. The measurement showed that w_p was always greater than w_d for the range of the height we investigated (h_d = 100–3000 μm) (Figure 2a). The magnified view of w_p against w_d for 100–500 μm is shown (Figure 2b). This observation reconfirmed that spreading inevitably happened laterally in PolyJet printing [1,2]. The time lag between the deposition of resin and curing by UV led to the reflow of resin and spreading, which increased the lateral dimensions of the base of the printed features with curved corners (Figure 1c). We note that the extent of spreading of resin (and the resulting increase in the width) was the same for all the designed dimensions. Therefore, we reported the deviation with respect to their original values, not by the percentage. The report by the absolute values would provide a clear understanding of the mechanism that led to the deviation.

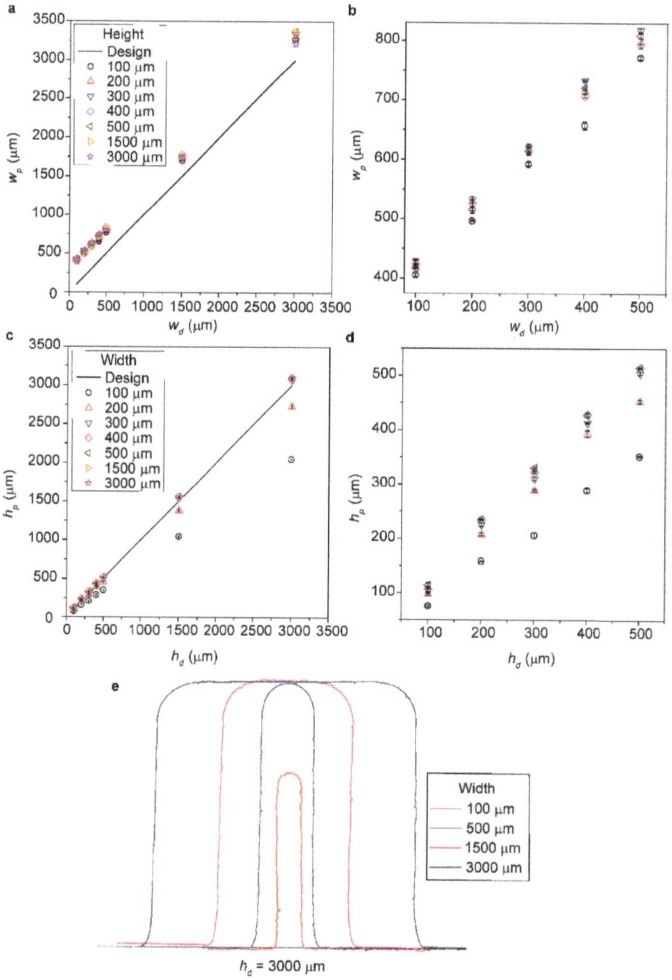

Figure 2. (**a**,**b**) Plot of wp against wd for h_d = 100–3000 μm, and a magnified view for w_d = 100–500 μm. (**c**,**d**) Plot of hp against hd for w_d = 100–3000 μm, and a magnified view for h_d = 100–500 μm. (**e**) Cross-sectional profile for h_d = 3000 μm with varying wd, suggesting the decrease in the printed height for w_d = 100 μm.

We then investigated the height of the printed features. Figure 2c shows the plot of h_p against h_d for a range of w_d. The magnified view of h_p against h_d for 100–500 μm is shown (Figure 2d). The measurement of h_p for HS mode showed that the actual height of the structures was not produced for $w_d \leq 200$ μm. It is plausible that the resin deposited over the narrow width ($w_d \leq 200$ μm) spread towards the base before UV curing without maintaining the volume of the resin to produce the intended height. The plot also suggested that the percentage of the decrease in the printed height was consistent for $w_d = 100$ μm (28.1% by average) and $w_d = 200$ μm (4.3% by average) regardless of h_d, which was indicated by the linear trend for each series. This observation suggested the decrease in the printed height consistently occurred in each cycle of printing. For $w_d \geq 300$ μm, the structures achieved the designed height (h_d) at the center of the features. We observed that for $w_d \geq 300$ μm, h_p obtained was slightly larger than h_d (~7.3% by average). This difference in h_p was systemic, and we attributed it to the calibration of the printer. Figure 2e shows the outline (i.e., upper surface) of the cross-section of the structures for varying w_d with the same height ($h_d = 3000$ μm). This illustration depicts that the structure printed with $w_d = 100$ μm was lower than the designed, while the structures with $w_d \geq 300$ μm reached h_d at the centerline. The inability to produce designed h_p for low w_d suggested that PolyJet printer is not suitable to produce micromolds to replicate narrow channels.

3.3. Influence of Layer Thickness on the Spreading of Resin

In the previous section, we discussed that the spreading of the resin caused the deviation in the printed dimensions in HS mode (layer thickness = 28 μm). Albeit discretely, PolyJet printing offered different settings for layer thickness. To understand the influence of the layer thickness in the attainable microstructures, we studied the printed dimensions obtained in three available printing modes: HQ (16 μm), HS (28 μm), and Draft (36 μm).

The plot of w_p against w_d depicts the deviation in dimensions of the printed structures from the designed dimensions (Figure 3a). The spreading of resin was more pronounced with the larger layer thickness due to the higher volume of resin deposited in each cycle. HQ mode (with the lowest layer thickness) exhibited the least deviation in dimension among the three modes investigated. Therefore, we concluded that layer thickness is a major factor that decided the extent of spreading of resin. Figure 3b shows the cross-sectional views at the center of the designed and the printed structures (replicated with PDMS) with $w_d = h_d = 300$ μm. We observed that w_p of the structures exceeded w_d for all the layer thickness, while the degree of spreading increased in the order of the layer height. Interestingly, however, the spreading of the resin for HS and Draft modes are comparable despite the difference in the layer heights (28 μm and 36 μm). We discuss this observation in the later sections by analyzing the volume of the resins deposited in each mode of printing.

3.4. Characterization of Printed Height for Varying Designed Widths

The previous section discussed the influence of the layer thickness on the width of the printed features (w_p). We also studied the effect of the layer thickness on the height of the printed features (h_p) for varying designed widths (w_d). The plot shows h_p of the structures obtained for $h_d = 500$, 1500, and 3000 μm against varying w_d (Figure 4a). As discussed earlier, h_d was attained only when the structures were designed above a particular threshold (w_d'). From the measurement of h_p, w_d' was identified to be 300 μm for HQ and HS modes, and 500 μm for Draft mode, respectively. Figure 4b depicts the cross-section of the designed and printed structures (replicated with PDMS) taken at the center of the microchannel. For $w_d = 300$ μm, h_d was attained in HQ and HS modes but not in Draft mode. These observations suggested that layer thickness was another key parameter (in addition to the dimension of the features) to determine the fidelity of printing in terms of both width and height. While the layer height was preset by the printer at discrete values, decreasing the layer height resulted in decreasing the lateral spreading (that increased the print fidelity in width) and the threshold width ($w_d{}'$) (that increased the print fidelity in height).

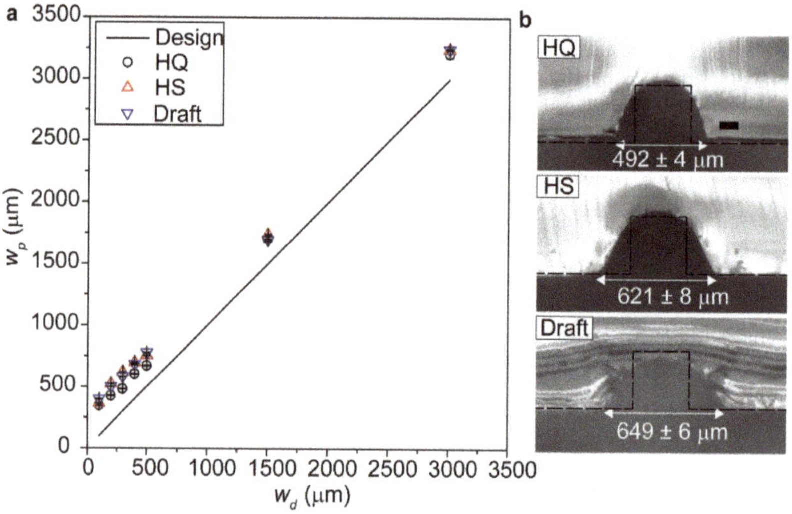

Figure 3. (**a**) Plot of the w_p against w_d obtained for three different layer thickness: HQ (16 µm), HS (28 µm), and Draft (36 µm). (**b**) Microscopic images showing the cross-section of the designed and printed cuboid-shaped structures replicated in PDMS for the three different modes of printing. The microstructures were with $w_d = h_d = 300$ µm. Scale bar = 100 µm.

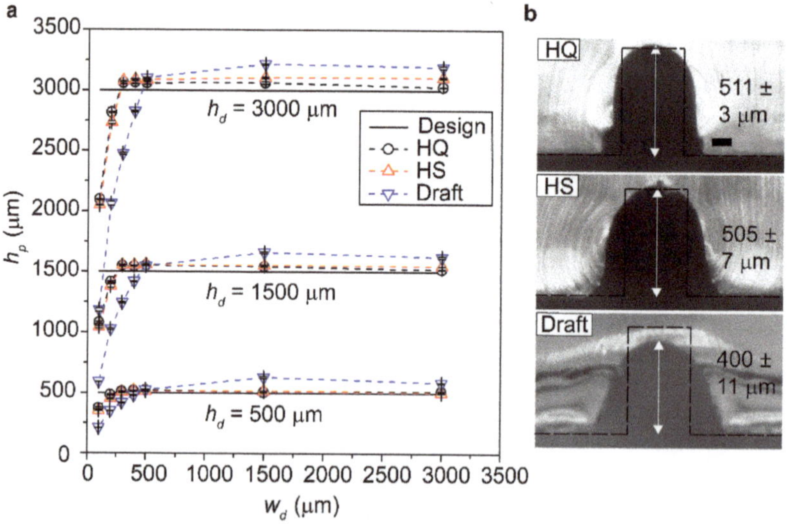

Figure 4. (**a**) Plot of h_p against w_d for the three different layer thickness: HQ (16 µm), HS (28 µm), and Draft (36 µm) for h_d = 500, 1500 and 3000 µm. HQ and HS modes required $w_d = 300$ µm to achieve $h_p > h_d$, while Draft mode required $w_d = 500$ µm to achieve $h_p > h_d$. (**b**) Microscopic images showing the cross-section of designed and printed cuboid-shaped structures replicated in PDMS for the three different modes of printing. The microstructures were with $w_d = 300$ µm and $h_d = 500$ µm. Scale bar = 100 µm.

3.5. Fabrication of Intended Height by HS Printing Mode

Our original hypothesis was that HS mode (with the intermediate layer thickness) would exhibit the degree of spreading between HQ and Draft modes. However, our study of the printed structures provided conflicting observations; Figure 3b suggested that the degree of spreading was similar between HS and Draft modes, while Figure 4a suggested the threshold width (w_d') was the same for HS and HQ modes. To explain these observations, we measured the cross-sectional area of the printed structures (which were the indication of the deposited volume of the resin) obtained for three printing modes. Figure 5 shows a bar graph of the cross-sectional area of the structures with w_d = 300 μm and h_d = 300, 400, and 500 μm. The graph revealed that, with the same design, the structures printed in HS mode were systemically larger (i.e., a higher volume of the resin was deposited) than the structures printed in HQ and Draft modes. Based on the image analysis, we concluded that HS mode deposited a larger amount of resin (by ~16.5% by average) than the other two modes. The additional resin in HS mode contributed to the lateral spreading of the resin equivalent to Draft mode, while it allowed printing the designed height for narrow features by maintaining a sufficient volume of resins after spreading.

The available print modes are predefined and cannot be altered. However, our experiment implied that the volume of the resin printed in each cycle is another key parameter to achieve intended microstructures. The increase in the resin volume, in principle, helps to achieve the intended height of the microstructures with the trade-off for the increased width of the printed features. Such understanding would help to calibrate the printer when open-source polymer-jetting printers become available.

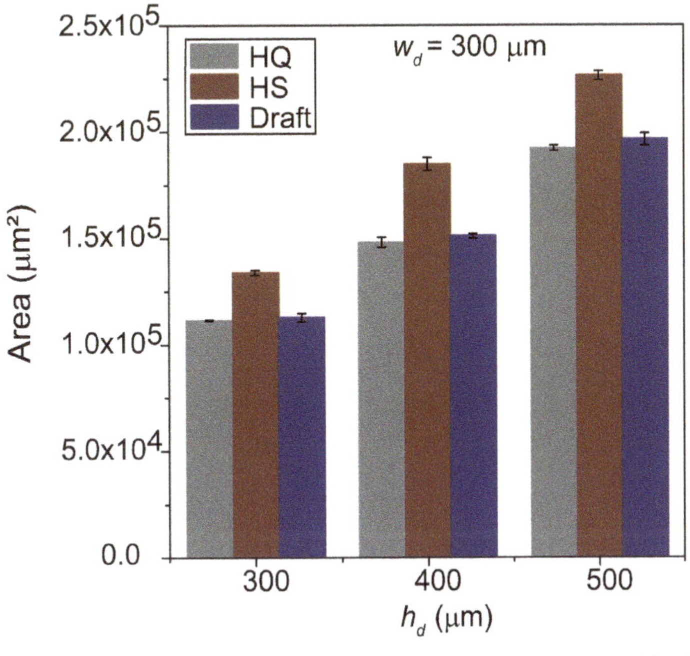

Figure 5. Bar graph showing the cross-sectional area of the structures obtained for the three different layer thickness: HQ (16 μm), HS (28 μm) and Draft (36 μm). The features were with w_d = 300 μm and h_d = 300, 400 and 500 μm.

3.6. Mathematical Modelling

Lastly, we fitted our experimental data obtained for HS mode on a surface using second-order polynomials. We developed the models for two cases—(1) region of low widths (100 μm, 200 μm) where the intended height was not achieved and (2) region of high widths (300–3000 μm) that was produced with intended height. Polynomial equations were obtained for the printed width and height of the two regimes of HS mode. The model equation is given as:

$$Z(x,y) = C_4x^2 + C_5y^2 + C_3xy + C_1x + C_2y + C_0 \quad (1)$$

where x and y denote the designed width and height, respectively. $Z(x,y)$ represent printed width or height estimated for the given designed width and height. Parameters of the model are tabulated (Table 1). Plots describing the fit of the experimental data with the modeled surfaces are provided (Figure 6). The coefficient of determination (R^2) gives the value of how close the experimental data lies to the modeled surface. We obtained a R^2 value > 96% for all our model equations. We tested the dimensions of 200 μm × 500 μm and 900 μm × 900 μm ($w_d \times h_d$) (called Case (1) and Case (2), respectively). The tested dimensions were excluded during the development of the model. The model predicted $h_p < h_d$ in Case (1) and $h_p \sim h_d$ in Case (2). The observation suggested that our model efficiently represented the actual feature dimensions of PolyJet printing. The predicted values of the two cases matched closely with the experimental data. The deviation between the predicted dimensions and experimental measurement was in the range of 0.1–4.5%. Thus, our model can be employed to predict the outcomes of PolyJet printing. Using this model, one can estimate the printed dimensions beforehand and account for appropriate corrections to the designed dimensions.

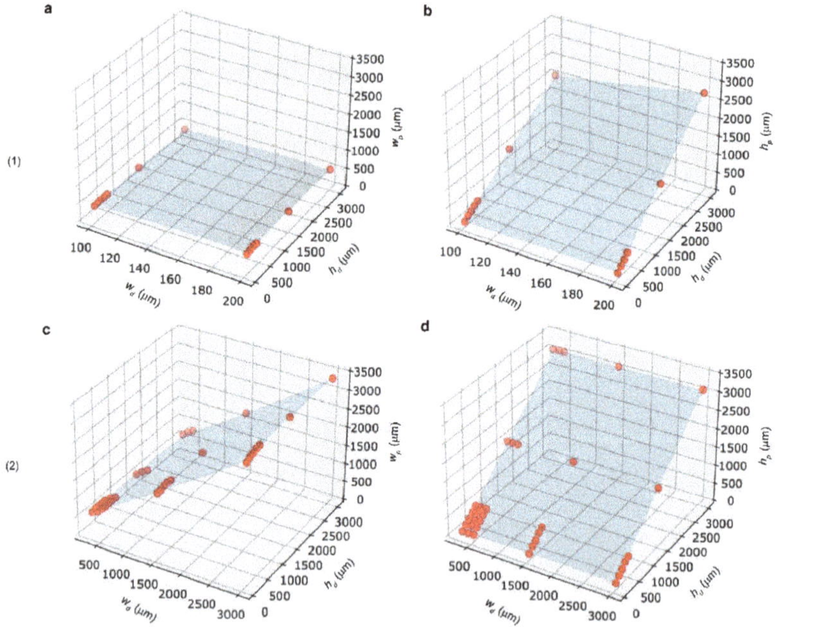

Figure 6. Plots showing the surface fitting of experimental data of HS mode. (**a**,**b**) represent the modeled surface for w_p and h_p of low widths. (**c**,**d**) represent the modeled surface for w_p and h_p of high widths.

Table 1. Parameters of the mathematical model to predict printed dimensions.

Printed Dimension	R^2	C_0	C_1	C_2	C_3	C_4	C_5
w_p (1)	0.9632	8.76×10^{-2}	5.84	-1.78×10^{-2}	8.77×10^{-5}	-1.63×10^{-2}	4.58×10^{-7}
h_p (1)	0.9999	1.23×10^{-3}	8.21×10^{-2}	4.73×10^{-1}	2.25×10^{-3}	-7.77×10^{-5}	-5.57×10^{-6}
w_p (2)	0.9995	3.41×10^{2}	8.54×10^{-1}	3.94×10^{-2}	7.62×10^{-6}	4.22×10^{-5}	-1.00×10^{-5}
h_p (2)	0.9999	6.58	1.62×10^{-1}	1.02	1.68×10^{-6}	-4.45×10^{-6}	1.49×10^{-6}

4. Conclusions

This paper discussed the characterization of the lateral and vertical dimensions of the cuboid-shaped microstructures printed by Objet30 Prime 3D printer. We made the following key observations that were not reported in the previous studies: (1) the spreading of resin affected both w_p and h_p of the structures; (2) the printed structures attained h_d only when w_d was above a particular threshold width (w_d'); (3) the layer thickness of printing was an important parameter to determine w_p and h_p; (4) an excess resin in each printing cycle allowed achieving an intended height with the trade-off for the increased width by spreading in HS mode. The fidelity of printing is determined by both the lateral and the vertical dimensions of the designed structures. The threshold width (w_d') to ensure the fidelity of the height was 300 µm for HQ and HS modes and 500 µm for Draft mode. Interestingly, our analysis suggested that HS mode deposited additional resin in each cycle of printing, which helped in forming the intended height while exhibiting spreading equivalent to Draft mode. In addition to the above experimental findings, we developed two mathematical models to predict the outcomes of PolyJet printing for the regions of low and high widths of HS mode. The models successfully predicted both the lateral and vertical dimensions based on the designed dimensions. The models represented the experimental data in the closeness of $R^2 > 96\%$. Validation of the models showed that the maximum deviation was less than 4.5% for both regimes. The prediction that the printed height did not reach the intended height for low-width features was a crucial contribution of our work. Our study would serve as a guideline to choose the appropriate dimensions of the features and layer thickness of printing to fabricate features with required fidelity.

This study contributes to establishing a practical understanding of the inherent limitations of PolyJet printers. The outcomes of our work would be beneficial to comprehend the influencing parameters of the printing. PolyJet printers are increasingly employed for fabricating molds for non-planar microfluidic devices. At present, the limitation of the resolution of printed microstructures prevents PolyJet printers to fabricate microstructures with high fidelity. The findings of our work can help in designing microstructures under the limitation of PolyJet printers, which can be useful in microfluidics, soft robotics, and other fields of engineering employing replica molding based on 3D-printed micromolds.

Author Contributions: S.V., P.P. and M.H. planned the study. S.V. performed the experiments and drafted the manuscript. M.H. supervised the work and edited the manuscript. All authors have read and agreed to the published version of the manuscript.

Funding: This research received no external funding.

Acknowledgments: S.V. thanks for the President's Graduate Fellowship awarded by the Ministry of Education (MOE), Singapore. M.H. thanks for financial supports from Start-up Research Grant at Singapore University of Technology and Design (SREP14088) and the 2nd A*STAR-AMED Joint Call (A19B9b0067).

Conflicts of Interest: The authors declare no conflict of interest.

References

1. Hwang, Y.; Paydar, O.H.; Candler, R.N. 3D printed molds for non-planar PDMS microfluidic channels. *Sens. Actuator A Phys.* **2015**, *226*, 137–142. [CrossRef]
2. Lee, J.M.; Zhang, M.; Yeong, W.Y. Characterization and evaluation of 3D printed microfluidic chip for cell processing. *Microfluid. Nanofluidics* **2016**, *20*, 1–15. [CrossRef]

3. Walczak, R.; Adamski, K. Inkjet 3D printing of microfluidic structures—On the selection of the printer towards printing your own microfluidic chips. *J. Micromech. Microeng.* **2015**, *25*, 85013. [CrossRef]
4. Whitesides, G.M. The origins and the future of microfluidics. *Nat. Cell Biol.* **2006**, *442*, 368–373. [CrossRef]
5. Weibel, D.B.; Whitesides, G.M. Applications of microfluidics in chemical biology. *Curr. Opin. Chem. Biol.* **2006**, *10*, 584–591. [CrossRef]
6. Kakaç, S.; Kosoy, B.; Li, D.; Pramuanjaroenkij, A. (Eds.) *Microfluidics Based Microsystems: Fundamentals and Applications*; Springer: Berlin/Heidelberg, Germany, 2010.
7. Nguyen, N.-T.; Wereley, S.T.; Shaegh, S.A.M. *Fundamentals and Applications of Microfluidics*; Artech House: Norwood, MA, USA, 2019.
8. Xu, Q.; Hashimoto, M.; Dang, T.T.; Hoare, T.; Kohane, D.S.; Whitesides, G.M.; Langer, R.; Anderson, D.G. Preparation of monodisperse biodegradable polymer microparticles using a microfluidic flow-focusing device for controlled drug delivery. *Small* **2009**, *5*, 1575–1581. [CrossRef] [PubMed]
9. Hashimoto, M.; Langer, R.; Kohane, D.S. Benchtop fabrication of microfluidic systems based on curable polymers with improved solvent compatibility. *Lab Chip* **2012**, *13*, 252–259. [CrossRef] [PubMed]
10. Terry, S.C.; Jerman, J.H.; Angell, J.B. A gas chromatographic air analyzer fabricated on a silicon wafer. *IEEE Trans. Electron. Devices* **1979**, *26*, 1880–1886. [CrossRef]
11. Kovacs, G.T.A.; Maluf, N.I.; Petersen, K.E. Bulk micromachining of silicon. *Proc. IEEE* **1998**, *86*, 1536–1551. [CrossRef]
12. Xia, Y.; Whitesides, G.M. Whitesides, Soft lithography. *Annu. Rev. Mater. Res.* **1998**, *28*, 153–184.
13. Duffy, D.C.; McDonald, J.C.; Schueller, O.J.; Whitesides, G.M. Rapid prototyping of microfluidic systems in poly (dimethylsiloxane). *Anal. Chem.* **1998**, *70*, 4974–4984. [CrossRef] [PubMed]
14. McDonald, J.C.; Duffy, D.C.; Anderson, J.R.; Chiu, D.T.; Wu, H.; Schueller, O.J.; Whitesides, G.M. Fabrication of microfluidic systems in poly (dimethylsiloxane). *Electrophoresis* **2000**, *21*, 27–40. [CrossRef]
15. Mukhopadhyay, R. When PDMS isn't the best. *Anal. Chem.* **2007**, *79*, 3248–3253. [CrossRef]
16. Chang, F.-C.; Su, Y.-C. Controlled double emulsification utilizing 3D PDMS microchannels. *J. Micromechanics Microengineering* **2008**, *18*, 065018. [CrossRef]
17. Riche, C.T.; Roberts, E.J.; Gupta, M.; Brutchey, R.L.; Malmstadt, E.J.R.M.G.R.L.B.N. Flow invariant droplet formation for stable parallel microreactors. *Nat. Commun.* **2016**, *7*, 10780. [CrossRef]
18. Lian, M.; Collier, C.P.; Doktycz, M.J.; Retterer, S.T. Monodisperse alginate microgel formation in a three-dimensional microfluidic droplet generator. *Biomicrofluidics* **2012**, *6*, 44108. [CrossRef]
19. Mohamed, M.G.A.; Kumar, H.; Wang, Z.; Martin, N.; Mills, B.; Kim, K. Rapid and inexpensive fabrication of multi-depth microfluidic device using high-resolution LCD stereolithographic 3D printing. *J. Manuf. Mater. Process.* **2019**, *3*, 26. [CrossRef]
20. Zhang, J.M.; Li, E.Q.; Aguirre-Pablo, A.A.; Thoroddsen, S.T. A simple and low-cost fully 3D-printed non-planar emulsion generator. *RSC Adv.* **2016**, *6*, 2793–2799. [CrossRef]
21. Rotem, A.; Abate, A.R.; Utada, A.S.; Van Steijn, V.; Weitz, D.A. Drop formation in non-planar microfluidic devices. *Lab Chip* **2012**, *12*, 4263–4268. [CrossRef]
22. Bonyár, A.; Sántha, H.; Ring, B.; Varga, M.; Kovács, J.G.; Harsányi, G. 3D Rapid Prototyping Technology (RPT) as a powerful tool in microfluidic development. *Procedia Eng.* **2010**, *5*, 291–294. [CrossRef]
23. Fuad, N.M.; Carve, M.; Kaslin, J.; Wlodkowic, D. Characterization of 3D-printed moulds for soft lithography of millifluidic devices. *Micromachines* **2018**, *9*, 116. [CrossRef]
24. Martinez, R.V.; Branch, J.L.; Fish, C.R.; Jin, L.; Shepherd, R.F.; Nunes, R.M.D.; Suo, Z.; Whitesides, G.M. Robotic tentacles with three-dimensional mobility based on flexible elastomers. *Adv. Mater.* **2013**, *25*, 205–212. [CrossRef]
25. Kitson, P.J.; Rosnes, M.H.; Sans, V.; Dragone, V.; Cronin, L. Configurable 3D-Printed millifluidic and microfluidic 'lab on a chip' reactionware devices. *Lab Chip* **2012**, *12*, 3267–3271. [CrossRef] [PubMed]
26. Au, A.K.; Huynh, W.; Horowitz, L.F.; Folch, A. 3D-Printed Microfluidics. *Angew. Chem. Int. Ed.* **2016**, *55*, 3862–3881. [CrossRef] [PubMed]
27. Amin, R.; Knowlton, S.; Hart, A.; Yenilmez, B.; Ghaderinezhad, F.; Katebifar, S.; Messina, M.; Khademhosseini, A.; Tasoglu, S. 3D-printed microfluidic devices. *Biofabrication* **2016**, *8*, 022001. [CrossRef] [PubMed]
28. Chan, H.N.; Chen, Y.; Shu, Y.; Chen, Y.; Tian, Q.; Wu, H. Direct, one-step molding of 3D-printed structures for convenient fabrication of truly 3D PDMS microfluidic chips. *Microfluid. Nanofluidics* **2015**, *19*, 9–18. [CrossRef]
29. Olanrewaju, A.O.; Robillard, A.; Dagher, M.; Juncker, D. Autonomous microfluidic capillaric circuits replicated from 3D-printed molds. *Lab Chip* **2016**, *16*, 3804–3814. [CrossRef] [PubMed]
30. Comina, G.; Suska, A.; Filippini, D. PDMS lab-on-a-chip fabrication using 3D printed templates. *Lab Chip* **2014**, *14*, 424–430. [CrossRef] [PubMed]
31. Parthiban, P.; Doyle, P.S.; Hashimoto, M. Self-assembly of droplets in three-dimensional microchannels. *Soft Matter* **2019**, *15*, 4244–4254. [CrossRef]
32. Glick, C.C.; Srimongkol, M.T.; Schwartz, A.J.; Zhuang, W.S.; Lin, J.C.; Warren, R.H.; Tekell, D.R.; Satamalee, P.A.; Lin, L. Rapid assembly of multilayer microfluidic structures via 3D-printed transfer molding and bonding. *Microsyst. Nanoeng.* **2016**, *2*, 16063. [CrossRef]

33. Kamei, K.-I.; Mashimo, Y.; Koyama, Y.; Fockenberg, C.; Nakashima, M.; Nakajima, M.; Li, J.; Chen, Y. 3D printing of soft lithography mold for rapid production of polydimethylsiloxane-based microfluidic devices for cell stimulation with concentration gradients. *Biomed. Microdevices* **2015**, *17*, 1–8. [CrossRef] [PubMed]
34. Hwang, Y.; Seo, D.; Roy, M.; Han, E.; Candler, R.N.; Seo, S. Capillary flow in PDMS cylindrical microfluidic channel using 3-D printed mold. *J. Microelectromechanical Syst.* **2016**, *25*, 238–240. [CrossRef]
35. Knowlton, S.; Yu, C.H.; Ersoy, F.; Emadi, S.; Khademhosseini, A.; Tasoglu, S. 3D-printed microfluidic chips with patterned, cell-laden hydrogel constructs. *Biofabrication* **2016**, *8*, 025019. [CrossRef]
36. Goh, W.H.; Hashimoto, M. Fabrication of 3D microfluidic channels and in-channel features using 3D printed, water-soluble sacrificial mold. *Macromol. Mater. Eng.* **2018**, *303*, 1700484. [CrossRef]
37. Shallan, A.I.; Smejkal, P.; Corban, M.; Guijt, R.M.; Breadmore, M.C. Cost-effective three-dimensional printing of visibly transparent microchips within minutes. *Anal. Chem.* **2014**, *86*, 3124–3130. [CrossRef]
38. Donvito, L.; Galluccio, L.; Lombardo, A.; Morabito, G.; Nicolosi, A.; Reno, M. Experimental validation of a simple, low-cost, T-junction droplet generator fabricated through 3D printing. *J. Micromechanics Microengineering* **2015**, *25*, 035013. [CrossRef]
39. Ching, T.; Li, Y.; Karyappa, R.; Ohno, A.; Toh, Y.-C.; Hashimoto, M. Fabrication of integrated microfluidic devices by direct ink writing (DIW) 3D printing. *Sens. Actuators B Chem.* **2019**, *297*, 126609. [CrossRef]
40. Macdonald, N.P.; Cabot, J.M.; Smejkal, P.; Guijt, R.M.; Paull, B.; Breadmore, M.C. Comparing microfluidic performance of three-dimensional (3D) printing platforms. *Anal. Chem.* **2017**, *89*, 3858–3866. [CrossRef] [PubMed]

micromachines

MDPI

Article

Development of Rapid and High-Precision Colorimetric Device for Organophosphorus Pesticide Detection Based on Microfluidic Mixer Chip

Jiaqing Xie [1], Haoran Pang [1], Ruqian Sun [1], Tao Wang [2,*], Xiaoyu Meng [1] and Zhikang Zhou [1]

1 College of Mechanical and Electronic Engineering, Northwest A&F University, Yangling 712100, China; xiejq@nwafu.edu.cn (J.X.); 18729071855@nwafu.edu.cn (H.P.); sunruqian@nwafu.edu.cn (R.S.); mxy917@nwafu.edu.cn (X.M.); zk_zhou@nwafu.edu.cn (Z.Z.)

2 Institute of Intelligent Manufacturing Technology, Shenzhen Polytechnic, Shenzhen 518055, China

* Correspondence: charlietree@szpt.edu.cn

Abstract: The excessive pesticide residues in cereals, fruit and vegetables is a big threat to human health, and it is necessary to develop a portable, low-cost and high-precision pesticide residue detection scheme to replace the large-scale laboratory testing equipment for rapid detection of pesticide residues. In this study, a colorimetric device for rapid detection of organophosphorus pesticide residues with high precision based on a microfluidic mixer chip was proposed. The microchannel structure with high mixing efficiency was determined by fluid dynamics simulation, while the corresponding microfluidic mixer chip was designed. The microfluidic mixer chip was prepared by a self-developed liquid crystal display (LCD) mask photo-curing machine. The influence of printing parameters on the accuracy of the prepared chip was investigated. The light source with the optimal wavelength of the device was determined by absorption spectrum measurement, and the relationship between the liquid reservoir depth and detection limit was studied by experiments. The correspondence between pesticide concentration and induced voltage was derived. The minimum detection concentration of the device could reach 0.045 $mg{\cdot}L^{-1}$ and the average detection time was reduced to 60 s. The results provide a theoretical and experimental basis for portable and high-precision detection of pesticide residues.

Keywords: colorimetric device; organophosphorus pesticide residues; microfluidics; LCD mask photo-curing

Citation: Xie, J.; Pang, H.; Sun, R.; Wang, T.; Meng, X.; Zhou, Z. Development of Rapid and High-Precision Colorimetric Device for Organophosphorus Pesticide Detection Based on Microfluidic Mixer Chip. *Micromachines* **2021**, *12*, 290. https://doi.org/10.3390/mi12030290

Academic Editor: Andrea Ehrmann

Received: 17 February 2021
Accepted: 8 March 2021
Published: 9 March 2021

1. Introduction

Organophosphorus pesticides are kinds of organic compound containing phosphorus, which is widely used in agricultural production, household health, garden management and other fields. Organophosphorus pesticides can effectively inhibit the activity of cholinesterase in the nervous system of animals and human beings, so that the acetylcholine decomposition process is suppressed, resulting in the accumulation of acetylcholine in nerve endings [1–3]. In recent years, the extensive use of organophosphorus pesticides has caused pollution in groundwater, surface water and soil, and its residues in food seriously threaten human health. The Codex Alimentarius Commission (CAC) has set strict standards of pesticide residues in food. For example, the maximum content of glufosinate (a typical organophosphorus pesticide) in most fruit and vegetables is 0.05 $mg{\cdot}L^{-1}$ [4–6]. At present, the detection methods of organophosphorus pesticide residues mainly include the precision instrumental analysis method and rapid detection method. The precision instrumental analysis method has superiority of high sensitivity and high selectivity, however, the shortcomings of a large volume, high cost, time-consuming duration and tedious pretreatment has limited the application of this method. Rapid detection methods mainly includes the biological detection method, immune method, enzyme inhibition method and biosensor method [7,8]. The enzyme inhibition method is based on the principle of colorimetry. The organophosphorus pesticides can inhibit the activity of acetylcholinesterase (AChE), slow down or stop the decomposition of acetylcholine. Therefore, acetylcholine, AChE and a chromogenic agent can be mixed to detect the sample concentration of pesticide residues. The presence of organophosphorus pesticides can be determined by the change of color or the change of physicochemical signal of enzyme reactions with a specific compound [9]. The rapid detection instrument and colorimetric instrument developed based on this principle has been able to realize the preliminary screening of pesticide residues in agricultural products. This method is currently facing many challenges, such as various accessories, inconvenience to carry, cumbersome operation, a lengthy process, poor accuracy, low sensitivity and repeatability. It is necessary to find a portable, low-cost and high-precision pesticide residue detection scheme to detect the concentration of organophosphorus pesticides.

Microfluidic chip technology integrates the processes of sample reaction, separation and detection involved in the fields of chemistry, biology and medicine into one chip, which makes the analysis equipment miniaturized and automated [10,11]. Jia [12] developed an impedance immunosensor based on a microfluidic chip for rapid detection of pesticide residues in vegetable samples. The microfluidic chip consisted of a detection microchamber inlet and outlet microchannel. A gold interdigitated array microelectrode (IDAM) was embedded in the microchannel of the microfluidic chip, which can be used for direct detection of practical samples. The research confirmed the value of microfluidic chips in pesticide detection, but the preparation of an immunosensor is difficult, and the detection time can be further compressed. Deng [13] developed a rapid semi-quantification detection method of trichlorfon residues by a microfluidic paper-based phosphorus-detection chip, the chip fabrication process was optimized. The author emphasizes the low cost of the chip, however, the durability of paper-based microfluidic chips was not investigated. Asghar [14] developed an innovative immuno-based microfluidic device that can rapidly detect and capture *Candida albicans* from phosphate-buffered saline (PBS) and human whole blood. The microchip technology showed an efficient capture of Candida albicans in PBS with an efficiency of 61–78% at various concentrations ranging from 10 to 10^5 colony-forming units per milliliter (cfu·mL^{-1}). The mixing efficiency of microfluidic chip is also an important factor of pesticide detection device. Through reasonable structure design, the chip channel with high mixing efficiency at low Reynolds number can be obtained [15]. Fan [16] developed a rapid microfluidic mixer utilizing sharp corner structures, it can be potentially used in the fluid mixing in variety of lab-on-a-chip applications. The preparation method of the chip also needs to be considered. Plevniak [17] developed a low-cost, smartphone-based, 3D-printed microfluidic chip system for rapid diagnosis of anemia in 60 s, and a 3D-printed chip with a 5 gel lens on camera was assembled for capturing and analyzing color-scale results from the chip view-window. The color-scale image capture and analysis app written in-house was developed to extract RGB (red, green, and blue) peak values in the region of interest. The 3D printing technology greatly reduces the cost of microfluidic chip preparation, while the optical observation technology can achieve semi-quantitative blood sample observation. However, it is difficult to realize accurate quantitative analysis for samples with no significant color change. Previous studies have confirmed the advantages of microfluidic chips in the field of detection, especially in improving the mixing efficiency and reducing the consumption of reagents [18,19]. Meanwhile, the photoelectric detection method has higher detection accuracy and lower detection concentration than the image method.

Based on the above analysis, a microfluidic mixer chip which can realize efficient and quick mixing was developed. The chip was successfully fabricated by a self-development LCD mask photo-curing machine, and the corresponding colorimetric device was developed to realize the rapid and accurate detection of organophosphorus pesticides. Firstly, two kinds of mixing microchannel were designed, the mixing efficiency was evaluated by simulation, and the optimal microchannel structure was confirmed. Secondly, the microfluidic mixer chip was fabricated by a self-development LCD mask photo-curing machine, the influence of printing parameters on the microchannel accuracy was investigated. Finally, a portable colorimetric device was developed, the corresponding relationship between pesticide concentration and induced voltage was constructed, and quantitative detection of organophosphorus pesticides was realized. The results indicated that the detection efficiency was increased and the detection sample was reduced.

2. Experimental

2.1. Basic Theory

In order to achieve high-precision detection of organophosphorus pesticides, related theories involved in enzyme reaction, microfluidic chip design theory, mixing index analysis and laminar flow simulation were adopted. In addition, the calculation method of detection limit was developed to evaluate the performance of the colorimetric device. The theoretical contents related to the research are listed as follows.

The color reaction was based on the Ellman method [20]. Under the catalysis of acetylcholinesterase (AChE), the thioacetylcholine iodide was hydrolyzed into thiocholine and acetic acid. The thiocholine reacted with 5, 5′-dithiobis (2-nitrobenzoic acid) to form a yellow product, as shown in Figure 1. The higher the AChE activity, the darker the color of the reagent after reaction. By contrast, AChE activity was inhibited and the color of reagent was lighter.

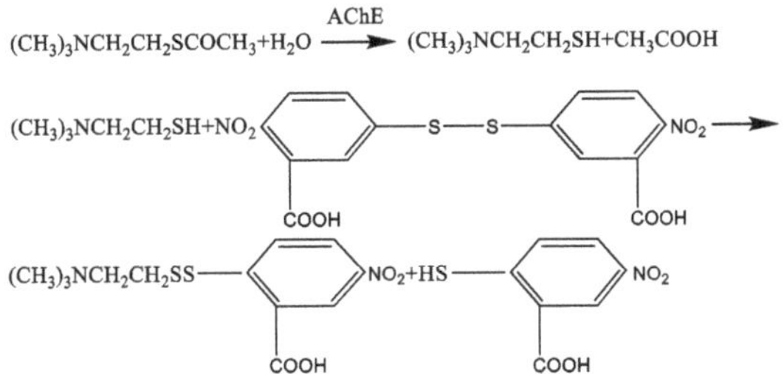

Figure 1. The principle of color reaction.

The diffusion coefficient D can be defined as [21],

$$D = \frac{kT}{6\pi\mu r} \tag{1}$$

where k denotes the Boltzmann constant, T gives the absolute temperature, μ is dynamic viscosity and r is molecular radius. The diffusion coefficient is inversely proportional to the dynamic viscosity of the solution at a certain temperature.

The cross-section structure of the channel fabricated on the chip was rectangular, and the expression of Reynolds number can be expressed as follows,

$$R_e = \frac{4\rho A v}{p\mu} \tag{2}$$

where A is the interface area, ρ is the liquid density, p is the wetting perimeter length, and v is the flow rate.

The relationship between transmitted light intensity I and incident light intensity I_0 can be expressed by the Lambert law,

$$I = I_0 10^{-\alpha \cdot l} \tag{3}$$

where l is the path length, α is the absorption coefficient describing the absorption capacity of a substance to light.

When the incident light is definite, the transmitted light is proportional to the concentration of the substance, and the above principle can be used to detect the concentration of pesticides.

The detection limit of the colorimetric device x_{LOD} is determined by the following formula [22–24],

$$x_{LOD} = \frac{2ts_y}{nt^2{s_y}^2 - Ar^2} \times (ts_y \sum x_i - \sqrt{\frac{A^2r^2}{k} + Ar^2 \sum {x_i}^2 - n\frac{A}{k}t^2{s_y}^2 - At^2{s_y}^2}) \quad (4)$$

where t is the Student's t-function parameter, usually assumed to be t = 3 for convenience [22], s_y is the standard deviation of the measurement, n is the tested number of concentrations, k designates the number of repeat measurements, r is the sensitivity as the slope of a linear fit, x_i represents the pesticide concentration of the i-th experiment, the determinant in the denominator A is given by:

$$A = n \sum {x_i}^2 - (\sum {x_i}^2)^2 \quad (5)$$

The Navier–Stokes equation [25] describing the behavior of incompressible fluid is used to simulate the mass and momentum transfer of fluid, which can be expressed as follows,

$$\frac{\partial}{\partial x_j}(\rho u_j) = 0 \quad (6)$$

$$\frac{\partial}{\partial x_i}(\rho u_i u_j) - \frac{\partial P}{\partial x_i} + \frac{\partial \tau_{ij}}{\partial x_i} \quad (7)$$

where ρ and u_j are density and velocity vector, respectively, u is velocity vector of fluid, P is pressure on fluid, τ_{ij} is stress tensor, respectively.

The mass flux is given by diffusion and convection, and the resulting mass balance is,

$$\nabla \cdot \left(-D\nabla c + c\vec{u}\right) = 0 \quad (8)$$

where c gives the concentration.

In substances involved in color reactions, the AChE has the largest molecular size and the minimum diffusion rate, so the standard deviation of the AChE concentration on the cross-section at different positions of the microchannel was used to measure the uniformity of the fluid distribution, as well as the mixing efficiency of the chip [26].

2.2. Materials and Methods

The chromogenic agent (0.75 $g \cdot L^{-1}$) was prepared by mixing 5,5-dithio-bis-2-nitro benzoic acid (DTNB) (Shanghai Chemical Reagent Co., Ltd. Shanghai, China) with phosphoric acid buffer (0.1 $mol \cdot L^{-1}$). The concentration of AChE (Merck Life Science Technology Co., Ltd., Darmstadt, Germany) solution was 150 $g \cdot L^{-1}$. Different concentrations of glufosinate-ammonium (Lear Chemical Co., Ltd., Shenzhen, China) standard solutions were prepared to test the inhibition on AChE.

A transparent photosensitive resin (Shenzhen Novartis Intelligent Technology Co., Ltd., Shenzhen, China) was used to fabricate microfluidic mixer chip, the parameters of photosensitive resin after ultraviolet curing is shown in Table 1.

Table 1. Photosensitive resin parameters after ultraviolet curing.

Parameter	Value
Density (g·cm^{-3})	1.05–1.25
Tensile modulus (GPa)	1.8–2.8
Tensile strength (MPa)	64–72
Bending modulus (GPa)	1.8–2.3
Heat distortion temperature (°C)	44–47
Elongation at break (%)	8–13

2.3. Characterization and Instruments

A self-developed LCD mask photo-curing machine was adopted for microfluidic mixer chip fabrication. The device adopted 405 nm ultraviolet light as the curing light source and the LCD panel as the selective light transmission plate to realize the curing of photosensitive resin for complex parts. Figure 2 is the schematic diagram of the LCD mask photo-curing machine: the machine includes platform, Z axis, material trough and LCD panel, and the incident position of the light source can be controlled by the LCD panel [27]. The shape of cured resin parts was observed by the laser microscope (VHX-1000, KEYENCE Co., Ltd., Osaka, Japan) with a display resolution of height 0.005 µm and width 0.01 µm.

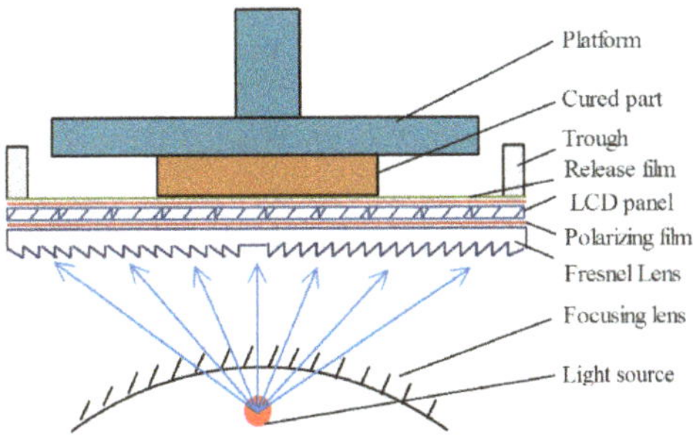

Figure 2. Schematic diagram of self-developed liquid crystal display (LCD) mask photo-curing machine.

The self-developed colorimetric device was adopted to detect the concentration of glufosinate-ammonium, a typical organophosphorus pesticide, the equipment included a light-emitting diode (LED) light source (Jinxin Photoelectric Technology Co., Ltd., Guangzhou, China), precision current source (Shenzhen wave particle Technology Co., Ltd., Shenzhen, China), precision voltage source (Shenzhen wave particle Technology Co., Ltd., Shenzhen, China), silicon photodetector (Shanghai Bose Intelligent Technology Co., Ltd., Shanghai, China) and Bluetooth voltage detection module. The schematic diagram of the equipment is shown in Figure 3a. Figure 3b is the colorimetric device after assembly: the LED light source is driven by a precision current source to ensure the stability of light intensity, and the customized narrow band filter with a diameter of 10 mm (Beijing Yongxing perception Instrument Co., Ltd., Beijing, China) was adopted to ensure that light of 385 nm to 430 nm wavelength could pass through, as shown in Figure 3c. The induced voltage of the silicon photodetector varies according to the intensity of the transmitted light, the induced voltage can be received by Bluetooth module and stored in the self-developed mobile client, and the resolution of silicon photodetector is 0.01 V, as shown in Figure 3d.

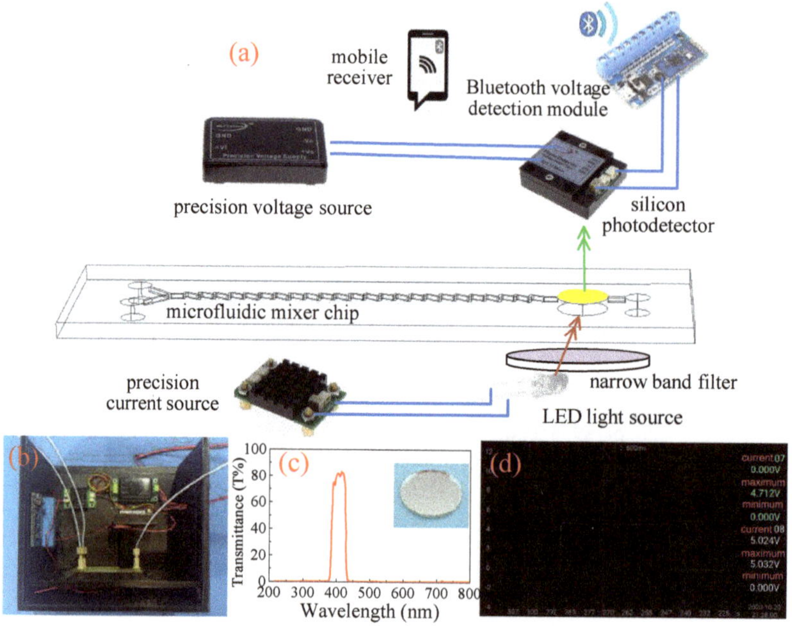

Figure 3. Schematic diagram of colorimetric device (**a**), colorimetric device after assembly (**b**), the absorption spectrogram of the narrow band filter (**c**), and self-developed mobile client (**d**).

In the experiment, the two solvents were pushed into the microfluidic chip channel through the syringes driven by a micro injection pump (XFP02-B, Suzhou iFLYTEK Scientific Instrument Co., Ltd., Suzhou, China), the liquid flow rate was controlled by micro injection pump. The ultraviolet-visible spectrophotometer (UV-2250, Shimadzu Instrument Equipment Co., Ltd., Kyoto, Japan) was adopted to obtain the absorption spectra of the solution after reaction. The three-dimensional structure of the chip was designed by SOLIDWORKS (Dassault Systèmes SOLIDWORKS Corp, Massachusetts, USA) software. The multi-physics coupling analysis software COMSOL Multiphysics 5.5 (COMSOL Inc., Stockholm, Sweden) was applied to simulate the mixing process. The free quadrilateral mesh was adopted in the model, and the wall condition was no sliding, the fully developed flow was applied in the fluid inflow model, and the inlet pressure was 0. The MATLAB 2018 (Mathworks Inc., Natick, MA, USA) software was used to process the simulation images to obtain the concentration standard deviation on the chip channel section.

3. Results and Discussion

3.1. Influence of Microchannel Shape on Mixing Efficiency

Among the substances involved in the color reaction, the molecular size of AChE is the largest, while the diffusion coefficient is the smallest, so the purpose of microchip channel design is to realize the rapid and complete mixing of AChE. The length, width and height of AChE were 98 nm, 79 nm and 3 nm, respectively [28]. The physical parameters of the mixed liquid are shown in Table 2. The diffusion coefficient of AChE can be obtained by substituting the parameters into Equation (1), and the calculated result is 0.22×10^{-11} [29]. In addition, the flow rate of the microfluidic mixer chip should be controlled at a low speed to ensure complete reaction after mixing. Therefore, the aim of microfluidic chip design is to realize the efficient mixing of AChE at low Reynolds number (10–82). The main principle is to make the fluid stretch, compress, fold and surround in the channel to realize the rapid mixing of the liquid involved in the reaction, which can be achieved by changing the channel width and bending the channel shape. The change of the channel width makes the liquid flow perpendicular to the fluid flow direction, and the bending of the microchannel makes the pressure difference inside the liquid. In this case, the method of channel bending and width variation were proposed to improve the mixing efficiency [30,31]. Based on the above principles, two types of microchannel structure were designed, as shown in Figure 4. The channel volumes of the two types of structure are kept consistent to ensure the same mixing time.

Table 2. Physical parameters of mixed liquid.

Parameter	Value
Boltzmann constant ($J{\cdot}K^{-1}$)	1.38×10^{-23}
Absolute temperature (K)	293.15
Dynamic viscosity (Pa·s)	10×10^{-3}
Water density ($Kg{\cdot}m^{-3}$)	1000

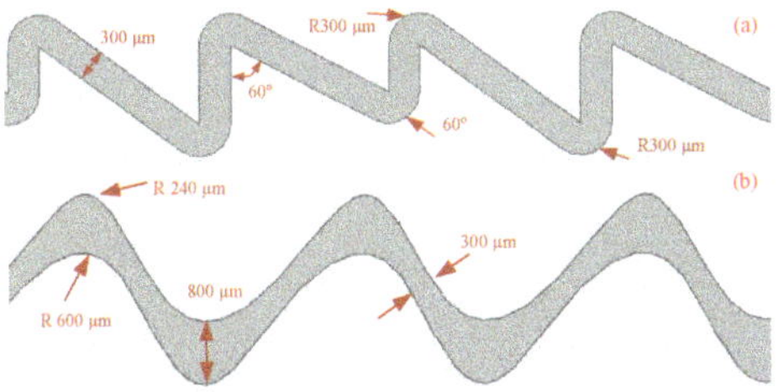

Figure 4. The designed two types of microstructures with curved channel (**a**), curved width variation channel (**b**).

The mixing process of the designed two kinds of microchannels were simulated, and the mixing index of AChE was compared, as shown in Figure 5a; the simulation results demonstrate that the mixing efficiency of channel structure (b) is significantly higher than that of channel structure (a). Meanwhile, the increase of channel flow has no significant effect on the improvement of mixing efficiency. By contrast, increasing the flow rate will reduce the mixing reaction time of the solution in the channel. Based on the above analysis, the flow rate of 50 $\mu L{\cdot}min^{-1}$ was selected in the experiment and the corresponding mixing time was 30 s; the simulated concentration distribution is shown in Figure 5b.

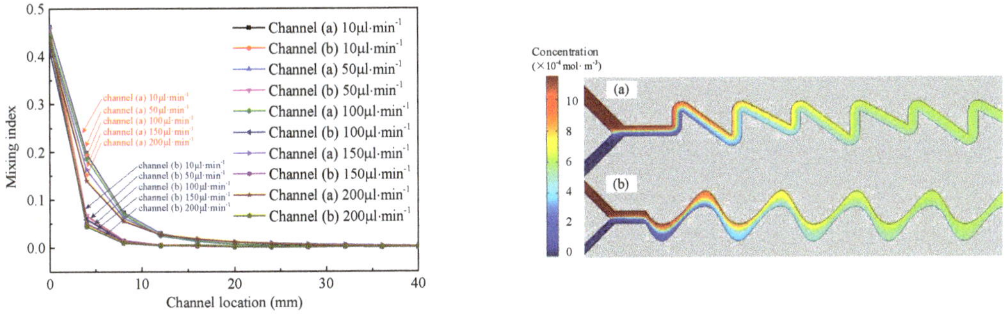

(**a**) (**b**)

Figure 5. The simulation result of mixing index of two kinds of channels (**a**) and the simulated concentration distribution at the speed of 50 $\mu L{\cdot}min^{-1}$ (**b**).

Based on the above analysis, a microfluidic mixer chip was designed, as shown in Figure 6, and contained two liquid inlets, a mixing channel, a liquid reservoir and an outlet.

During the detection, the mixture of acetylcholine and the tested sample was injected into inlet 1, and AChE and chromogenic agent were injected into inlet 2.

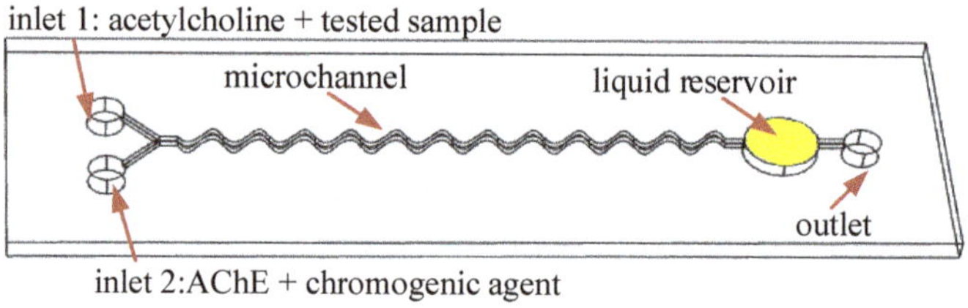

Figure 6. The diagram of the designed microfluidic mixer chip.

3.2. Influence of Printing Parameters on Chip Fabrication Accuracy

The curing roughness and shape accuracy of the LCD mask photo-curing machine was investigated before preparing the microfluidic mixer chip. The relationship between UV exposure time and surface roughness of cured parts is shown in Figure 7. The suitable exposure time is between 2–6 s. When the exposure time is 6 s, the surface roughness decreases to the minimum of 0.17 μm.

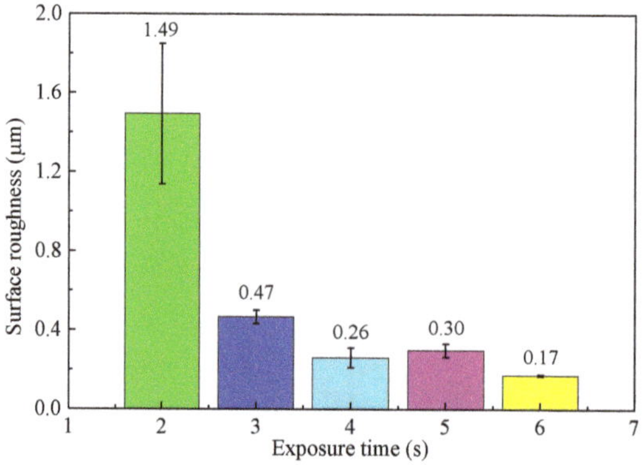

Figure 7. Relationship between ultraviolet (UV) exposure time and surface roughness.

Due to the small size of the rectangular microgroove, there will certain deviation between the designed size and the actual preparation size during the curing process. In order to accurately prepare the rectangular microgroove with the target size, the influence of single layer printing thickness on channel size was investigated. A microchannel structure with a height and width of 300 μm was designed, the effect of printing layer thickness on dimensional accuracy was analyzed, and the exposure time is 6 s. The single layer curing thickness with minimum curing error appears at 30 μm, as shown in Figure 8a, the cross-section shape of the cured microchannel is shown in Figure 8b.

According to the above research, a rectangular cross-section microfluidic chip was prepared. The photograph of the fabricated microfluidic chip is shown in Figure 9a, The scanning electron microscope (SEM) photograph shows that the obtained structure has good size consistency, as shown in Figure 9b.

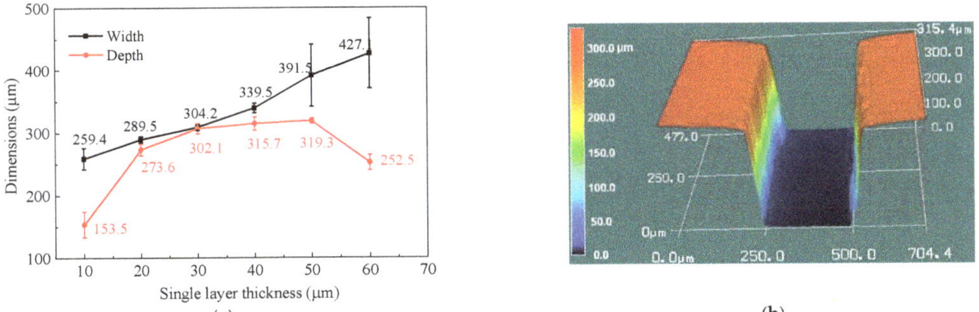

Figure 8. The influence of single layer curing thickness on channel size (**a**), and the cross-section shape of cured microchannel (**b**).

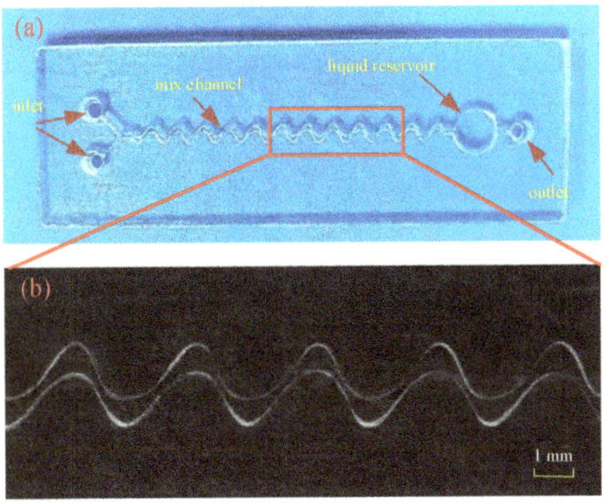

Figure 9. The microfluidic mixer chip fabricated by liquid crystal display (LCD) mask UV curing method: (**a**) overall structure and (**b**) scanning electron microscope (SEM) photograph of microchannel structure.

The chip was directly sealed with high transparency semi-solid acrylic material. The adhesive force between the semi-solid acrylic and the cured resin is large, and it will lose the bonding ability when contacting with water. In addition, the contact angle between acrylic and water is 57.5°, which is close to the cured resin (53.3°).

3.3. Relationshiop between Pesticide Concentration and Detection Voltage

The absorption spectrum measurement results for 0, 1 and 2 mg·L^{-1} concentrations of organophosphorus pesticide are shown in Figure 10. Theoretically, the sample without pesticide has the deepest color, corresponding to the highest light absorption value. In this test, whether the application of pesticide will change the absorption wavelength of the sample was observed. Therefore, pure water was selected as the contrast sample during the test. The experimental results indicated that the maximum absorption wavelength is 407 nm, it can be determined that the largest voltage difference would be obtained with 407 nm light source. In addition, the application of pesticides inhibited the color reaction and reduced the absorption peak.

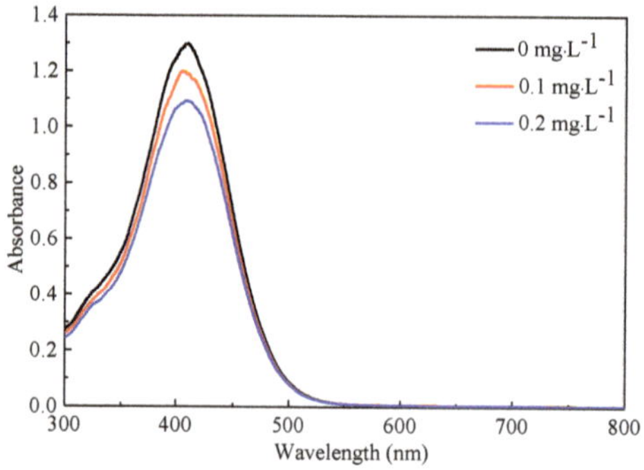

Figure 10. Absorption spectrum measurement results for 0, 1 and 2 mg·L^{-1} concentrations.

The microfluidic chips with liquid reservoir depth of 540 μm, 720 μm, 900 μm, 1080 μm, and 1260 μm were separately prepared. The voltage difference between the pesticide concentration of 0 and 2 mg·L^{-1}. The influence of liquid reservoir depth on voltage difference is shown in Figure 11, the voltage difference increases linearly with the liquid reservoir depth, the test error decreases with the liquid reservoir depth.

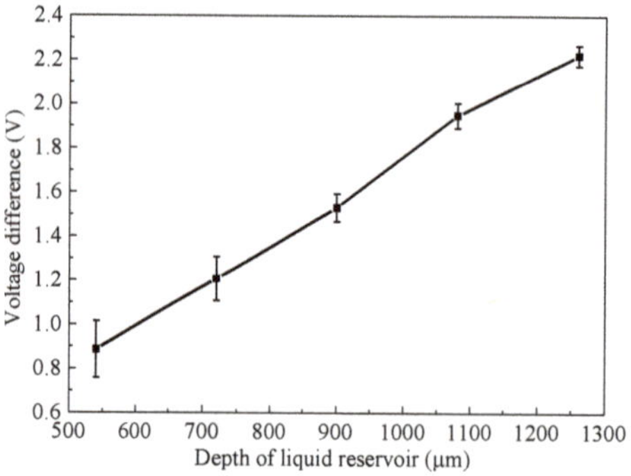

Figure 11. Influence of liquid reservoir depth on voltage difference.

Based on Equation (4), the detection limit of the device at different liquid reservoir depths was calculated, and each test was repeated 10 times; the test results are shown in Figure 12. The detection limit of the device decreases with the increasing of liquid reservoir depth. When the depth is greater than 900 μm, the decrease trend becomes slow. The reason is that the injection sample easily produces micro bubbles in the filling process when the liquid reservoir depth is too large, which affects the stability of the device. In addition, the injection time and the required sample volume increase with the liquid reservoir depth. After comprehensive analysis, the microfluidic chip with a liquid reservoir depth of 900 μm was selected as the mixing microfluidic chip of the device.

The relationship between induced voltage and pesticide concentration is shown in Figure 13. The test at each standard concentration was repeated 3 times, and the average values of each test were used for linear fitting. The fitted curve indicated that with the increase of glufosinate-ammonium content, the AChE activity was gradually suppressed, and the color reaction was inhibited, resulting in the increase of induced voltage. The linear relationship can be expressed as $y = 2.27 + 0.801x$, the linearity expression R^2 can reach 0.985, while the detection limit is 0.045 mg·L^{-1}.

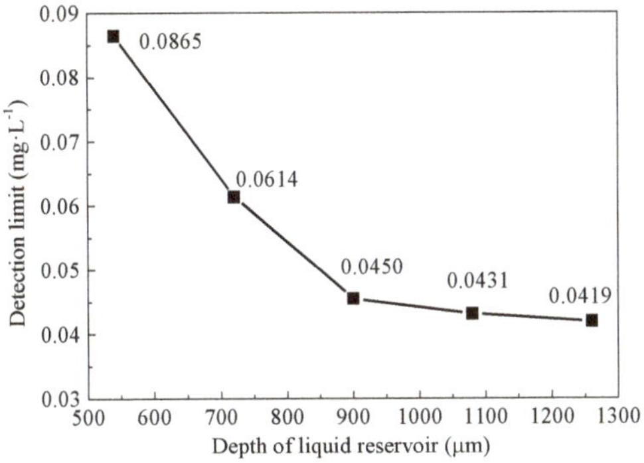

Figure 12. Relationship between liquid reservoir depth and detection limit.

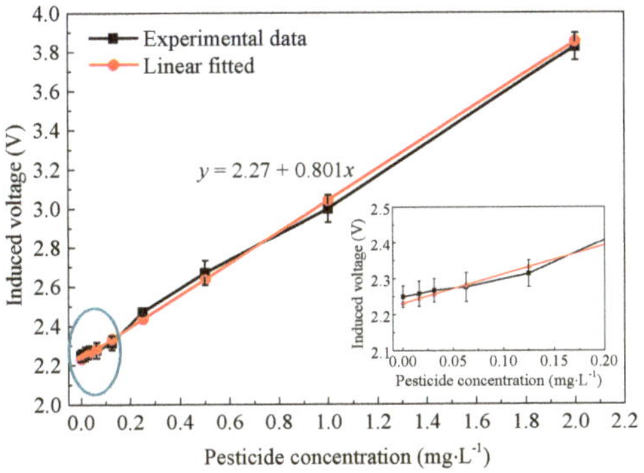

Figure 13. Relationship between induced voltage and pesticide concentration.

Figure 14 shows the color reaction process of the mixed liquid with time observed under microscope; after the reaction liquid mixed through the microchannel and reaches the liquid reservoir, it takes up to 30 s to achieve stable color reaction. The total test time is the sum of mixing time and reaction time, which can be reduced to 60 s.

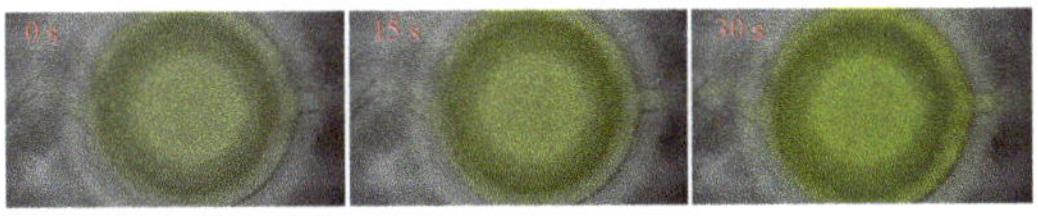

Figure 14. Color change of mixed liquid with time.

Based on the above analysis, the accumulation time required for liquid mixing and reaction is 60 s at a flow rate of 50 $\mu L \cdot min^{-1}$ with a detection sample content of 25.2 μL. Compared with the existing pesticide residue detector, the detection time is reduced from more than 10 min to one minute, and the required sample content is very low. In addition, the detection limit (0.045 $mg \cdot L^{-1}$) is lower than the traditional pesticide detector (0.05 $mg \cdot L^{-1}$), which can meet the demand of pesticide residue detection with high precision [32,33]. However, all the experiments in this study were carried out at room temperature. In the follow-up study, an added temperature control function was proposed to further improve the device's detection efficiency.

4. Conclusions

In this study, a colorimetric device for rapid detection of organophosphorus pesticide residues with high precision based on a microfluidic mixer chip was proposed. The detection system includes a microfluidic mixer chip and a silicon photodetector. The microfluidic simulation indicated that the microchannel with the characteristics of width variation and shape bending had better mixing efficiency. The optimal preparation parameters of the LCD mask photo-curing process were that the exposure time was 6 s and single layer thickness was 30 μm. The reaction liquid had the maximum absorption at 407 nm, and there was a linear relationship between pesticide concentration and induced voltage. The linearity expression R^2 could reach 0.985, the minimum detection concentration of the system reached 0.045 $mg \cdot L^{-1}$ and the average detection time reduced to 60 s. The results provide a theoretical and experimental basis for portable and high-precision detection of pesticide residues.

Author Contributions: Formal analysis, R.S. and Z.Z.; Funding acquisition, J.X.; Methodology, H.P.; Writing—original draft, J.X.; Writing—review and editing, T.W. and X.M. All authors have read and agreed to the published version of the manuscript.

Funding: This work has been financed by the China Postdoctoral Science Foundation (No. 2019M653761), and Provincial Innovation and Entrepreneurship Training Program of Northwest A&F University (S201910712099).

Institutional Review Board Statement: Not applicable.

Informed Consent Statement: Not applicable.

Conflicts of Interest: The authors declare no conflict of interest.

References

1. Curl, C.L.; Fenske, R.A.; Elgethun, K. Organophosphorus pesticide exposure of urban and suburban preschool children with organic and conventional diets. *Environ. Health Perspect.* **2003**, *111*, 377–382. [CrossRef] [PubMed]
2. Jariyal, M.; Jindal, V.; Mandal, K.; Gupta, K.V.; Singh, B. Bioremediation of organophosphorus pesticide phorate in soil by microbial consortia. *Ecotoxicol. Environ. Saf.* **2018**, *159*, 310–316. [CrossRef]
3. Yassine, A.; Omar, B.; Karine, G.S. Study of the degradation of an organophosphorus pesticide using electrogenerated hydroxyl radicals or heat-activated persulfate. *Sep. Purif. Technol.* **2018**, *208*, 27–33.
4. Zhe, Z.; Godefroy, S.B.; Hanyang, L.; Baoguo, S.; Yongxiang, F. Transformation of china's food safety standard setting system—Review of 50 years of change, opportunities and challenges ahead. *Food Control* **2018**, *93*, 106–111.
5. Chen, X.; Liu, X.; Dong, B.; Hu, J. Simultaneous determination of pyridaben, dinotefuran, dn and uf in eggplant ecosystem under open-field conditions: Dissipation behaviour and residue distribution. *Chemosphere* **2018**, *195*, 245–251. [CrossRef]
6. Lara-Ortega, F.J.; Robles-Molina, J.; Brandt, S.; Alexander, S.; Franzke, J. Use of dielectric barrier discharge ionization to minimize matrix effects and expand coverage in pesticide residue analysis by liquid chromatography-mass spectrometry. *Anal. Chim. Acta* **2018**, *1020*, 76–85. [CrossRef]
7. Siyuan, B.I.; Zhu, Z.; Chi, W.; Chen, M.; Wang, X.; Zhou, S. Determination of eight organophosphorus pesticide residues in pepper by quechers-gas chromatography. *Agric. Biotechnol.* **2018**, *7*, 126–128.
8. Li, X.Z.; Yu, Z.; Yang, T.Y.; Ding, J.H. Detection of organophosphorus pesticide residue on the surface of apples using sers. *Spectrosc. Spectr. Anal.* **2013**, *33*, 2711–2720.
9. Andreas, H.A.; Tunemalm, K.; Fredrik, E. Crystal structures of acetylcholinesterase in complex with organophosphorus compounds suggest that the acyl pocket modulates the aging reaction by precluding the formation of the trigonal bipyramidal transition state. *Biochemistry* **2015**, *46*, 4815–4825.
10. Wirdatmadja, S.A.; Moltchanov, D.; Balasubramaniam, S.; Koucheryavy, Y. Microuidic system protocols for integrated on-chip communications and cooling. *IEEE Access* **2017**, *5*, 2417–2429. [CrossRef]
11. Agostini, M.; Greco, G.; Cecchini, M. Full-saw microfluidics-based lab-on-a-chip for biosensing. *IEEE Access* **2019**, *7*, 70901–70909. [CrossRef]
12. Jia, H.; Guo, Y.; Sun, X.; Wang, X. An Electrochemical Immunosensor Based on Microfluidic Chip for Detection of Chlorpyrifos. *Int. J. Electrochem. Sci.* **2015**, *10*, 8750–8758.
13. Deng, S.; Yang, T.; Zhang, W.; Ren, C.; Zhang, J.; Zhang, Y. Rapid detection of trichlorfon residues by a microfluidic paper-based phosphorus-detection chip (μppc). *New J. Chem.* **2019**, *43*, 7194–7197. [CrossRef]
14. Asghar, W.; Sher, M.; Khan, N.S.; Vyas, J.M.; Demirci, U. Microfluidic Chip for Detection of Fungal Infections. *ACS Omega* **2019**, *4*, 7474–7481. [CrossRef]
15. Suh, Y.K.; Kang, S. A review on mixing in microfluidics. *Micromachines* **2010**, *1*, 82–111. [CrossRef]

16. Fan, L.L.; Zhu, X.L.; Zhao, H.; Zhe, J.; Zhao, L. Rapid microfluidic mixer utilizing sharp corner structures. *Microfluid. Nanofluid.* **2017**, *21*, 36–47. [CrossRef]
17. Plevniak, K.; Campbell, M.; Myers, T.; Hodges, A.; He, M. 3d Printed Auto-Mixing Chip Enables Rapid Smartphone Diagnosis of Anemia. *Biomicrofluidics* **2016**, *10*, 54113–54123. [CrossRef]
18. Xu, B.; Guo, J.; Fu, Y.; Chen, X.; Guo, J. A Review on Microfluidics in the Detection of Food Pesticide Residues. *Electrophoresis* **2020**, *41*, 821–832. [CrossRef]
19. Sinko, G.; Calic, M.; Bosak, A.; Kovarik, Z. Limitation of the Ellman Method: Cholinesterase Activity Measurement in the Presence of Oximes. *Anal. Biochem.* **2007**, *370*, 223–227. [CrossRef] [PubMed]
20. Neves, D.A.; Paulo, A.A.; Eliane, N.S.; Paulo, S.L. Microcalorimetric Study of Acetylcholine and Acetylthiocholine Hydrolysis by Acetylcholinesterase. *Adv. Enzyme Res.* **2017**, *5*, 1–12. [CrossRef]
21. Roeser, H.P.; Bohr, A.; Haslam, D.T.; López, J.S.; Stepper, M.; Nikoghosyan, A.S. Size Quantization in High-Temperature Superconducting Cuprates and a Link to Einstein's Diffusion Law. *Acta Astronaut.* **2012**, *76*, 37–41. [CrossRef]
22. Ernesto, B. Limit of detection and limit of quantification determination in gas chromatography. In *Advances in Gas Chromatography*; INTECH: London, UK, 2014. Available online: https://www.intechopen.com/books/advances-in-gas-chromatography/limit-of-detection-and-limit-of-quantification-determination-in-gas-chromatography (accessed on 10 January 2021).
23. Loock, H.-P.; Wentzell, P.D. Detection limits of chemical sensors: Applications and misapplications. *Sens. Actuators B Chem.* **2012**, *173*, 157–163. [CrossRef]
24. Armbruster, D.A.; Pry, T. Limit of blank, limit of detection and limit of quantitation. *Clin. Biochem. Rev.* **2008**, *29*, S49–S54. [PubMed]
25. Taylor, C.; Hood, P. A numerical solution of the navier-stokes equations using the finite element technique. *Comput. Fluids* **2017**, *1*, 73–100. [CrossRef]
26. Liu, R.H.; Stremler, M.A.; Sharp, K.V.; Olsen, M.G.; Santiago, J.G.; Adrian, R.J. Passive mixing in a three-dimensional serpentine microchannel. *J. Microelectromech. Syst.* **2000**, *9*, 190–197. [CrossRef]
27. Xie, J.; He, Y.; Ma, W.; Liu, T.; Chen, J. Study on the Liquid Crystal Display Mask Photo-Curing of Photosensitive Resin Reinforced with Graphene Oxide. *J. Appl. Polym. Sci.* **2020**, *137*, 49538–49546. [CrossRef]
28. Zhu, B.; Cai, Y.; Wu, Z.; Niu, F.; Yang, H. Dielectrophoretic microfluidic chip integrated with liquid metal electrode for red blood cell stretching manipulation. *IEEE Access* **2019**, *7*, 152224–152232. [CrossRef]
29. Dvir, H.; Silman, I.; Harel, M.; Rosenberry, T.L.; Sussman, J.L. Acetylcholinesterase: From 3d Structure to Function. *Chem. Biol. Interact.* **2010**, *187*, 10–22. [CrossRef] [PubMed]
30. Qi, J.; Li, W.; Chu, W.; Yu, J.; Wu, M.; Liang, Y.; Yin, D.; Wang, P.; Wang, Z.; Wang, M.; et al. A Microfluidic Mixer of High Throughput Fabricated in Glass Using Femtosecond Laser Micromachining Combined with Glass Bonding. *Micromachines* **2020**, *11*, 213. [CrossRef] [PubMed]
31. Jung, S.Y.; Park, J.E.; Kang, T.G.; Ahn, K.H. Design Optimization for a Microfluidic Crossflow Filtration System Incorporating a Micromixer. *Micromachines* **2019**, *10*, 836. [CrossRef]
32. Yong, H.; Yan, W.; Fu, J.Z.; Wu, W.B. Fabrication of paper-based microfluidic analysis devices: A review. *RSC Adv.* **2015**, *5*, 78109–78127.
33. Bhakta, S.A.; Borba, R.; Toba, M., Jr.; Garcia, C.D.; Carrilho, E. Determination of nitrite in saliva using microfluidic paper-based analytical devices. *Anal. Chim. Acta* **2014**, *809*, 117–122. [CrossRef] [PubMed]

micromachines MDPI

Article

An Ultra-High-Q Lithium Niobate Microresonator Integrated with a Silicon Nitride Waveguide in the Vertical Configuration for Evanescent Light Coupling

Jianhao Zhang [1,2], Rongbo Wu [1,2], Min Wang [3,4,*], Youting Liang [3,4], Junxia Zhou [3,4], Miao Wu [3,4], Zhiwei Fang [3,4], Wei Chu [4,*] and Ya Cheng [1,3,4,5,6,7,8,*]

1 State Key Laboratory of High Field Laser Physics and CAS Center for Excellence in Ultra-Intense Laser Science, Shanghai Institute of Optics and Fine Mechanics, Chinese Academy of Sciences, Shanghai 201800, China; jhzhang@siom.ac.cn (J.Z.); rbwu@siom.ac.cn (R.W.)
2 Center of Materials Science and Optoelectronics Engineering, University of Chinese Academy of Sciences, Beijing 100049, China
3 State Key Laboratory of Precision Spectroscopy, East China Normal University, Shanghai 200241, China; 15253172638@163.com (Y.L.); 52180920026@stu.ecnu.edu.cn (J.Z.); wumiao1993@126.com (M.W.); zwfang@phy.ecnu.edu.cn (Z.F.)
4 The Extreme Optoelectromechanics Laboratory (XXL), School of Physics and Electronic Science, East China Normal University, Shanghai 200241, China
5 Collaborative Innovation Center of Extreme Optics, Shanxi University, Taiyuan 030006, China
6 Collaborative Innovation Center of Light Manipulations and Applications, Shandong Normal University, Jinan 250358, China
7 CAS Center for Excellence in Ultra-Intense Laser Science, Shanghai 201800, China
8 Shanghai Research Center for Quantum Sciences, Shanghai 201315, China
* Correspondence: mwang@phy.ecnu.edu.cn (M.W.); wchu@phy.ecnu.edu.cn (W.C.); ya.cheng@siom.ac.cn (Y.C.)

Citation: Zhang, J.; Wu, R.; Wang, M.; Liang, Y.; Zhou, J.; Wu, M.; Fang, Z.; Chu, W.; Cheng, Y. An Ultra-High-Q Lithium Niobate Microresonator Integrated with a Silicon Nitride Waveguide in the Vertical Configuration for Evanescent Light Coupling. *Micromachines* **2021**, *12*, 235. https://doi.org/10.3390/mi12030235

Academic Editor: Andrea Ehrmann

Received: 1 February 2021
Accepted: 21 February 2021
Published: 25 February 2021

Abstract: We demonstrate the hybrid integration of a lithium niobate microring resonator with a silicon nitride waveguide in the vertical configuration to achieve efficient light coupling. The microring resonator is fabricated on a lithium niobate on insulator (LNOI) substrate using photolithography assisted chemo-mechanical etching (PLACE). A fused silica cladding layer is deposited on the LNOI ring resonator. The silicon nitride waveguide is further produced on the fused silica cladding layer by first fabricating a trench in the fused silica while using focused ion beam (FIB) etching for facilitating the evanescent coupling, followed by the formation of the silicon nitride waveguide on the bottom of the trench. The FIB etching ensures the required high positioning accuracy between the waveguide and ring resonator. We achieve Q-factors as high as 1.4×10^7 with the vertically integrated device.

Keywords: lithium niobate microring resonator; silicon nitride waveguide; photolithography assisted chemo-mechanical etching

1. Introduction

Lithium niobate (LN) has long been recognized as an important material platform for integrated photonic devices because of its wide transparent window, high nonlinear coefficient, and excellent electro-optical property [1]. In particular, the latest advancement of the fabrication technologies of high-quality photonic micro- and nanostructures on lithium niobate on insulator (LNOI) has further promoted the development of integrated photonics on the LNOI platform. The building block photonic structures, such as microresonator and optical waveguide, are typically fabricated on LNOI substrate using either maskless focused ion beam (FIB) milling or lithographic processing involving argon ion milling [2–11]. Meanwhile, a chemo-mechanical etching process (termed photolithography assisted chemo-mechanical etching (PLACE) hereafter) has been developed to achieve ultra-high surface smoothness, resulting in high-quality (high-Q) microdisk resonator with a Q-factor of 4×10^7 and ultra-low-loss optical waveguide with a propagation loss of 0.03 dB/cm [12–14]. So far, a broad range of nonlinear optical processes have been demonstrated with the ultra-high-Q LN microresonators that were fabricated using the PLACE technology, ranging from optomechanics [15] and optical frequency comb [16] to nonlinear frequency conversion [17] and on-chip micro-disk lasing.

Light must be efficiently coupled into the microresonaotors using either a fiber taper [18] or an integrated optical waveguide [19] in order to excite the nonlinear optical effects. The on-chip integration of the microresonator and the coupling optical waveguide provides an efficient means for up scaling of the photonic integration circuits (PICs), which is critical for some applications, such as photonic computation and quantum information processing [20], etc. However, in the PLACE scheme, the low-loss optical waveguides and high Q microresonators are both generated using the chemical-mechanical polishing (CMP) technique. In the CMP process, it is required that the distance between the closely located photonic structures should be on the micrometer scale, but not the nanometer scale; otherwise, the LNOI in the narrow gap between the neighboring structures cannot be efficiently removed by polishing. In this case, the lateral evanescent coupling between a microresonator and a waveguide is difficult to achieve, simply because of the fact that the evanescent coupling in the visible and near infrared ranges requires the gap width to be in the order of a few hundred nanometers. For this reason, on-chip evanescent coupling has not been realized between an optical waveguide and a microresonator fabricated using the PLACE technique.

Here, we overcome the difficulty by utilizing a vertical coupling scheme between a crystalline LN microring resonators and a silicon nitride (Si_3N_4) waveguide. Si_3N_4 is also considered to be an attractive candidate for monolithic integration of photonic circuits because of its low propagation loss. Importantly, Si_3N_4 has a refractive index that is close to that of LN, which makes it easy to fulfill the phase matching condition between the Si_3N_4 waveguide and LN microresonator. We characterized the integrated device by measuring the Q-factor of the fabricated LN microresonator, and demonstrated the coupling control by varying the thickness of the SiO_2 cladding layer.

2. Materials and Methods

The microring resonator is fabricated on a commercially available x-cut LNOI wafer with a thickness of 900 nm (NANOLN, Jinan Jingzheng Electronics Co., Ltd., Jinan, Shandong, China). The LN thin film is bonded to a 2 μm-thick SiO_2 layer supported by a 500-μm-thick LN substrate. Figure 1a depicts the configuration of the LNOI wafer, followed by the schematic of process flow, as shown in Figure 1b–k. In general, the fabrication procedures include: (1) the deposition of a thin layer of chromium (Cr) with a thickness of 400 nm on the surface of the LNOI by magnetron sputtering (Figure 1b); (2) space-selective ablation of a Cr layer coated on top of the LNOI to generate the pattern of the microring resonator using a focused femtosecond laser beam (Figure 1c). In this step, the femtosecond laser ablation was conducted by a commercial laser system (Pharos, Light-Conversion, Lithuania) at a repetition rate of 500 kHz and a scan speed of 40 mm/s. The center wavelength of the femtosecond laser was 1030 nm, and the pulse width was set to be ~270 fs. A 100× objective lens (M Plan Apo NIR, Mitutoyo, Japan) with a numerical aperture (NA) of 0.7 was employed to pattern the Cr layer in order to obtain a high ablation resolution. Femtosecond laser ablation was carried out by translating the sample with a three-dimensional (3D) motion stage (ABL1500-ANT130, Aerotech Inc., USA); (3) etching of the LNOI layer by CMP (Figure 1d). In this step, the LN without being covered by the Cr mask will be completely removed, while the LN protected by Cr mask will survive from the CMP because of the high hardness of Cr; (4) removal of the residual Cr mask left on the surface of LNOI by chemical wet etching, and further eliminate the roughness by a second CMP process (Figure 1e); (5) the deposition of the SiO_2 film on the LNOI waveguide to form the cladding layer by plasma enhanced chemical vapor deposition (PECVD) (Figure 1f); (6) polishing the surface of SiO_2 cladding layer with the third CMP

(Figure 1g); (7) patterning of the SiO_2 layer while using focused ion beam (FIB) etching (Figure 1h). In particular, the depth of the etched trench can be controlled with an accuracy of ~1 nm using the FIB etching; (8) the deposition of a Si_3N_4 film on the SiO_2 layer by PECVD to fill the trench fabricated in the SiO_2 cladding layer (Figure 1i); (9) removing the Si_3N_4 above the SiO_2 layer with the fourth CMP (Figure 1j); and, (10) patterning of the Si_3N_4 film in the trench using FIB etching to form the waveguide (Figure 1k). More details of the femtosecond laser micromachining of Cr, the CMP processing, and the FIB etching can be found elsewhere [12–14]. Figure 1l shows a schematic 3D view of the hybrid LN and Si_3N_4 coupling structure.

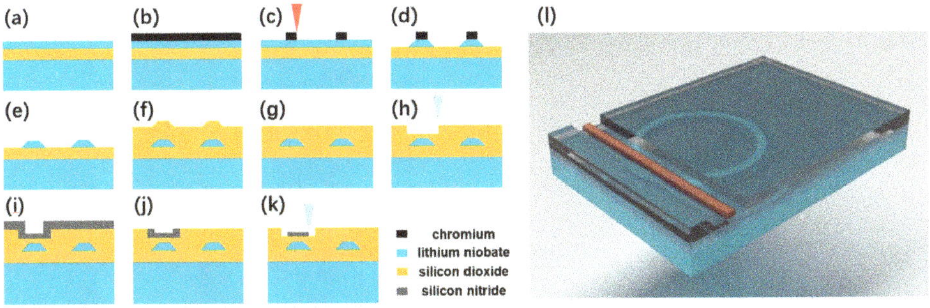

Figure 1. (**a**–**k**) Process flow of fabricating a lithium niobate (LN) microring resonator coupling with a silicon nitride (Si_3N_4) waveguide in the vertical configuration. (**l**) Three-dimensional (3D) diagram of the coupling structure.

3. Results

Figure 2a shows the top-view scanning electron micrograph (SEM) of the vertically coupled LN microring and Si_3N_4 waveguide. The profiles of the microring buried beneath SiO_2 layer can be distinguished in the SEM image. The radius of the LN microring is 50 µm. The cross section of the LN microring, as indicated by the blue dashed area in Figure 2a, is shown in Figure 2b. In the current design, the LN microring has a trapezoidal cross-section with a top width of 2.5 µm, a bottom width of 7.5 µm, and a height of 800 nm, which is covered by a 1.5-µm-thick SiO_2 cladding layer. The cross section of the coupling area (yellow dashed area in Figure 2a) is shown in Figure 2c. The three-layer structure, i.e., LN, SiO_2 and Si_3N_4 from bottom to top, can be clearly seen. The thickness of the PECVD SiO_2 layer was reduced from 1.5 µm to 600 nm by FIB etching for evanescent light coupling. The rib Si_3N_4 waveguide was fabricated by FIB etching with a width of 3.5 µm and a rib height of 550 nm.

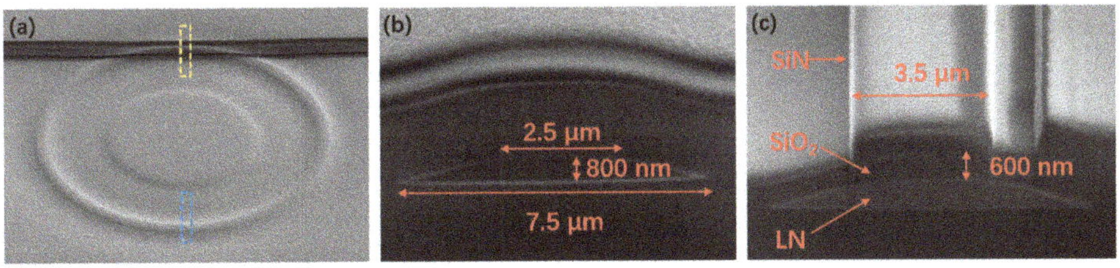

Figure 2. (**a**) Top view scanning electron micrograph (SEM) image of LN microring resonator and Si_3N_4 waveguide. (**b**) Sectional view SEM image of the structure at the location of the bule dashed box in (**a**). (**c**) Sectional view SEM image of the structure at the location of the yellow dashed box in (**a**).

Furthermore, we measured the surface roughness for both the LN microring resonator fabricated by CMP and the trench in SiO_2 layer (i.e., the bottom of the trench) generated by FIB etching that are critical in obtaining high Q factors. Figure 3a,b present the SEM images of fabricated LN microring after CMP (step e in Figure 1) and the etched trench of the SiO_2 layer using FIB etching (step h in Figure 1). Using an atomic force microscope (AFM), we measured the surface root-mean-square roughness (Rq) in the areas that are indicated by the red squares in Figure 3a,b respectively. An ultralow surface roughness of Rq ~ 0.45 nm can be achieved by the CMP processing, while a slightly higher roughness of Rq ~ 1.22 nm was achieved after FIB etching. Based on the smooth surface morphology, we can expect a high-Q LN microring resonator.

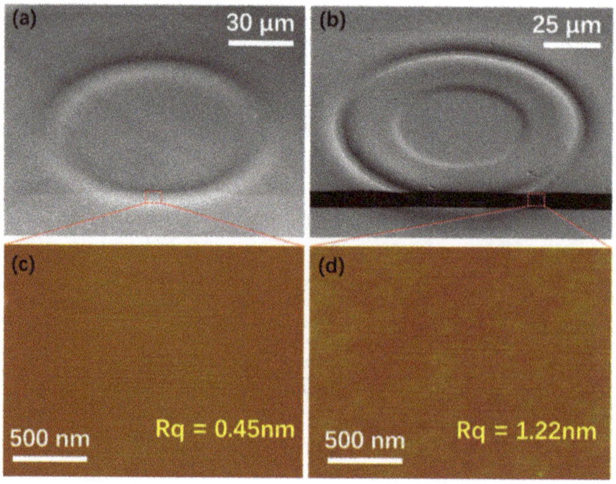

Figure 3. (**a**) Top view SEM image of the fabricated LN microring resonator. (**b**) Top view SEM image of the trench fabricated in the SiO_2 cladding layer. (**c**,**d**) Atomic force microscope (AFM) images of the surfaces of the microring in (**a**) and the trench in (**b**) respectively.

We used an experimental setup to examine the coupling effect of the configuration and characterize the optical mode structure of the LN ring resonator, as schematically shown in Figure 4. A tunable laser (TLB 6728, New Focus Inc., San Jose, CA, USA) was employed to couple light into and out of Si_3N_4 waveguide through the lensed fiber with a taper angle of 90°. In order to enhance the detection signal, the tunable laser was boosted by an erbium-ytterbium-doped fiber amplifier (EYDFA, Golight, Inc., Culbertson, NE, USA) before coupling into the Si_3N_4 waveguide. The linewidth of the tunable laser is 200 kHz. The polarization of the pump laser was adjusted by an in-line fiber polarization controller. A photodetector measured the transmission of resonant mode (New focus 1811-FC-AC, Newport Inc., Irvine, CA, USA). We used an arbitrary waveform generator (AFG3052C, Tektronix Inc., Beaverton, DC, USA) to synchronize the tunable laser and oscilloscope signals.

Figure 5a shows the transmission spectrum for the wavelength range from 1537 nm to 1562 nm. The free spectral range (FSR) of the microresonator was measured to be 3.34 nm. A pair of the splitting whispering-gallery modes at the resonant wavelength around 1543.52 nm was chosen for the measurement of the Q-factor by fitting with a Lorentz function. The Q factors were measured to be 1.49×10^7 and 1.09×10^7, respectively, as indicated by the Lorentz curves shown in Figure 5b. The Q factors vary from 2.9×10^6 to 1.49×10^7, which is mainly caused by the different coupling condition of the whispering gallery modes. The high Q-factor of the LN microresonator indicates that the fabricated device with the vertical integration configuration functions effectively for evanescent light coupling between the LN microring and the Si_3N_4 waveguide.

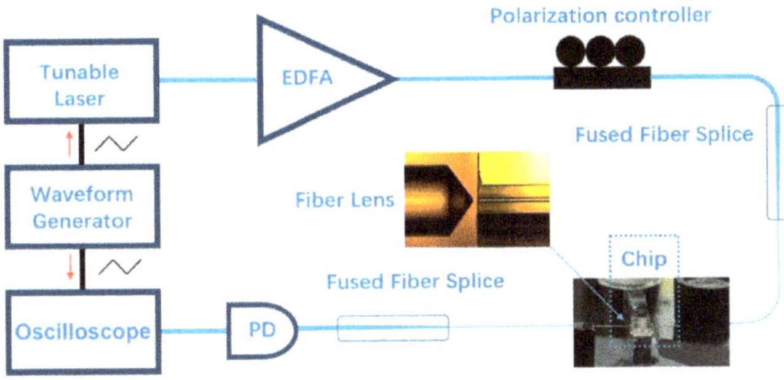

Figure 4. Schematic of experimental testing setup. Inset: optical micrograph of the fiber lens coupling with the Si_3N_4 waveguide.

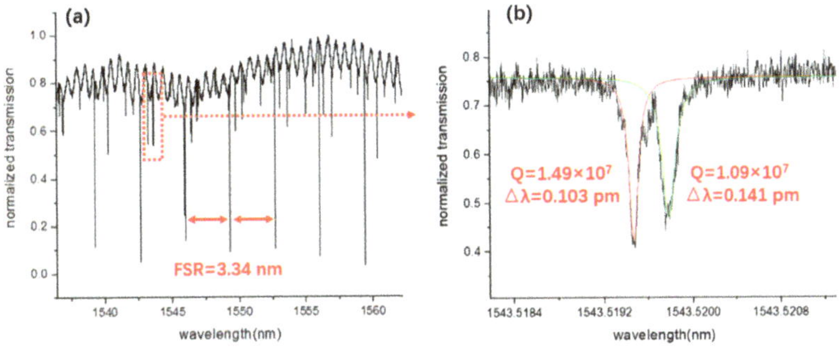

Figure 5. (**a**) Transmission spectrum of the LN microring resonator. (**b**) The Lorentz fitting of the splitting modes at the location of the red dotted box in Figure 4a reveals a Q-factor of 1.49×10^7 (red solid line) and 1.09×10^7 (green solid line), respectively.

Finally, we demonstrated that, in our scheme, the coupling efficiency as well as Q-factor can be tuned by changing the thickness of the SiO_2 cladding layer, i.e., the vertical distance between the Si_3N_4 waveguide and the LN microring. The distance was adjusted by precisely controlling the FIB etching depth of the PECVD SiO_2 layer. When the thickness of the SiO_2 cladding layer was set to 1100 nm, the coupling efficiency was relatively low, leading to the undercoupling between the Si_3N_4 waveguide and the LN microring and giving rise to a high Q-factor of 6.6×10^6, as illustrated in Figure 6a,d. Subsequently, we adjusted the distance to 600 nm, as shown in Figure 6b. The transmission loss of the light in the microring was close to the coupling loss between the waveguide and microring, which indicated that the critical coupling condition was reached. The deepest dip can be observed in the transmission curve presented in Figure 6e, and the Q-factor was 4.5×10^6. In general, the Q factor should be higher for the critical coupling condition than that obtained in the undercoupling condition. Here, the slightly lower Q-factor measured in the critical coupling condition can be attributed to various imperfections in the fabrication process, which could influence the intrinsic Q of the microring itself beneath the coupling Si_3N_4 waveguide. When the distance was further reduced to 100 nm, a higher coupling efficiency was reached at the strong over-coupling, whilst the Q factor decreased to 2.6×10^6, as illustrated in Figure 6c,f. The Q factor that we mentioned here is loaded Q factor.

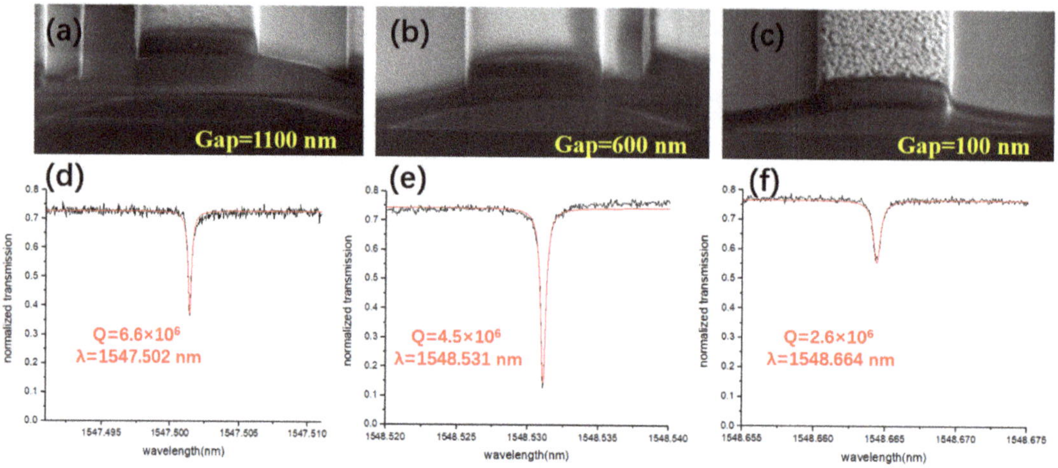

Figure 6. Sectional view SEM images of the coupling structures when the thickness of the SiO_2 cladding layer are 1100 nm (**a**), 600 nm (**b**), and 100 nm (**c**), respectively. The Lorentz fitting of the modes of the structures in (**a–c**) are illustrated in (**d–f**), respectively.

4. Conclusions

To conclude, we have demonstrated efficient evanescent coupling between the crystalline LN microring resonator fabricated by PLACE and the Si_3N_4 waveguide fabricated by FIB with a vertical configuration. By controlling the distance between the waveguide and microresonator, nearly critical coupling condition has been achieved with ultra-high Q factors, i.e., the Q-factor of the fabricated LN microresonator was measured to be 1.49×10^7. Furthermore, we demonstrated that the coupling efficiency can be continuously tuned upon demand by varying the thickness of the SiO_2 cladding layer. We should point out that the coupling efficiency can also be tuned by changing the relative position between the waveguide and the microring resonator in the horizontal plane. The scheme proposed in this work is also beneficial for large-scale PIC integration, as multiple microresonators can be remotely connected on a single chip using the same waveguide and the coupling efficiency can be individually tuned, as reasoned above. Thus, the scheme provides a promising photonic integration solution widely adopted by a broad range of LNOI photonic applications, which range from micro/nano-nonlinear optics and optical interconnect to on-chip artificial intelligence demonstration, etc.

Author Contributions: Conceptualization, Y.C.; methodology, Y.C. and W.C.; validation, J.Z. (Jianhao Zhang), M.W. (Ming Wang) and R.W.; formal analysis, J.Z. (Jianhao Zhang), R.W., W.C. and M.W. (Ming Wang); investigation, J.Z. (Jianhao Zhang), R.W., Z.F. and W.C.; resources, J.Z.(Jianhao Zhang), M.W. (Ming Wang), Z.F., R.W. and J.Z. (Junxia Zhou); data analyze, J.Z. (Jianhao Zhang), Y.L., W.C. and M.W. (Miao Wu); draft preparation, Y.C., W.C and J.Z. (Jianhao Zhang); supervision, W.C. and Y.C.; funding acquisition, Y.C. and W.C. All authors have read and agreed to the published version of the manuscript.

Funding: We acknowledge supports from National Key R&D Program of China (2019YFA0705000), National Natural Science Foundation of China (11674340, 11734009, 11874154, 11874375, 61761136006, 61590934), the Strategic Priority Research Program of CAS (XDB16030300), Shanghai Municipal Science and Technology Major Project (2019SHZDZX01), and Key Research Program of Frontier Sciences, CAS (QYZDJ-SSW-SLH010).

Conflicts of Interest: The authors declare no conflict of interest.

References

1. Boes, A.; Corcoran, B.; Chang, L.; Bowers, J.; Mitchell, A. Status and potential of lithium niobate on insulator (LNOI) for photonic integrated circuits. *Laser Photon. Rev.* **2018**, *12*, 1700256. [CrossRef]
2. Diziain, S.; Geiss, R.; Zilk, M.; Schrempel, F.; Kley, E.-B.; Tünnermann, A.; Pertsch, T. Mode analysis of photonic crystal L3 cavities in self-suspended lithium niobate membranes. *Appl. Phys. Lett.* **2013**, *103*, 251101. [CrossRef]
3. Zhang, M.; Wang, C.; Cheng, R.; Shams-Ansari, A.; Lončar, M. Monolithic ultra-high-Q lithium niobate microring resonator. *Optica* **2017**, *4*, 1536–1537. [CrossRef]
4. Hu, H.; Yang, J.; Gui, L.; Sohler, W. Lithium niobate-on-insulator (LNOI): Status and perspectives. *Proc. SPIE* **2012**, *8431*, 84311D.
5. Geiss, R.; Saravi, S.; Sergeyev, A.; Diziain, S.; Setzpfandt, F.; Schrempel, F.; Grange, R.; Kley, E.-B.; Tünnermann, A.; Pertsch, T. Fabrication of nanoscale lithium niobate waveguides for second-harmonic generation. *Opt. Lett.* **2015**, *40*, 2715–2718. [CrossRef]
6. Luo, R.; He, Y.; Liang, H.; Li, M.; Lin, Q. Semi-nonlinear nanophotonic waveguides for highly efficient second-harmonic generation. *Laser Photon. Rev.* **2019**, *13*, 1800288. [CrossRef]
7. Krasnokutska, I.; Tambasco, J.L.J.; Li, X.J.; Peruzzo, A. Ultra-low loss photonic circuits in lithium niobate on insulator. *Opt. Express* **2018**, *26*, 887–894. [CrossRef]
8. Lu, J.; Surya, J.B.; Liu, X.; Xu, Y.; Tang, H.X. Octave-spanning supercontinuum generation in nanoscale lithium niobate waveguides. *Opt. Lett.* **2019**, *44*, 1492–1495. [CrossRef]
9. He, M.; Xu, M.; Ren, Y.; Jian, J.; Ruan, Z.; Xu, Y.; Gao, S.; Sun, S.; Wen, X.; Zhou, L.; et al. High-performance hybrid silicon and lithium niobate Mach–Zehnder modulators for 100 Gbit s−1 and beyond. *Nat. Photon.* **2019**, *13*, 359–364. [CrossRef]
10. Chen, J.-Y.; Ma, Z.-H.; Sua, Y.M.; Li, Z.; Tang, C.; Huang, Y.-P. Ultra-efficient frequency conversion in quasi-phase-matched lithium niobate microrings. *Optica* **2019**, *6*, 1244–1245. [CrossRef]
11. Luo, R.; He, Y.; Liang, H.; Li, M.; Lin, Q. Highly tunable efficient second-harmonic generation in a lithium niobate nanophotonic waveguide. *Optica* **2018**, *5*, 1006–1011. [CrossRef]
12. Wu, R.; Zhang, J.; Yao, N.; Fang, W.; Qiao, L.; Chai, Z.; Lin, J.; Cheng, Y. Lithium niobate micro-disk resonators of quality factors above 107. *Opt. Lett.* **2018**, *43*, 4116–4119. [CrossRef] [PubMed]
13. Zhang, J.; Fang, Z.; Lin, J.; Zhou, J.; Wang, M.; Wu, R.; Gao, R.; Cheng, Y. Fabrication of crystalline microresonators of high quality factors with a controllable wedge angle on lithium niobate on insulator. *Nanomaterials* **2019**, *9*, 1218. [CrossRef] [PubMed]
14. Wu, R.B.; Wang, M.; Xu, J.; Qi, J.; Chu, W.; Fang, Z.W.; Zhang, J.H.; Zhou, J.X.; Qiao, L.L.; Chai, Z.F.; et al. Long low-loss-litium niobate on insulator waveguides with sub-nanometer surface roughness. *Nanomaterials* **2018**, *8*, 910. [CrossRef] [PubMed]
15. Jiang, W.C.; Lin, Q. Chip-scale cavity optomechanics in lithium niobate. *Sci. Rep.* **2016**, 36920. [CrossRef] [PubMed]
16. Zhang, M.; Buscaino, B.; Wang, C.; Shams-Ansari, A.; Reimer, C.; Zhu, R.; Kahn, J.M.; Lončar, M. Broadband electro-optic frequency comb generation in a lithium niobate microring resonator. *Nature* **2019**, *568*, 373–377. [CrossRef] [PubMed]
17. Fang, Z.; Haque, S.; Farajollahi, S.; Luo, H.; Lin, J.; Wu, R.; Zhang, J.; Wang, Z.; Wang, M.; Cheng, Y.; et al. Polygon coherent modes in a weakly perturbed whispering gallery microresonator for efficient second harmonic, optomechanical, and frequency comb generations. *Phys. Rev. Lett.* **2020**, *125*, 173901. [CrossRef] [PubMed]
18. Wang, L.; Wang, C.; Wang, J.; Bo, F.; Zhang, M.; Gong, Q.; Lončar, M.; Xiao, Y.-F. High-Q chaotic lithium niobate microdisk cavity. *Opt. Lett.* **2018**, *43*, 2917. [CrossRef] [PubMed]
19. Wolf, R.; Breunig, I.; Zappe, H.; Buse, K. Scattering-loss reduction of ridge waveguides by sidewall polishing Opt. *Express* **2018**, *26*, 19815–19820. [CrossRef]
20. Pant, M.; Towsley, D.; Englund, D.; Guha, S. Percolation thresholds for photonic quantum computing. *Nat. Commun.* **2019**, *10*, 1070. [CrossRef] [PubMed]

micromachines

MDPI

Article

Synthesis of Printable Polyvinyl Alcohol for Aerosol Jet and Inkjet Printing Technology

Mahmuda Akter Monne [1], Chandan Qumar Howlader [1], Bhagyashree Mishra [1] and Maggie Yihong Chen [1,2,*]

1 Materials Science, Engineering, and Commercialization Program, Texas State University, San Marcos, TX 78666, USA; mam638@txstate.edu (M.A.M.); chandan.howlader@txstate.edu (C.Q.H.); b_m415@txstate.edu (B.M.)
2 Ingram School of Engineering, Texas State University, San Marcos, TX 78666, USA
* Correspondence: yc12@txstate.edu

Abstract: Polyvinyl Alcohol (PVA) is a promising polymer due to its high solubility with water, availability in low molecular weight, having short polymer chain, and cost-effectiveness in processing. Printed technology is gaining popularity to utilize processible solution materials at low/room temperature. This work demonstrates the synthesis of PVA solution for 2.5% *w*/*w*, 4.5% *w*/*w*, 6.5% *w*/*w*, 8.5% *w*/*w* and 10.5% *w*/*w* aqueous solution was formulated. Then the properties of the ink, such as viscosity, contact angle, surface tension, and printability by inkjet and aerosol jet printing, were investigated. The wettability of the ink was investigated on flexible (Kapton) and non-flexible (Silicon) substrates. Both were identified as suitable substrates for all concentrations of PVA. Additionally, we have shown aerosol jet printing (AJP) and inkjet printing (IJP) can produce multi-layer PVA structures. Finally, we have demonstrated the use of PVA as sacrificial material for micro-electro-mechanical-system (MEMS) device fabrication. The dielectric constant of printed PVA is 168 at 100 kHz, which shows an excellent candidate material for printed or traditional transistor fabrication.

Keywords: aerosol jet printing; dielectric; etching; inkjet printing; polyvinyl alcohol (PVA), polymers; sacrificial material; micro-electro-mechanical-system (MEMS)

Citation: Monne, M.A.; Howlader, C.Q.; Mishra, B.; Chen, M.Y. Synthesis of Printable Polyvinyl Alcohol for Aerosol Jet and Inkjet Printing Technology. *Micromachines* **2021**, *12*, 220. https://doi.org/10.3390/mi12020220

Academic Editor: Andrea Ehrmann

Received: 20 January 2021
Accepted: 12 February 2021
Published: 22 February 2021

1. Introduction

In 1915, Poly(vinyl alcohol) was discovered by F. Klatte from the precursor Poly(vinyl acetate) [1]. After that, the preparation of PVA was firstly described by W. O. Herrmann and W. Haehnel in 1924 [2–4]. However, it cannot be prepared with traditional polymerization because its monomer vinyl alcohol is not stable and rearranges readily to acetaldehyde. It is usually manufactured from hydrolysis of polyvinyl acetate, which involves the partial or total replacement of the ester groups of vinyl acetate by hydroxyl groups under defined conditions. Later, the PVA is precipitated, washed, and finally dried. The PVA properties depend on the polymer chain's length or the degree of polymerization and the degree of hydrolysis [5,6]. PVA is an example of a water-soluble semi-crystalline synthetic polymer, and it is also slightly soluble in ethanol. It is important to mention that the higher the degree of hydrolysis and polymerization, the lower the solubility of PVA in cold water. Polyvinyl alcohol is a semi-crystalline, non-toxic, water-soluble, and biodegradable polymer. It is also biocompatible with human tissues and has excellent gas-barrier properties [7]. PVA has many applications, including but not limited to the manufacturing of cleaning and detergent products, in the food packaging industry, water treatment, textile, agriculture, and construction [7–11]. It also has recently attracted an increasing amount of attention for pharmaceutical uses (i.e., drug delivery) and in medical applications (e.g., wound dressing, soft contact lenses, and eye drops) [12–15]. PVA films can be produced in either a melt or solution form. Melt processing is compatible only with low hydrolysis or heavily plasticized PVA. At the same time, a PVA film can be deposited from the solution form through drop casting, spin coating, and electrospinning, etc. [16–18]. However, these methods have a few limitations in terms of waste of material. In case of spin coating, it has been reported that about 95% of material is wasted with no design/features/patterning capability [19–21]. Our work reports the synthesis of PVA and the potential of aerosol jet and inkjet printing technologies as a mean to provide a novel platform to produce multilayer structures. This phenomenon will have a substantial impact on both the material processing and application perspectives.

Recent developments in the thin-film deposition sector have focused on cheap, simple, eco-friendly, and energy-saving processes. Printing is a modern fabrication process that fits perfectly within this framework. The significant advantages of printing technology are material utilization efficiency, mask-free and additive patterning, large area capability, compatibility with many substrates, and the low-cost fabrication process [22–27]. Printing technology is used to successfully deposit conductive and non-conductive nanomaterial, polymers, ceramic, 1D/2D materials, dielectric, biological, and pharmaceutical-based materials [28–31].

With the advancement of flexible electronics, low molecular weight and water-soluble elastic polymers as well as high permittivity dielectrics are in demand for MEMS device and flexible transistor fabrication [32–37]. Many polymers like polyvinylpyrrolidone (PVP), polymethylmethacrylate (PMMA), and PVA have been studied for their water solubility, electrical, and dielectric properties [38]. However, PVA is the most studied polymer due to its versatile properties. It is highly water-soluble, has a low cost, and has suitable film-forming properties.

The selection of appropriate sacrificial material is a crucial part of the MEMS fabrication process. The sacrificial material must be deposited and finally removed without disturbing other layers. Along with developing printed technology, researchers have been working on selecting appropriate sacrificial materials for years. In [39], the use of sacrificial layer technique is discussed to create multilayer metalized structure using SU-8 dielectric. Multilayer suspended structure was successfully fabricated by adopting sputter coating, which is another form of solution processible technique.

Additionally, the MEMS fabrication process is less standardized than other microelectronics fabrication processes. 3D printed MEMS is a potential candidate among researchers for a selection of less expensive materials and producing new structures for future applications [40]. However, it may be difficult to achieve sufficient resolution only with this technology compared to traditional MEMS fabrication technology [36,37,41]. A fundamental study has been conveyed to fabricate graphene or graphite-based flexible MEMS devices which enables future researchers to fabricate 2D material-based MEMS devices [42,43].

This study aims to present the synthesis of PVA solution for printing technology (aerosol jet and inkjet printing), characterize the solution, and present the application of PVA for MEMS and transistor device fabrication.

2. Experiments

Ink Formulation: The PVA solution is synthesized by mixing low molecular weight (<400 kDa) PVA crystal with DI water and heating with a microwave. Different percentage of w/w PVA solutions were prepared for two printing techniques. Viscosity of the PVA solution was varied by changing the percent of PVA powder. It is observed that 0.5%–4.5% w/w PVA is good for inkjet printing technology and aerosol jet technology with the ultrasonic atomizer. The solution with >4.5% w/w PVA is suitable for aerosol jet printing with the pneumatic atomizer and other printing methods such as screen printing, spray coating, bar coating, blade coating, etc. To prepare a 2% w/w 10ml PVA solution, 10 mL of DI water is mixed with 0.2 g of low molecular weight PVA crystal and heated with a microwave. The microwave's power setting is kept at 1000 watts, and the solution is microwaved for 180 s until a homogeneous solution is achieved. Later, the homogeneous solution is centrifuged for 30 min at 5000 rpm to eliminate the big and clustered polymer chains. Finally, the solution is filtered using a 0.25–0.45 μm filter before loading it to the printer reservoir. Figure 1 shows the 98% hydrolyzed PVA crystal from Sigma Aldrich (St. Louis, MO 63178, USA) and the in-lab prepared PVA solution.

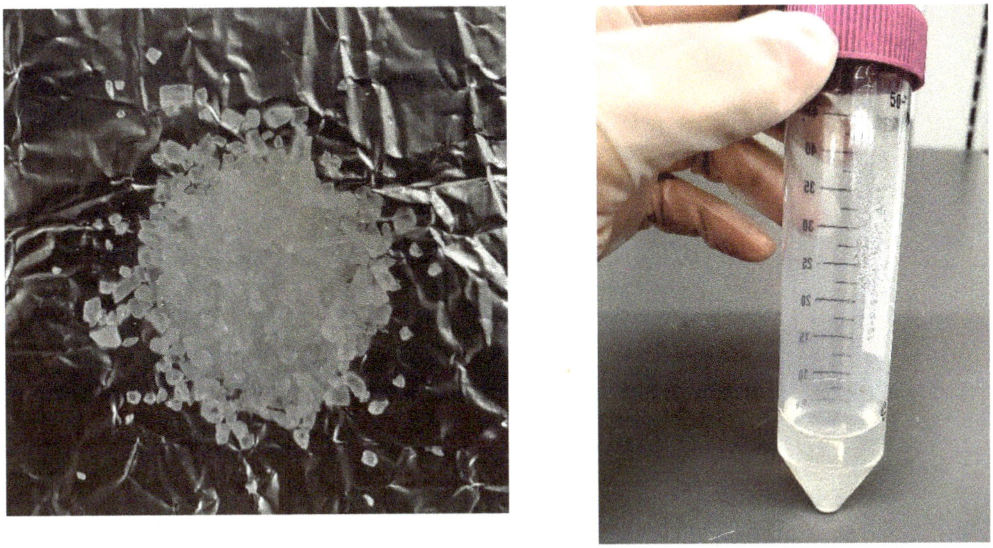

(a) (b)

Figure 1. (**a**) The 98% hydrolyzed PVA crystal. (**b**) In-lab prepared PVA solution.

Ink Characterization (Viscosity): Ink was prepared with 5 different concentrations; 2.5% *w*/*w*, 4.5% *w*/*w*, 6.5% *w*/*w*, 8.5% *w*/*w*, and 10.5% *w*/*w* PVA in DI water. An AMETEK Brookfield (DV2T) (Chandler, AZ 85225, USA) viscometer is used to measure the viscosity of each solution. Since the instrument gives multiple measured data for the viscosity measurement for different shear rates, so standard deviation of viscosity is added to Table 1. The standard deviation in Table 1 column 6 is among all the values of viscosity (at different set of shear rates) including minimum and maximum viscosity.

Table 1. Viscosity measurement of all five PVA solution.

PVA Percentage (%)	Spindle Speed (rpm)	Shear Rate (1/s)	Viscosity Minimum (cP)	Viscosity Maximum (cP)	Standard Deviation for Viscosity (cP)
2.5	50–200	375–1500	4.93	5.00	0.03
4.5	25–100	187.5–750	8.80	9.00	0.08
6.5	15–50	112.5–375	24.2	25.5	0.5
8.5	10–25	75–187.5	34.8	35.3	0.3
10.5	10–25	75–187.5	81	82	1

Viscosity was measured at 25 °C, with different spindle speeds and shear rate. Multiple shear rates were used due to various concentrations of PVA, which results in different viscosity. Usually, low-viscous inks need higher spindle speed and shear rate to get an accurate result. Table 1 shows a summary of viscosity measurement.

Contact Angle and Surface Tension (on Kapton and Silicon): Appropriate surface cleaning is a crucial step to acquire a good quality print. Substrates are cleaned in an air plasma for 15 mins to remove organic contaminants and physical ablation. The plasma cleaning process introduces chemical functional groups (carbonyl, carboxyl, hydroxyl) on the surface to make it hydrophilic. This paper also investigates the effect of plasma cleaning on contact angle. Since printable PVA is a water-based solution, a hydrophilic surface is essential to achieve a good quality print. The analysis of surface tension and contact angle is carried out to present the degree of hydrophilicity before and after the plasma cleaning process, which ultimately helped us decide hydrophilicity or hydrophobicity and the wettability of the printable PVA solution. The literature found that the surface tension in the range of (25–75) mN/m is adequate to achieve a good liquid tension toward the surface [40]. Through experimentation the surface tension is found to be between 30 mN/m to 50 mN/m for both the substrates. Since the expected surface tension is obtained without plasma treatment, thus plasma treatment is not carried out to measure the surface tension.

On the other hand, contact angle presents the wettability or surface coverage of a solution. The contact angle between 0°–90° is preferable to achieve a good quality print. However, a contact angle toward 0° will have high wettability, which means it will spread out easily after printing process. This solution is good to cover large area printing. In addition, the contact angle toward 90° will have less spreading issue and higher possibility of holding any structure.

Surface tension and contact angle has been measured using a KRUSS Drop Shape Analyzer (DSA 1000) (Gulf Coast Region, TX 78373, USA). The contact angle and surface tension measurements were done at 20 °C and with a 2 mm syringe tip size. Two different substrates, Kapton and Silicon, were used to test for all five concentrations of PVA. In printed electronics, it is essential to check surface tension and contact angle for two reasons; (i) determine the bonding between solution and substrate material and (ii) determine if the solution can hold the structure. Figure 2 shows the contact angle of 2.5% *w*/*w* PVA on Kapton without and with plasma treatment. The contact angle is found to be 73.5°–74.4° for without O_2 plasma-treated surface and 129°–130° for the plasma-treated one. It is to be noted that the substrate wettability is useful when the contact angle is between 0°- 90°. So, the Kapton substrates do not need plasma treatment to achieve good quality printing with PVA.

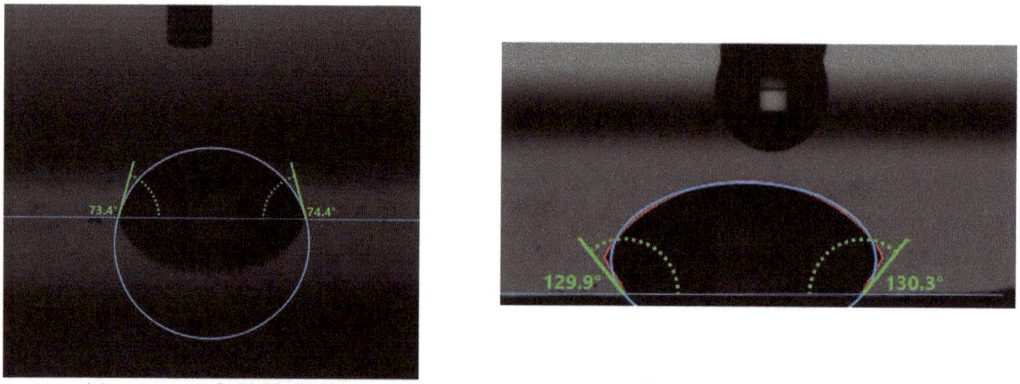

(**a**) contact angle 73.5°–74.4° (**b**) contact angle 129°–130°

Figure 2. The contact angle of 2.5% PVA on Kapton (**a**) without plasma treatment and (**b**) with plasma treatment.

Figure 3 shows the contact angle of 2.5% PVA on Silicon, and the angle is found to be 44.3°–44.4° without O_2 plasma-treated surface and 62.7°–69.6° for the plasma-treated one. From these two analyses, it is observed that the O_2 plasma improves the contact angle of the substrate significantly.

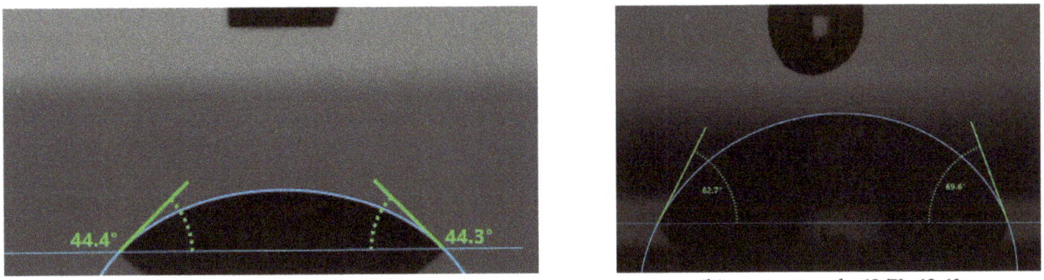

(**a**) contact angle 44.3°–44.4° (**b**) contact angle 62.7°–69.6°

Figure 3. The contact angle of 2.5% PVA on Silicon (**a**) without plasma treatment and (**b**) with plasma treatment.

Figure 4 shows the surface tension of 2.5% PVA on Kapton and Silicon. The surface tension is found to be 42.03 mN/m for Kapton and 32.51 mN/m for Silicon substrate. To achieve a good inkjet printing with any material, the surface tension has to be in between (25–75) mN/m and viscosity has to be in between (1–8)cP [44].

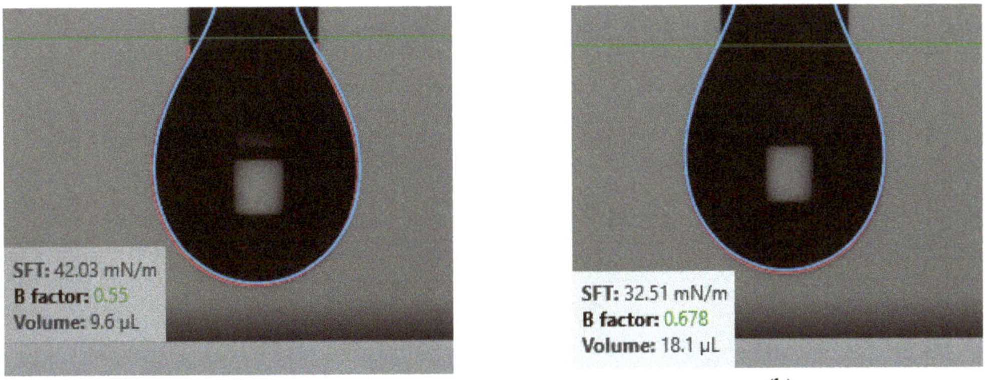

(**a**) (**b**)

Figure 4. The surface tension of 2.5% PVA on (**a**) Kapton and (**b**) Silicon.

Table 2 shows the summary of contact angles on Kapton and Silicon for all five concentrations of PVA. It is observed that both substrates show higher surface tension for higher concentrated or higher viscous PVA. In addition, the contact angle does not vary much for the Silicon substrate for all concentrated PVA while Kapton has the contact angle between 0°–90° without a plasma-treated surface, and above 100° for a plasma-treated one. It is possible to achieve good printing with both the substrates, only Kapton does not need plasma treatment.

Table 2. The summary of the contact angle on Kapton and Silicon without surface treatment.

PVA Concentration (%)	Contact Angle (Kapton)	Contact Angle (Silicon)	Surface Tension (Kapton) (mN/m)	Standard Deviation for Surface Tension (Kapton) (mN/m)	Surface Tension (Silicon) (mN/m)	Standard Deviation for Surface Tension (Silicon) (mN/m)
2.5	73.5°–74.4°	44.3°–44.4°	42.0	0.6	32.5	0.6
4.5	74.5°–75°	44.3°–45.2°	45.5	0.3	39	2
6.5	77.6°–78.5°	45.3°–45.5°	46	1	44.9	0.9
8.5	80°–82.3°	45.4°–45.9°	48.8	0.8	47	2
10.5	85.2°–85.5°	46.3°–46.6°	50.4	0.8	49.6	0.9

Aerosol Jet Printing of PVA: An aerosol jet AJ 300 system from the company Optomec was used to print a 5 mm by 5 mm pattern. The reason for printing a large pattern is to show the print-quality for large scale printing area. In this work, two different printing methods have been adopted to print a large area with AJP and small area with IJP. Furthermore, by developing an appropriate ink waveform and printing parameters, it is also possible to achieve both large and small area printing with both the deposition system. The aerosol jet system has two different atomizers called ultrasonic atomizer and pneumatic atomizer. The ultrasonic atomizer is used for the solutions that has a viscosity range of (1–5) cP, and the pneumatic atomizer is used for all of the kinds of solution that have a viscosity from (1–1000) cP. This article attempts to develop appropriate printing parameters and recipe to print with ultrasonic atomizer.

During the ultrasonic atomization, the solution is agitated by a pressure wave of appropriate frequency and amplitude. When a suitable pressure wave is applied, a capillary wave structure forms at the liquid interface with a desire aerosol droplet dimension of 3 μm to 5 μm. This capillary wave looks like a standing wave, where crests of these waves break off to form the aerosol droplets. Different sizes of nozzles (e.g., 100 μm, 150 μm, 200 μm, 250 μm, and 300 μm) can be used with AJ 300 system. The polymer chains are tends to amalgamate, which is why, the 200 μm nozzle is used to deposit the film. The other printing parameters used during the deposition process are: printing speed 1 mm/sec, sheath gas flow 5 sccm, atomizer gas flow 50 sccm, atomizer current 627 mA, platen temperature 45 °C, and curing temperature 80–90 °C for 15 min. Table 3 shows the summary of the printing parameters for PVA deposition with ultrasonic atomizer.

Table 3. The summary of the printing parameters for PVA deposition with ultrasonic atomizer.

Atomizer	Nozzle Size (μm)	Printing Speed (mm/sec)	Sheath Gas Flow (SCCM)	Atomizer Gas Flow (SCCM)	Atomizer Current (mA)	Platen Temperature (°C)	Curing Temperature (°C)
Ultrasonic	200	1	5	50	627	45	80–90

A test pattern size of 5 mm by 5 mm was designed using AutoCAD 2019 (Autodesk, San Rafael, CA 94903, USA). The file was saved as .dxf format. Then, using manufacturer provided plug-in, VMTools, the .dxf file was converted to .prg file. The .prg file is the only file format that is supported by the AJ 300 deposition system. A multi-layer structure is printed on a Silicon substrate. Figure 5 shows the two layers and four layers of PVA printed patterns on the substrate. Although, the target length and width are 5 mm by 5 mm, however, after printing the pattern size is 5.62 mm by 5.68 mm, and 5.93 mm by 5.93 mm for two and four PVA layers, respectively. It is to be noted that, by appropriate pattern modification, it is possible to achieve the target pattern size. The maximum size of pattern that can be printed by an AJ 300 is 300 mm by 300 mm with the highest in-situ platen temperature of 120 °C. For large sized or multi-layer pattern printing, the ink spreading will increase along with time.

2.5% PVA was also printed on a Kapton substrate. Figure 6 shows the 5 mm by 5mm pattern printed on both Kapton and Silicon substrate. Both have three printed layers, but after printing, 6.17 mm by 6.14 mm was found for Kapton, and 5.83 mm by 5.83 mm was found for the Silicon substrate. So, it can be concluded that the Silicon substrate is superior to the Kapton substrate from the print-quality perspective.

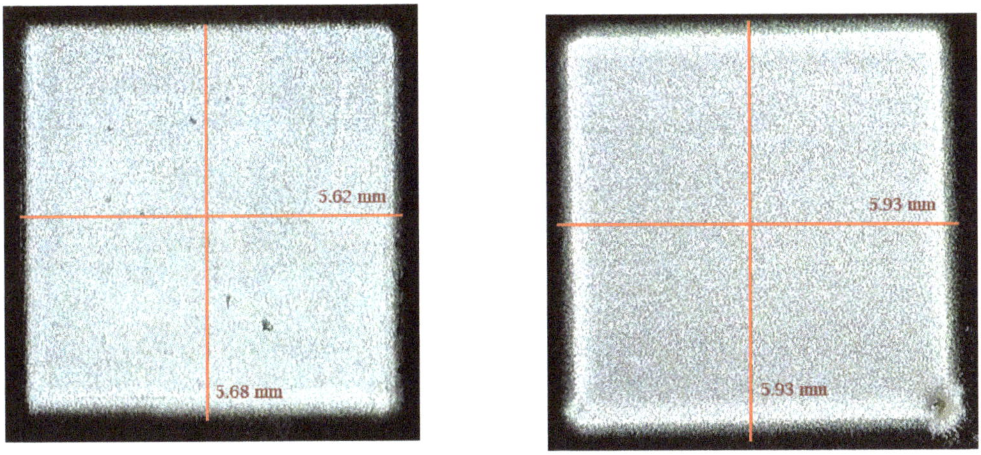

(**a**) (**b**)

Figure 5. 2.5% PVA printed on Silicon substrate: (**a**) Two printed layers; (**b**) four printed layers.

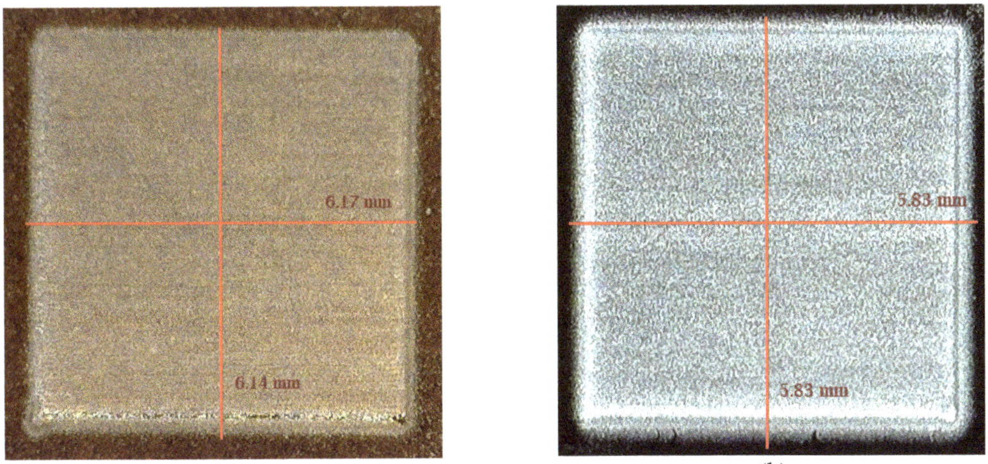

(**a**) (**b**)

Figure 6. 2.5% PVA printed on Kapton and Silicon. (**a**) 5 mm × 5 mm targeted pattern printed on Kapton; (**b**) 5 mm × 5 mm targeted pattern printed on Silicon.

Thickness Profile of Printed PVA: KLA Tencor D-300 (Milpitas, CA 95035, USA) profiler was used to measure the thickness and roughness profile of the printed patterns. Figure 7 shows the thickness and roughness profile of four printed layers of 2.5% PVA on the Silicon substrate.

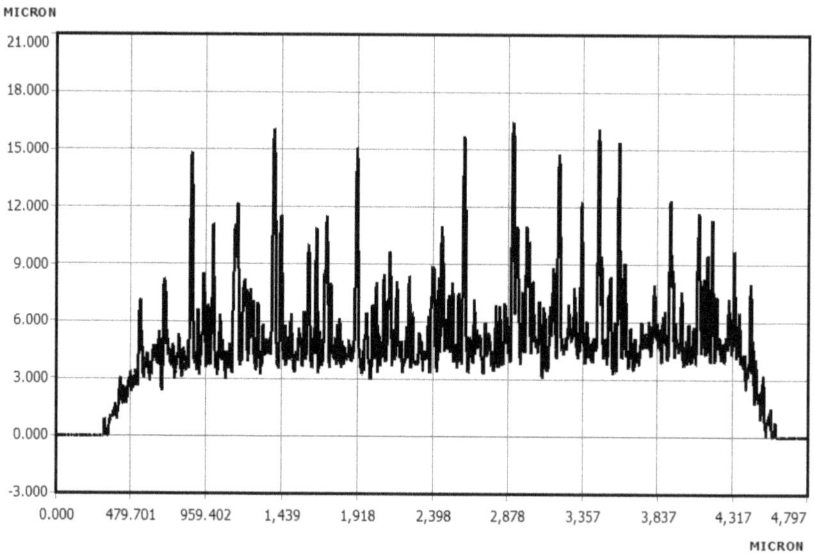

Figure 7. The thickness and roughness profile of 2.5%, and four printed layers of PVA.

The average thickness and roughness were found to be 5.7 μm and 1.8 μm, respectively. Table 4 shows the summary of thickness and roughness profiles for different printed layers on Silicon. It is observed from the table that the spreading and roughness both increased with the increment of printed layers.

Table 4. Thickness and roughness profiles for different printed layers of PVA.

Printed Layers (on Silicon)	Targeted Pattern (mm × mm)	Achieved Pattern (mm × mm)	Thickness (μm)	Roughness (μm)
1	5 × 5	5.44 × 5.44	1.8	0.45
2	5 × 5	5.62 × 5.68	2.53	0.7
3	5 × 5	5.83 × 5.83	3.73	0.86
4	5 × 5	5.93 × 5.93	5.68	1.8

Inkjet Printing of PVA: The Fujifilm Dimatix DMP-2800 (Santa Clara, CA 95050, USA) deposition system allows fluidic materials on 8 by 11-inch substrate with a maximum substrate height of 25 mm. The deposition head utilizes a disposable piezo inkjet cartridge. The platen of the system is vacuum controlled with an adjustable temperature up to 60 °C. The ink-dispensing mechanism by piezo inkjet cartridge head is activated by four different voltage phases, which can be adjusted up to 40 V to achieve good jetting. The voltage phases break down a solution into droplets to start printing/jetting process. It is very difficult to handle long chain polymers with inkjet printing because the highest voltage of DMP-2800 cannot break the long polymer chains to make it into droplets form. However, PVA is a short chain polymer, hence it is possible to break it's polymer chain with the voltage. Although, due to the amalgamation tendency, the nozzles tend to get clogged very frequently. PVA test print is carried out for two different types of pattern: For smaller area coverage and for larger area coverage, (i) 5000 μm × 150 μm line and (ii) 5000 μm × 5000 μm square, respectively.

Figure 8 shows the inkjet printed PVA on a Silicon substrate. Figure 8a is the single-layer printed line with an average thickness of 3.5 μm and roughness (Ra) of 0.84 μm (Figure 8b). Additionally, Figure 8c shows the multilayer printed PVA with an average thickness of 5.5 μm and roughness (Ra) of 1.14 μm (Figure 8d). It is observed that the printed PVA has perfect coverage for the smaller area with low surface roughness, and it has some dewetting and increased roughness issues for larger area coverage.

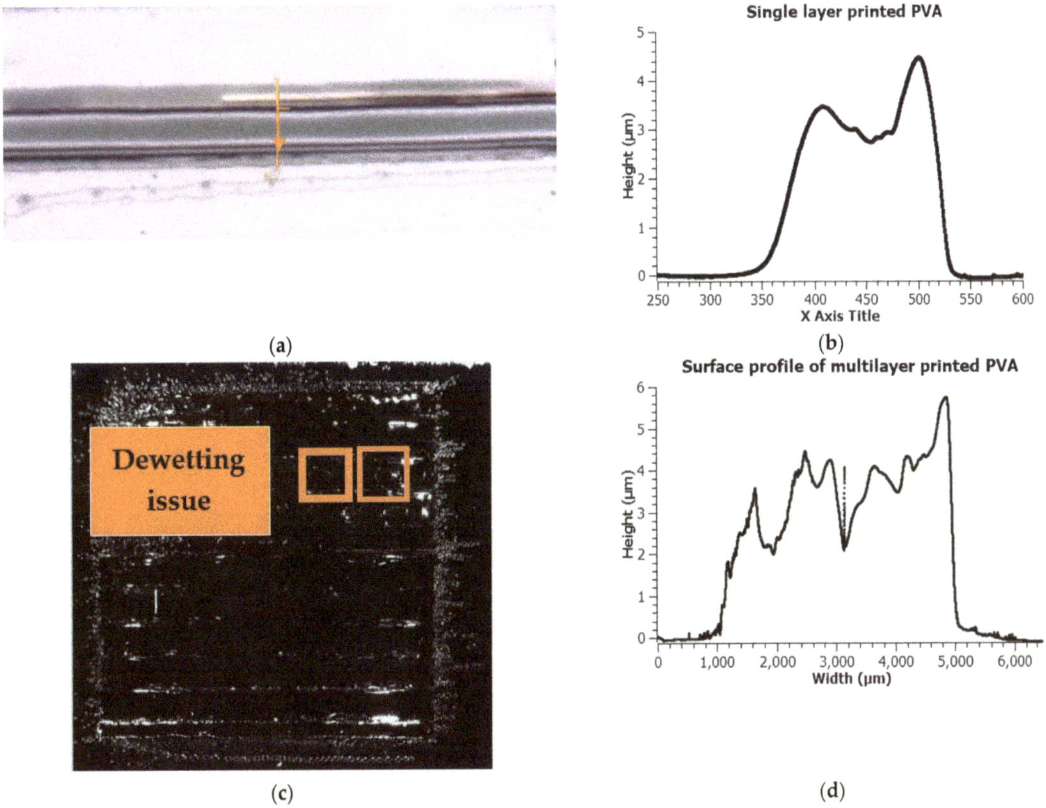

Figure 8. Printed PVA on Silicon substrate (**a**) (ii) 5000 μm × 150 μm line. (**b**) Thickness of the line; (**c**) 5000 μm × 5000 μm square. (**d**) Thickness of the square.

The recipe used to start jetting the PVA is the default Fujifilm Dimatix jetting waveform. Figure 9 shows the jetting condition of PVA solution with the drop watcher of DMP-2800. The nozzles showed here are from 5 to 10 with a jetting velocity of about 4 m/sec. In Table 5, it is seen that the thickness of the single layer deposited PVA is about 3.5 μm. We have used only a single printed layer of 2.5% PVA solution to create a suspended structure for MEMS device application. Typically for a standard MEMS device, the distance between the top and bottom electrode should be (0.5–4) μm. Ideally, the lower the gap between top and bottom electrode, the lower the actuation voltage. In this work, with 2.5% PVA solution one printed layer of PVA gives about 3.5 μm thickness, and that is why the print is limited to single layer only. For the concentrations > 2.5%, the thickness is assumed to be > 4 μm. Thus, higher concentration of PVA is not taken into consideration to create the suspended structure for the proposed application.

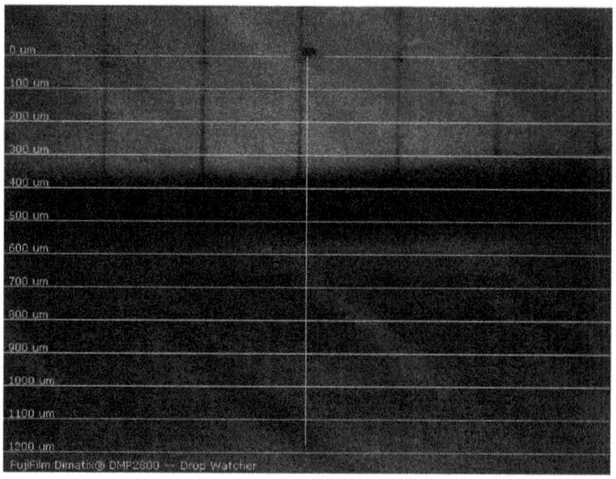

Figure 9. Jetting condition of PVA from the nozzles 5 to 10.

Table 5. Summary table of characterized printed PVA.

Targeted Pattern	Achieved	Printed Layers	Thickness (μm)	Standard Deviation for Thickness (μm)
5000 μm × 150 μm line	5250 μm × 190 μm line	1	3.5	0.8
5000 μm × 5000 μm square	5255 μm × 4060 μm square	2	5.5	1

The other parameters used to achieve good jetting is presented in Table 6. The parameter values for nozzle size, jetting frequency, nozzle voltage, platen temperature, cartridge temperature, and meniscus is 10 pL, 2 kHz, 40 V, 50 °C, 30 °C, and 3, respectively.

Table 6. Additional inkjet printing parameters to print PVA.

Nozzle Size	Jetting Frequency	Nozzle Voltage	Platen Temperature	Cartridge Temperature	Meniscus
10 pL	2 kHz	40 V	50 °C	32 °C	3

PVA as Sacrificial Material for MEMS Device Fabrication: Figure 10 shows the schematic of a typical double clamped MEMS device where (a) shows the device schematic with sacrificial layer and (b) is the final device after removing the sacrificial layer. The sacrificial layer is an essential part of MEMS device fabrication and it is not a part of the final device. This layer is used to make a suspended MEMS cantilever structure, then removed by an etching process. In this work, printed PVA is introduced as a good alternative to other sacrificial materials to make a suspended structure for the MEMS cantilever application.

It is to be noted that, this work does not present a complete MEMS device structure. It only shows how PVA can be used as a material that can help create a suspended structure using a safe and simple etching process.

Figure 11a shows the top view and the schematic that is used to make a suspended structure. The cross-section region is illustrated in Figure 11b, while Figure 11c,d show the elaborated cross-section view of the structure before and after the removal of sacrificial layer. It is important to highlight that this structure is not a complete MEMS structure; here the printed PVA is portrayed as a sacrificial material which can be used in MEMS device fabrication in future work.

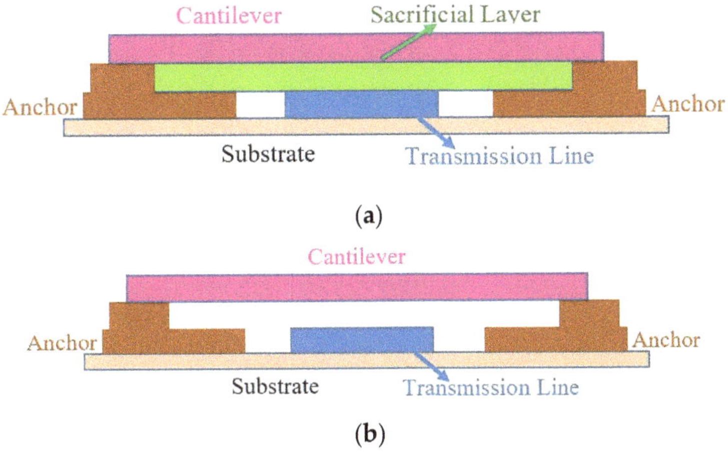

Figure 10. The schematic of a typical double clamped micro-electro-mechanical-system (MEMS) device structure. (**a**) MEMS device structure before the removal of sacrificial layer; (**b**) MEMS device structure after the removal of sacrificial layer.

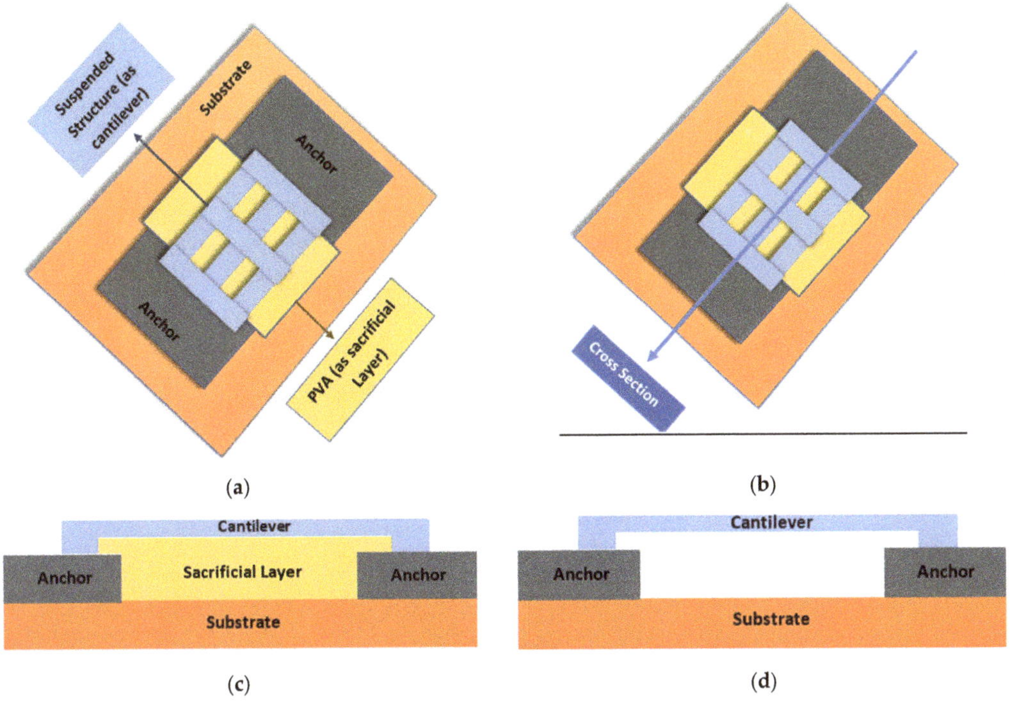

Figure 11. Schematic of a suspended structure where printed PVA used as a sacrificial material. (**a**) schematic and top view of the suspended structure; (**b**) the region from where the cross-section was taken, (**c**) structure before the final removal of the PVA sacrificial layer and (**d**)final suspended structure after the removal of sacrificial layer.

3. Results

Etching of PVA Layer: A DMP-2800 deposition system can handle materials with viscosity up to 12 cP, which is why 2.5% PVA solution is chosen to deposit sacrificial layer. A single layer of PVA with a thickness of 3.5 μm is deposited in between a top mesh suspended structure (as cantilever) and bottom anchors as clamper. Later, the PVA layer is removed using only hot DI water (80 °C–90 °C) without disrupting Ag layers. Table 7 shows the etching time and rate of 2.5% PVA layers with different thicknesses. It is to be noted that, with the increase of thickness, etching time increased and etching rate decreased due to the creation of strong polymeric bonding.

Table 7. Etching rate for PVA layers of different thicknesses.

PVA Layer Thickness (μm)	Etching Solution	Etching Time	Etching Rate
1	DI water with 110 °C	40 s	0.025 μm/s
4.5	DI water with 110 °C	~7 min	0.714 μm/min
9.5	DI water with 110 °C	~15 min	0.66 μm/min
15	DI water with 110 °C	~25 min	0.6 μm/min
23	DI water with 110 °C	~37 min	0.55 μm/min

PVA is used as a sacrificial material to create a free-standing mesh structure. Figure 12 shows the printed mesh structure (a) before and (b) after the removal of middle PVA layer. The FEI Helios NanoLab 400 Dual Beam system (Hillsboro, OR 97124-5793, USA) with a fully digital Field Emission Scanning Electron Microscope (FE SEM) is used to take the imaging. An Everhart–Thornley detector with a voltage, current, and magnification of 8.00 kV, 0.17 nA, and 122 x, respectively, is used during imaging. A higher current and exciting voltage would improve the imaging quality; however, a low exciting current is used since PVA is a non-conductive polymer and higher current will destroy the sample.

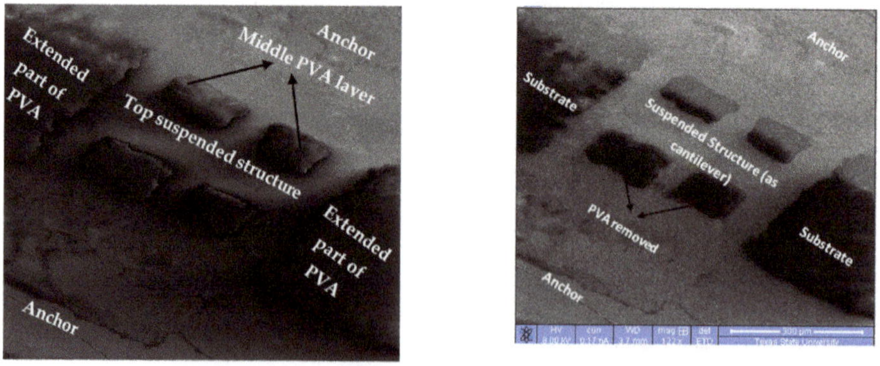

Figure 12. Mesh Cantilever structure (**a**) before PVA removal and (**b**) after PVA removal.

The SEM image shows a mesh double clamped suspended cantilever structure where printed PVA is sandwiched between top mesh structure and bottom double clamped anchors. In Figure 12a, the printed PVA is between mesh structure and anchors. In this image, PVA is deposited between top mesh structure (where mesh structure deposited by solution processible silver ink) and bottom anchor structure (anchors also deposited by silver ink). The substrate is not visible at all, because extended part of deposited PVA is covering the substrate region. It is important to emphasize that the four windows are not precise enough, not because of the non-accurate deposition of PVA layer, but because the windows are designed too small (50 μm by 50 μm) and it is very challenging to fabricate these small features with printing technique. The spreading of printed silver is responsible for the non-uniform squares. On the other hand, Figure 12b shows the suspended and free-standing mesh structure after the fully removal of PVA sacrificial layer. Additionally, in this image, substrate is also fully visible as the PVA layer is completely gone due to water etching.

It is to be noted that the mechanical structure or the cantilever of MEMS devices must be free-standing as this will be a movable part. PMMA is the most widely used sacrificial material for traditional MEMS fabrication techniques. However, PMMA is a long-chain polymer, and it is not suitable for printing technology. PVA is a short-chain polymer with low molecular weight; thus, PVA is used and suggested as a good alternative of PMMA as sacrificial material for printing technology.

PVA as Dielectric Material for Transistor Fabrication: A Hermes Mercury Probe 802–150 (Materials Development Corporation, Chatsworth, CA 91311, USA) is used to measure the dataset. The equipment measures capacitance of thin film on Si wafer using a mercury dot to form another electrode. It creates a MOS capacitor. From there, the dielectric constant is calculated using the following formula,

$$C = \epsilon_0 \epsilon_r Area / Thickness \tag{1}$$

where ϵ_r is the dielectric constant.

The printed thin film of PVA on Si wafer acts as metal-oxide-semiconductor (MOS) capacitor when attached to Hermes equipment's probe and back contact. When the DC voltage sweep is applied during CV measurement at low frequency, the MOS capacitor has depletion, accumulation, and an inversion zone.

The dielectric constant of printed PVA has been calculated from capacitance vs. DC bias voltage measurement. Figure 13 shows the capacitance vs. DC bias voltage at 100 kHz. The DC bias voltage is applied from −20 V to 20 V, and the highest capacitance measured is 4×10^{-10} F. From that measurement, the dielectric constant of printed PVA is calculated to be 168.

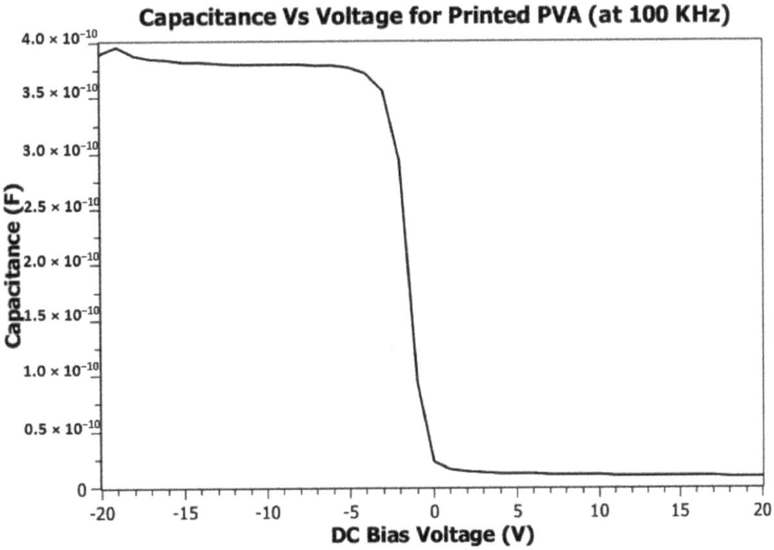

Figure 13. Capacitance Vs. bias voltage measurement.

4. Conclusions

We have demonstrated the synthesis, characterization, and application of printable PVA by Aerosol Jet Printing (AJP) and Inkjet Printing (IJP) technologies. Five different percentage of PVA of 2.5% *w*/*w*, 4.5% *w*/*w*, 6.5% *w*/*w*, 8.5% *w*/*w*, and 10.5% *w*/*w* aqueous solution were formulated and characterized with respect to viscosity, contact angle, surface tension, and printability. We have demonstrated that the PVA solution can be used to produce a multi-layer structure with patterning by AJP and IJP printing technology. This polymer is also promising to use as a sacrificial material for MEMS device fabrication. Additionally, the dielectric constant of printed PVA film is 168 at 100 kHz, which means it can be an excellent candidate material for printed or traditional transistor fabrication.

Author Contributions: Conceptualization, M.Y.C. and M.A.M.; methodology, M.A.M.; software, M.A.M. and M.Y.C.; validation, M.A.M., M.Y.C., C.Q.H. and B.M.; formal analysis, M.A.M.; investigation, M.A.M.; resources, M.Y.C.; data curation, M.A.M., C.Q.H. and B.M.; writing—original draft preparation, M.A.M.; writing—review and editing, M.Y.C., M.A.M., C.Q.H. and B.M.; visualization, M.Y.C. and M.A.M.; supervision, M.Y.C.; project administration, M.Y.C.; funding acquisition, M.Y.C. All authors have read and agreed to the published version of the manuscript.

Funding: This research received no external funding.

Conflicts of Interest: The authors declare no conflict of interest.

References

1. Premraj, R.; Doble, M. Biodegradation of polymers. *Indian J. Biotechnol.* **2005**, *4*, 186–193.
2. Qian, D.; Du, G.; Chen, J. Isolation and culture characterization of a new polyvinyl alcohol-degrading strain: Penicillium sp. WSH02-21. *World J. Microbiol. Biotechnol.* **2004**, *20*, 587–591. [CrossRef]
3. Khanna, P.K.; Singh, N.; Charan, S.; Subbarao, V.V.V.S.; Gokhale, R.; Mulik, U.P. Synthesis and characterization of Ag/PVA nanocomposite by chemical reduction method. *Mater. Chem. Phys.* **2005**, *93*, 117–121. [CrossRef]
4. Bachtsi, A.R.; Kiparissides, C. Synthesis and release studies of oil-containing poly(vinyl alcohol) microcapsules prepared by coacervation. *J. Control. Release* **1996**, *38*, 49–58. [CrossRef]
5. Yamagata, S.; Handa, H.; Taki, W.; Yonekawa, Y.; Ikada, Y.; Iwata, H. Nonsuture microvascular anastomosis. Experimental arterial end-to-end anastomosis using plastic adhesive and a soluble PVA tube. *Neurol. Surg.* **1979**, *7*, 1067–1073.
6. Duta-Capra, A.; Kadas-Iluna, I. Structural modifications in polyvinylalcohol. *Mater. Sci. Forum* **1994**, *166*, 761–766. [CrossRef]
7. Kenawy, E.R.; Kamoun, E.A.; Eldin, M.S.M.; El-Meligy, M.A. Physically crosslinked poly(vinyl alcohol)-hydroxyethyl starch blend hydrogel membranes: Synthesis and characterization for biomedical applications. *Arab. J. Chem.* **2014**, *7*, 372–380. [CrossRef]
8. Jiang, S.; Liu, S.; Feng, W. PVA hydrogel properties for biomedical application. *J. Mech. Behav. Biomed. Mater.* **2011**, *4*, 1228–1233. [CrossRef] [PubMed]
9. Yang, D.; Li, Y.; Nie, J. Preparation of gelatin/PVA nanofibers and their potential application in controlled release of drugs. *Carbohydr. Polym.* **2007**, *69*, 538–543. [CrossRef]
10. Thong, C.C.; Teo, D.C.L.; Ng, C.K. Application of polyvinyl alcohol (PVA) in cement-based composite materials: A review of its engineering properties and microstructure behavior. *Constr. Build. Mater.* **2016**, *107*, 172–180. [CrossRef]
11. Kamoun, E.A.; Kenawy, E.R.S.; Tamer, T.M.; El-Meligy, M.A.; Eldin, M.S.M. Poly(vinyl alcohol)-alginate physically crosslinked hydrogel membranes for wound dressing applications: Characterization and bio-evaluation. *Arab. J. Chem.* **2015**, *8*, 38–47. [CrossRef]
12. Wei, L.; Cai, C.; Lin, J.; Chen, T. Dual-drug delivery system based on hydrogel/micelle composites. *Biomaterials* **2009**, *30*, 2606–2613. [CrossRef]
13. Li, J.K.; Wang, N.; Wu, X.S. Poly(vinyl alcohol) nanoparticles prepared by freezing-thawing process for protein/peptide drug delivery. *J. Control. Release* **1998**, *56*, 117–126. [CrossRef]
14. Hua, S.; Ma, H.; Li, X.; Yang, H.; Wang, A. pH-sensitive sodium alginate/poly(vinyl alcohol) hydrogel beads prepared by combined Ca2+ crosslinking and freeze-thawing cycles for controlled release of diclofenac sodium. *Int. J. Biol. Macromol.* **2010**, *46*, 517–523. [CrossRef]
15. Taepaiboon, P.; Rungsardthong, U.; Supaphol, P. Drug-loaded electrospun mats of poly(vinyl alcohol) fibres and their release characteristics of four model drugs. *Nanotechnology* **2006**, *17*, 2317. [CrossRef]
16. Hashizume, M.; Kunitake, T. Preparation of Self-Supporting Ultrathin Films of Titania by Spin Coating. *Langmuir* **2003**, *19*, 10172–10178. [CrossRef]
17. Pérez, N.; Bartolomé, F.; García, L.M.; Bartolomé, J.; Morales, M.P.; Serna, C.J.; Labarta, A.; Batlle, X. Nanostructural origin of the spin and orbital contribution to the magnetic moment in Fe3-x O4 magnetite nanoparticles. *Appl. Phys. Lett.* **2009**, *94*, 093108. [CrossRef]

18. Rahimi, M.; Kameli, P.; Ranjbar, M.; Salamati, H. The effect of polyvinyl alcohol (PVA) coating on structural, magnetic properties and spin dynamics of Ni0.3Zn0.7Fe 2O4 ferrite nanoparticles. *J. Magn. Magn. Mater.* **2013**, *347*, 139–145. [CrossRef]
19. University of Louisville. Spin Coating Theory. 2013. Available online: https://louisville.edu/micronano/files/documents/standard-operating-procedures/SpinCoatingInfo.pdf/view (accessed on 24 December 2020).
20. Ding, I.K.; Melas-Kyriazi, J.; Cevey-Ha, N.L.; Chittibabu, K.G.; Zakeeruddin, S.M.; Grätzel, M.; McGehee, M.D. Deposition of hole-transport materials in solid-state dye-sensitized solar cells by doctor-blading. *Org. Electron.* **2010**, *11*, 1217–1222. [CrossRef]
21. Birnie, D.P. Spin Coating Technique. In *Sol-Gel Technologies for Glass Producers and Users*; Springer: Berlin, Germany, 2004.
22. Patra, S.; Young, V. A Review of 3D Printing Techniques and the Future in Biofabrication of Bioprinted Tissue. *Cell Biochem. Biophys.* **2016**, *74*, 93–98. [CrossRef]
23. Wang, Q.; Sun, J.; Yao, Q.; Ji, C.; Liu, J.; Zhu, Q. *3D Printing with Cellulose Materials*; Springer: Berlin, Germany; Cellulose: Madison, WI, USA, 2018.
24. Al-Halhouli, A.; Qitouqa, H.; Alashqar, A.; Abu-Khalaf, J. Inkjet printing for the fabrication of flexible/stretchable wearable electronic devices and sensors. *Sens. Rev.* **2018**, *38*, 438–452. [CrossRef]
25. Genina, N.; Kolakovic, R.; Palo, M.; Fors, D.; Juvonen, H.; Ihalainen, P.; Peltonen, J.; Sandler, N. Fabrication of printed drug-delivery systems. In *International Conference on Digital Printing Technologies*; Society for Imaging Science and Technology: Scottsdale, AZ, USA, 2013.
26. Delaney, J.T.; Liberski, A.R.; Perelaer, J.; Schubert, U.S. Reactive inkjet printing of calcium alginate hydrogel porogens—A new strategy to open-pore structured matrices with controlled geometry. *Soft Matter* **2010**, *6*, 866–869. [CrossRef]
27. Ning, H.; Tao, R.; Fang, Z.; Cai, W.; Chen, J.; Zhou, Y.; Zhu, Z.; Zheng, Z.; Yao, R.; Xu, M.; et al. Direct patterning of silver electrodes with 2.4 μm channel length by piezoelectric inkjet printing. *J. Colloid Interface Sci.* **2017**, *487*, 68–72. [CrossRef] [PubMed]
28. Kalsoom, U.; Nesterenko, P.N.; Paull, B. Recent developments in 3D printable composite materials. *RSC Adv.* **2016**, *6*, 60355–60371. [CrossRef]
29. Tao, H.; Marelli, B.; Yang, M.; An, B.; Onses, M.S.; Rogers, J.A.; Kaplan, D.L.; Omenetto, F.G. Inkjet Printing of Regenerated Silk Fibroin: From Printable Forms to Printable Functions. *Adv. Mater.* **2015**, *27*, 4273–4279. [CrossRef]
30. Rim, Y.S.; Bae, S.H.; Chen, H.; de Marco, N.; Yang, Y. Recent Progress in Materials and Devices toward Printable and Flexible Sensors. *Adv. Mater.* **2016**, *28*, 4415–4440. [CrossRef] [PubMed]
31. Truby, R.L.; Lewis, J.A. Printing soft matter in three dimensions. *Nature* **2016**, *540*, 371–378. [CrossRef] [PubMed]
32. Jewel, M.U.; Mahmud, M.S.; Monne, M.A.; Zakhidov, A.; Chen, M.Y. Low temperature atomic layer deposition of zirconium oxide for inkjet printed transistor applications. *RSC Adv.* **2019**, *9*, 1841–1848. [CrossRef]
33. Monne, M.A.; Zaid, A.; Mia, D.; Khanal, J.; Zakhidov, A.; Chen, M.Y. Anti-Reflective Coating for Flexible Devices Using Plasma Enhanced Chemical Vapor Deposition Technique. In Proceedings of the 2018 International Conference on Optical MEMS and Nanophotonics, Lausanne, Switzerland, 29 July–2 August 2018.
34. Monne, M.A.; Grubb, P.M.; Stern, H.; Subbaraman, H.; Chen, R.T.; Chen, M.Y. Inkjet-Printed Graphene-Based 1 × 2 Phased Array Antenna. *Micromachines* **2020**, *11*, 863. [CrossRef] [PubMed]
35. Monne, M.A.; Jewel, M.; Wang, Z.; Chen, M.Y. Graphene based 3D printed single patch antenna. *SPIE Proc. Low-Dimens. Mater. Devices* **2018**, *10725*, 107250C.
36. Monne, M.A.; Enuka, E.; Wang, Z.; Chen, M.Y. Inkjet printed graphene-based field-effect transistors on flexible substrate. *SPIE Proc. Low-Dimens. Mater. Devices* **2017**, *10349*, 1034905.
37. Monne, M.A.; Lan, X.; Zhang, C.; Chen, M.Y. Inkjet-Printed Flexible MEMS Switches for Phased-Array Antennas. *Int. J. Antennas Propag.* **2018**, *2018*, 4517848. [CrossRef]
38. Gaevski, M.; Mollah, S.; Hussain, K.; Letton, J.; Mamun, A.; Jewel, M.U.; Chandrashekhar, M.V.S.; Simin, G.; Khan, A. Ultrawide bandgap Al xGa1-xN channel heterostructure field transistors with drain currents exceeding 1.3 A mm-1. *Appl. Phys. Express* **2020**, *13*, 094002. [CrossRef]
39. Tatikonda, A.; Jokinen, V.P.; Evard, H.; Franssila, S. Sacrificial layer technique for releasing metallized multilayer SU-8 devices. *Micromachines* **2018**, *9*, 673. [CrossRef]
40. Blachowicz, T.; Ehrmann, A. 3D printed MEMS technology-recent developments and applications. *Micromachines* **2020**, *11*, 434. [CrossRef]
41. Monne, M.A.; Lan, X.; Chen, M.Y. Material Selection and Fabrication Processes for Flexible Conformal Antennas. *Int. J. Antennas Propag.* **2018**, *2018*, 9815631. [CrossRef]
42. Wang, Z.; Liu, Z.; Monne, M.A.; Wang, S.; Yu, Q.; Chen, M.Y. Interfacial separation and electrochemical delamination of CVD grown multilayer graphene for recyclable use of Cu powder. *RSC Adv.* **2016**, *6*, 24865–24870. [CrossRef]
43. Sone, J.; Murakami, M.; Tatami, A. Fundamental study for a graphite-based microelectromechanical system. *Micromachines* **2018**, *9*, 64. [CrossRef] [PubMed]
44. Krainer, S.; Smit, C.; Hirn, U. The effect of viscosity and surface tension on inkjet printed picoliter dots. *RSC Adv.* **2019**, *9*, 31708–31719. [CrossRef]

micromachines

MDPI

Article

Thermopneumatic Soft Micro Bellows Actuator for Standalone Operation

Seongbeom Ahn, Woojun Jung, Kyungho Ko, Yeongchan Lee, Chanju Lee and Yongha Hwang *

Department of Electro-Mechanical Systems Engineering, Korea University, Sejong 30019, Korea; ahnsb98@korea.ac.kr (S.A.); wjkst2010@korea.ac.kr (W.J.); gokyungho123@korea.ac.kr (K.K.); liws214@korea.ac.kr (Y.L.); cjl1226@korea.ac.kr (C.L.)
* Correspondence: hwangyongha@korea.ac.kr

Abstract: Typical pneumatic soft micro actuators can be manufactured without using heavy driving components such as pumps and power supplies by adopting an independent battery-powered mechanism. In this study, a thermopneumatically operated soft micro bellows actuator was manufactured, and the standalone operation of the actuator was experimentally validated. Thermopneumatic actuation is based on heating a sealed cavity inside the elastomer of the actuator to raise the pressure, leading to deflection of the elastomer. The bellows actuator was fabricated by casting polydimethylsiloxane (PDMS) using the 3D-printed soluble mold technique to prevent leakage, which is inherent in conventional soft lithography due to the bonding of individual layers. The heater, manufactured separately using winding copper wire, was inserted into the cavity of the bellows actuator, which together formed the thermopneumatic actuator. The 3D coil heater and bellows allowed immediate heat transfer and free movement in the intended direction, which is unachievable for conventional microfabrication. The fabricated actuator produced a stroke of 2184 μm, equivalent to 62% of the body, and exerted a force of 90.2 mN at a voltage of 0.55 V. A system in which the thermopneumatic actuator was driven by alkaline batteries and a control circuit also demonstrated a repetitive standalone operation.

Keywords: thermopneumatic; micro actuator; soft bellows actuator; independent actuation; polydimethylsiloxane; 3D printing; standalone

Citation: Ahn, S.; Jung, W.; Ko, K.; Lee, Y.; Lee, C.; Hwang, Y. Thermopneumatic Soft Micro Bellows Actuator for Standalone Operation. *Micromachines* **2021**, *12*, 46. https://doi.org/10.3390/mi12010046

Received: 30 November 2020
Accepted: 29 December 2020
Published: 1 January 2021

1. Introduction

Soft actuators are manufactured from soft materials with high compliance and degrees of freedom to handle unpredictable and dynamic tasks in unstructured environments [1–3]. The introduction of soft materials such as polydimethylsiloxane (PDMS) [4], polyurethane (PU) [5], and thermoplastic polyurethane (TPU) [6], which have the advantages of elasticity, easy processing (e.g., injection or casting), and low cost, inherently provides greater flexibility in comparison to the typical robot, which is made of a rigid material [7]. Fang et al. produced an assistive device that helped knee rehabilitation training using TPU fabric. Calderón et al. developed a soft robot composed of a silicone elastomer (Ecoflex 00-50) by mimicking an earthworm [8,9]. Wang et al. reported a soft grasping robot with a hinge using PDMS. In particular, PDMS has been used in biomedical applications that include lab-on-a-chip that deals with biological specifications and wearable devices attached to the skin, thereby proving its biocompatibility over the past decades [10–13].

Various driving methods have been developed for soft actuators, which include pneumatic pressure, cable-driven tendon, and electroactive polymer (EAP) [14]. Pneumatic actuation applies air pressure to inflate a membrane with a low Young's modulus; therefore, it requires an external drive source such as a compressor, pump, or syringe [15,16]. The cable-driven actuation produces a bend in the actuator when pulling the bendable cable embedded in the soft actuator by an external motor [17]. EAP is a soft material that generates mechanical deformation when stimulated by an electric field. However, to produce a noticeable deformation, a high voltage of several kV is required, for which a bulky power supply is required [18]. Thermopneumatic soft actuators are being developed that increase pressure by heating the internal air or fluid with built-in heaters, thereby switching to mechanical strokes [19]. Here, the thermopneumatic actuators can be configured by heaters and power supplies so that they offer favorable alternatives for miniaturization and standalone operation to relieve the payload without the need for additional devices for driving.

In this study, the micro thermopneumatic bellows actuators with standalone drives were designed, constructed, and evaluated by miniaturizing the 3D bellows actuators developed primarily for pneumatic driving and inserting 3D heaters for low voltage operation. The heater supplies heat energy by using Joule heating when the current passes through a conductive wire. The thermal energy supplied to the sealed internal air of the bellows actuator increases pressure and exerts force from the inside of the bellows actuator, generating movement in the intended direction as the folded bellows expands.

To manufacture the micro bellows actuator, the 3D-printed soluble mold technique was used to realize the fully-3D micro body beyond the limited 3D (2.5D) [20] structure of the conventional soft lithography technique [21,22]. Because the process of stacking separate thin structures is excluded, this fabrication technique, which uses sacrificial molds printed with build and support materials to cast liquid polymer, has the ability to minimize unnecessary bonding while also achieving the desired 3D structure.

It is desirable that the microheater takes a 3D shape so that it can extend the heating area to effectively heat the inner space of the micro bellows actuators. Therefore, the thermopneumatic soft micro bellows actuator was implemented by inserting a heater manufactured by winding a copper wire in the form of a 3D coil having a low resistivity for low-voltage driving. To analyze the thermodynamic and mechanical responses of the actuator, the temperature of the internal cavity was estimated by measuring the resistance change of the heater. The fabricated thermopneumatic bellows actuator was operated by an electric current to the heater using a direct current (DC) power supply, and the displacement and force were measured according to the rise in temperature of the sealed cavity. Finally, the standalone operation of the proposed thermopneumatic bellows actuator was demonstrated using an embedded control circuit driven by batteries.

2. Design and Fabrication

2.1. Design

To fully describe the design and operational principle of the proposed device, Figure 1 shows the structure of the thermopneumatic bellows actuator used in this work with the design parameters. Among the design parameters of the cross-section, as shown in Figure 1a, the bellows angle (θ) was determined to be 20° with the largest displacement ratio estimated by a numerical simulation performed by varying the bellows angle between 10° and 30°, as shown in the previous study [23]. The bellows thickness (t) for thermopneumatic actuator was 250 μm, considering the empirical limitations of the 3D printing and casting process. Polyamide-imide enameled copper wire of a diameter (d) of 200 μm was used for coil heaters. The coil heater diameter (D) was set to 1000 μm by the wire diameter and winding method of the copper wire, which is discussed in Section 2.2. Finally, the bellows inner diameter (w) was designed to be 1100 μm to prevent interference caused by contact with the heater during operation. The bellows actuator had five repeated bellows.

The working principle of the thermopneumatic actuation is that the internal air of the bellows actuator expands when heated by a heater, causing deflection of the bellows. The thermodynamic relation between the absolute temperature T, pressure P and volume V of a system is given by the well-known ideal gas equation:

$$PV = nRT \tag{1}$$

where n is the number of moles of the gas and R is the gas constant. The variation of volume and pressure by changing the temperature is then given by:

$$\frac{P_0 V_0}{T_0} = \frac{P_1 V_1}{T_1} \tag{2}$$

where V_1 and P_1 are the volume and pressure at a set temperature T_1, whereas V_0 and P_0 are the volume and pressure at room temperature T_0, which is 293.15 K. As the temperature increases, the volume of the air expands and the pressure of the air increases causing a deformation in the actuator. Therefore, the force F exerted by the bellows actuator is expressed as:

$$F = (P_1 - P_0)S = \left(\frac{P_0 V_0 T_1}{V_1 T_0} - P_0\right)S \tag{3}$$

where S is the effective area where the bellows head contacts the object. Since PDMS is hyperelastic, the pressure of the cavity according to the ideal gas law does not accurately account for the changes in PDMS features when it expands [24].

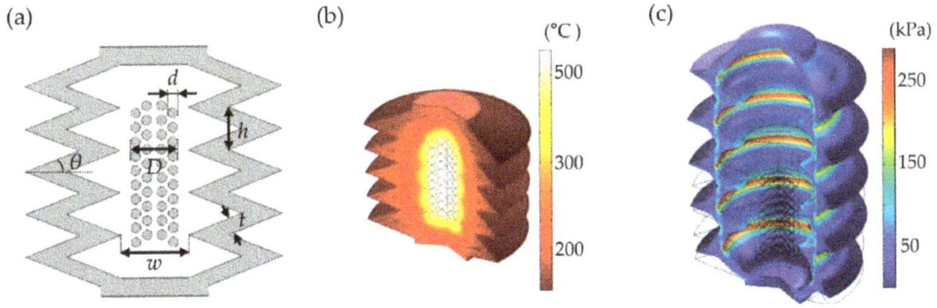

Figure 1. Concept of thermopneumatic bellows actuator. (**a**) Schematic cross-section showing the main design parameters of bellows angle (θ = 20°), bellows thickness (t = 250 µm), bellows height (h = 655 µm), wire diameter (d = 200 µm), coil heater diameter (D = 1000 µm), and bellows inner diameter (w = 1100 µm). The coil is double wound inside the bellows. (**b**) Temperature distribution of thermopneumatic bellows actuator using conjugate heat transfer module. The results are obtained by setting the power of the heater to 0.98 W, ambient temperature to 20 °C, and convection heat transfer coefficient to 5 W/(m^2·K). (**c**) Simulated deflection plots of the thermopneumatic bellows actuator using solid mechanics module. Wire frame outlines the resting actuator. Deflected actuator is shown with a color map depicting von Mises stress. (Red and blue indicate 280 kPa and 0 Pa, respectively.) The second-order Ogden model was used to evaluate the mechanical motion of PDMS bellows.

For numerical analysis of the motion of the designed thermopneumatic bellows actuator, COMSOL Multiphysics® (Stockholm, Sweden) was used to calculate the heat distribution of the entire actuator by the heater using the conjugate heat transfer module. The mechanical motion according to the thermal expansion of the air was sequentially calculated using the solid mechanical module given in Figure 1b–c. Using a simplified heater with an axisymmetric structure, the movement of the actuator was recorded as the maximum displacement when it reached the steady state which was 60 s after supplying power to the Joule heater through a time-dependent study. In the heat distribution analysis, the heat transfer coefficient of the convective heat flux generated on the bellows surface was set to 5 W/(m^2·K), and the ambient temperature was set to room temperature (20 °C).

When heated at high speed, PDMS was reported to be thermally decomposed at 530 °C [25]. When the heater applied 0.98 W of power to produce the highest heat within the limit of the PDMS deformation, the maximum temperature of the heater and the inner PDMS surface of the bellows were estimated to be 541.87 °C and 421.77 °C, respectively. In addition, the average temperature of the heated air of the entire cavity was 324.99 °C, resulting in a pressure of 12.62 kPa, which was consistent with previously reported work [23]. As shown in Figure 1c, the pressure generated by thermal expansion caused the soft PDMS bellows to unfold. In other words, the von Mises stress of 281.50 kPa was applied to the edge of the folded bellows, operating vertically against the direction in which the bellows were folded, and consequently resulting in the intended stroke of 1976 μm.

2.2. Fabrication

The thermopneumatic bellows actuator was fabricated using a technique of chemically removing the mold after casting PDMS, an elastomer [26]. As shown in Figure 2a, after designing with 3D CAD (Inventor®, Autodesk, Mill Valley, CA, USA) the mold was printed with a 3D printer (3Z STUDIO, Solidscape, Merrimack, NH, USA) using a build material and a support material that filled the empty space of the mold to support the structure during stacking of each printed layer. The printed output was dipped in a dewaxing solvent (BIOACT® VSO, Petroferm, Warren, NJ, USA) that flowed at 400 rpm at 50 °C and selectively removed the support material for 18 h, after which it was dried for 3 h in a vacuum chamber as shown in Figure 2b. There were 20 etch holes connecting the outer walls of the hexahedron mold with the edges of the inner bellows. The etch hole was designed to be 400 μm × 200 μm to help with the inflow of the dewaxing solvent but to not interfere with the appearance of the bellows. In Figure 2b, the central part was connected and fixed to the outer shell structure by supporting bridges (not represented in the figure) located at the bottom of the mold.

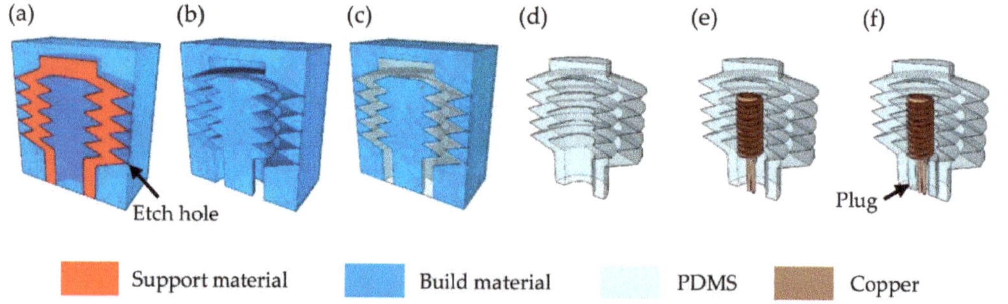

Figure 2. Fabrication process of cross-sectional thermopneumatic bellows actuator. (**a**) The actuator was printed through a 3D printer as a mold composed of build and support material. (**b**) Support material was dissolved through etch holes placed around the edge. (**c**) Polydimethylsiloxane prepolymer was poured and cured after vacuum. (**d**) PDMS bellows actuator after dissolving the mold. (**e**) 3D coil heater implanted inside the bellows actuator. (**f**) A thermopneumatic actuator after sealing with the PDMS plug.

A PDMS prepolymer (Sylgard 184, Dow Corning, Midland, MI, USA), which was mixed with a curing agent in a 15:1 weight ratio, was placed into a vacuum chamber to remove air bubbles, and then poured into the empty space of the mold. The mold filled with PDMS was degassed again inside the vacuum chamber for 2 h in order that PDMS could be fully filled inside the mold, and then it was cured in an oven at 75 °C for 24 h (Figure 2c). Next, the mold was placed in acetone for 3 h to dissolve, followed by cleaning with deionized (DI) water, after which a single body bellows actuator consisting of PDMS was formed (Figure 2d). It should be noted that the 3D-printed soluble mold technique fabricated 3D elements in a single body by excluding aligning and bonding steps of soft lithography, which were required for multiple separate layers. In addition, the use of 3D printers allowed faster prototyping than the typical microfabrication based on the soft lithography.

The Joule heater located inside the actuator was made of copper wire covered with enamel. Although several types of commercially available copper wires were tested, copper wires with a diameter of 0.15 mm or less, heated up to approximately 300 °C and broke. The Joule heater made of a 0.2 mm diameter copper wire maintained the mechanical stiffness for uniform heat transfer at the center of the bellows and retained a 3D coil shape without bending. It was also experimentally confirmed that the heater was reusable after the electric power was applied at the required temperature for the actuator. After the copper wire with a diameter of 0.2 mm was cut to a length of 50 mm to maintain a resistance of 50 ± 1 mΩ, the coil heater was formed by double winding the copper wire around a 0.2 mm diameter stainless steel pole, with an outer coil diameter of 1.0 mm. The manufactured heater was used as the actuator element only if the resistance measured using a milliohmmeter (MO-2012, LUTRON, Taipei, Taiwan) was uniformly within ±2%.

The 3D heater supplying the heat energy to the cavity inside the bellows actuator was inserted 5.0 mm from the bellows entrance and placed in the center of the cavity (Figure 2e). To tightly block the bellows entrance after the heater was inserted, a cylinder-shaped PDMS plug with an outer diameter equal to the inner diameter of the actuator entrance was manufactured separately and inserted into the bellows entrance. The plug had an oval center hole with a 0.4 mm long axis that allowed the heater to be connected externally. The liquid PDMS was covered with the gap between the actuator and the plug and cured in an oven at 75 °C for 30 min to seal the cavity of the actuator (Figure 2f).

The fabricated thermopneumatic bellows actuator using the 3D-printed soluble mold technique is shown in Figure 3. The micro thermopneumatic bellows actuator has a volume of 26.4 mm^3 and mass of 42 mg, which could operate as a thermopneumatic mechanism by applying low power at the level supplied by a battery. By cutting the fabricated actuator, as shown in Figure 3b into a plane perpendicular to the entrance, it was confirmed that the heater did not physically interfere with the bellows edges.

(a)

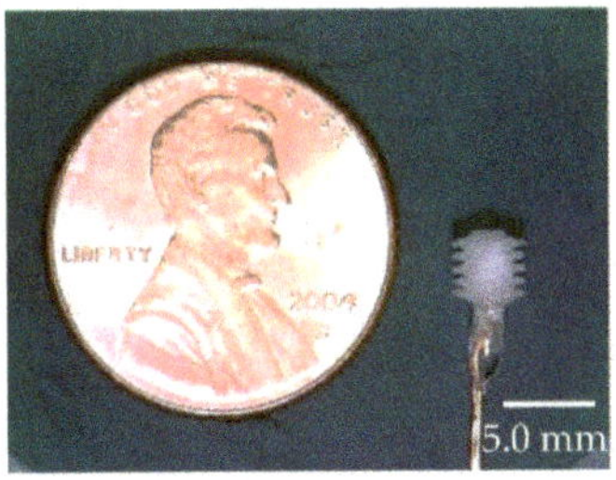

(b)

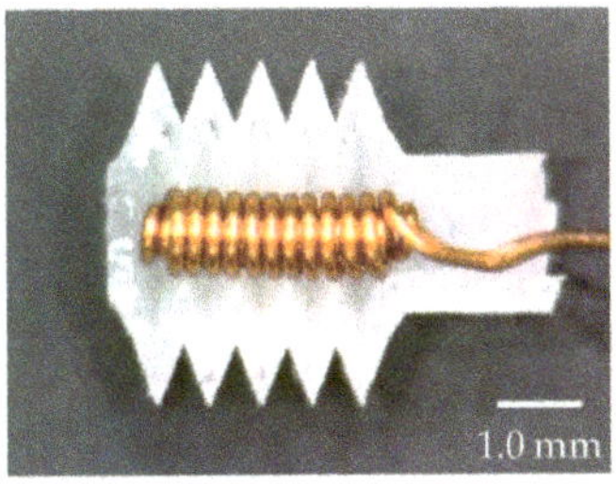

Figure 3. (**a**) Fabricated thermopneumatic bellows actuator. (**b**) Cross-section of the bellows actuator with the coil heater.

3. Experimental Results

3.1. Temperature Characteristics

Methods for temperature detection are divided into contact and non-contact types [27]. The non-contact temperature sensors, such as infrared (IR) thermometers, can only measure the surface temperature of an object. Because the cavity inside the thermopneumatic bellows actuator is surrounded by PDMS, the IR thermometer has a limitation in accurately measuring the temperature of the closed air in the cavity. On the other hand, contact temperature sensors such as thermocouples, resistance temperature detectors, and thermistors that require insertion inside the actuator cause the size of the actuator to increase owing to the space occupied by the sensor, and further require additional equipment to read the sensor. Alternatively, the temperature of the cavity with a volume of 12.7 mm^3 inside the micro actuator was analyzed by measuring the temperature of the heater itself because the cavity was heated to the same level as the heater in 1.52 s. To measure the temperature of the heater, it was experimentally matched with the relationship of the resistance and temperature of the heater before inserting the heater into the actuator. R-T (resistance-temperature) relationship of the heater was measured by employing a linear regression on the collected data from an IR camera (Seek Thermal Compact PRO, Santa Barbara, CA, USA) for the temperature and a digital multimeter (TX3 True RMS Digital Multimeter, Tektronix, Beaverton, OR, USA) for the resistance outside the bellows actuator (Figure 4a).

The resistance of the heater was 48 mΩ at room temperature (20 °C) and 96 mΩ at 300 °C. Consequently, the heater temperature above 300 °C, which was outside the measurement range of the IR camera, was estimated using the experimental relationship between the resistance and the temperature of the heater given R = 0.1715 T + 44.796. After inserting the heater into the bellows actuator, it was possible to measure the internal temperature of the bellows in real time through the resistance of the heater, as shown in Figure 4b. When 0.50 V was applied to the heater, the heater inside the bellows actuator was heated to 492 °C, and 12.6 s was taken to reach a steady state. The resistance of the heater according to the temperature was used to analyze the mechanical motion of the actuator using COMSOL® and to reduce the discrepancy with the measurement result.

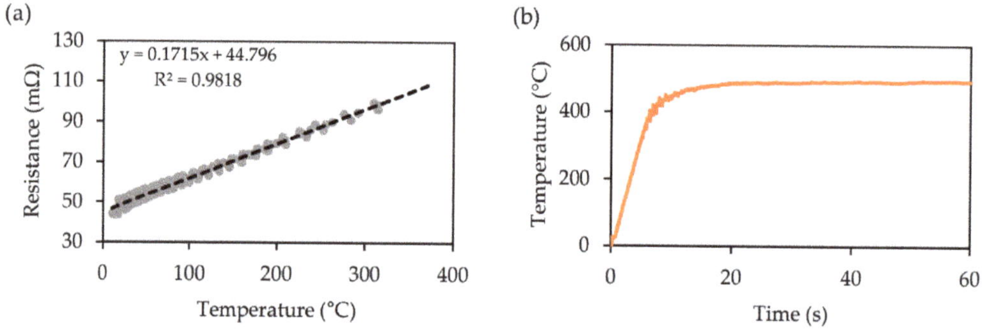

Figure 4. Temperature characteristics of a microheater. (**a**) Calibration of microheater through linear regression. R-squared of the linear regression is 0.9818. (**b**) Temperature–time curves under voltages of 0.5 V, measured based on the relationship between the resistance and temperature of the heater.

3.2. Displacement Characteristics

Figure 5 shows the increased displacement of the thermopneumatic bellows actuator by applying a voltage of 0.55 V at 0.05 V intervals. It was observed when a voltage of 0.60 V or more was applied to the heater, a pressure exceeding the ultimate tensile strength of PDMS was generated inside the actuator, resulting in damage to the bellows actuator. The displacement, therefore, was measured for a voltage less than 0.60 V. The displacement was measured perpendicular to the plane from the head of the bellows using an optical microscope (UM12, Microlinks Technology, Kaohsiung, Taiwan), and then the recorded image frames were analyzed using a custom-built Python-based image processing tool. The expected and measured displacements as a function of the supplied voltage are plotted in Figure 5b. When a voltage of 0.55 V was applied, the actuator showed a displacement of average of 2184 μm from the five samples. There was an average error of 16.2% compared to the predicted value. This is most likely because the measured performance of the actuator is affected by the mixing ratio of the base and agent mixture of PDMS. Also, the elastic properties of the fabricated PDMS have a discrepancy owing to the inconsistency with the parameters of the hyperelastic Ogden model for the numerical analysis [28,29].

The movement of the actuator reached a steady state after 15.24 s of applying a constant voltage. Therefore, in order to verify the repeated operation of the thermopneumatic bellows actuator, the displacement was measured by applying a pulse of 0.50 V with a 33% duty ratio with a heating time of 60 s (including a retention driving) and a cooling time of 120 s (Figure 5c). The displacement took 2.69 s longer to reach the steady-state response compared to the temperature of the inner heater using the heater resistance, as shown in Figure 4b. This is consistent with the calculated result that it took an average of 1.52 s for the air in the cavity to heat up to the same temperature as the heater. Therefore, it was confirmed that the increased temperature of the heater immediately heats the air in the actuator cavity, resulting in the mechanical motion of the actuator. It was observed that the actuator showed a 9.4% decrease in displacement after 10 repetitive operations, and a 12.8% decrease after 15 repetitions. The thermopneumatic driving method inherently increases the temperature of the PDMS bellows, causing a decrease in the ultimate tensile strength [30]. Thus, it is most likely that degradation by heat caused a decrease in performance of the actuator.

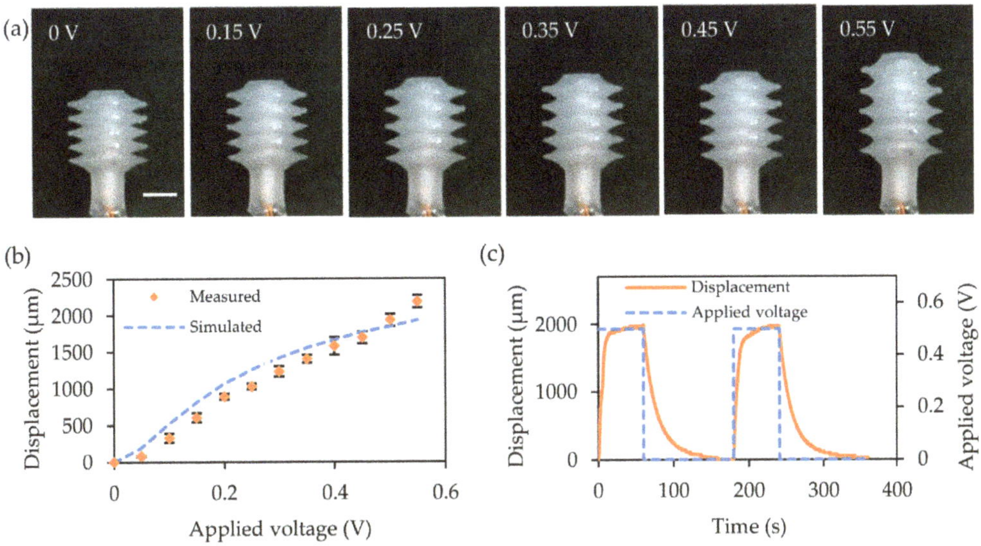

Figure 5. Displacement response generated by Joule heating. (**a**) Images of a thermopneumatic bellows actuator in stop motion due to a voltage applied at 0.10 V intervals from 0 V to 0.55 V. The scale bar is 2.0 mm. (**b**) Steady-state displacement of thermopneumatic bellows actuator versus applied voltage to the heater. Voltage was applied to the heater at intervals of 0.05 V up to the maximum operating voltage of 0.55 V. (**c**) Displacement of thermopneumatic bellows actuator when two cycles of 0.50 V pulse voltage with 180 s, duty ratio of 33% were applied.

In this paper, we built the bellows actuators based on PDMS, which has been verified for decades for biomedical applications. However, considering only thermal degradation, materials that are more thermally stable than PDMS such as α, ω-Trimethylsiloxy-poly(dimethyl-methylchloromethyl) siloxane (PDCMS) and polydimethylsiloxane/zeolitic imidazolate framework (PDMS/ZIF) can be applied for the thermopneumatic bellows actuators [31,32]. Changes in the design parameters of the actuator can also result in thermally stable behavior of a specific stroke. As the number of bellows increases, the displacement of the bellows actuator increases. Thus, by driving at a voltage lower than the voltage that could cause the thermal degradation and also increasing the number of bellows sufficient to achieve the required stroke, the exposure to high temperatures in PDMS can be reduced, thereby preventing the reduction of the ultimate tensile strength in PDMS. More importantly, the 3D-printed soluble mold technique, which is developed to produce a bellows actuator, readily enables a variety of soft polymer actuators or design changes without fabrication process revision, except for selecting the appropriate dissolving solvent [23].

Several studies have reported the PDMS membrane with thickness ranging from tens of μm to hundreds of μm to pneumatic actuation without side effects from gas permeation [10,33]. In particular, according to S Sawano, the 110 μm thick PDMS membrane pump exhibits only about 2% deflection change over 1 h and about 13% over 10 h due to the gas leakage [33]. Since the bellows actuator in this paper consists of a PDMS membrane (250 μm in thickness) that is more than twice as thick, it is relatively insensitive to the gas leakage caused by the pores inside the PDMS. In addition, the bellows actuator maintains the internal cavity pressure continuously by the supply of the electrical power for the required displacement. After the driving power is switched off, the pressure inside the cavity is lowered by air temperature cooling, not by the leakage, which in turn the actuator is restored to its original state. As a result, the operation of the bellows actuator is governed by the pressure formed by the electrical power supplied, while the gas leakage is negligible.

3.3. Force Characteristics

As shown in Figure 6, the force exerted by the actuator in the stroke direction according to the driving voltage was compared with the expected result. The tail of the actuator was fixed to the clamp, and the head, which was the part where the actuator deformed, was positioned such that it touched an electronic scale. The measured weight according to the force applied vertically by the actuator's expanding head was calculated as the exerted force by multiplying with the acceleration due to gravity. As the internally applied pneumatic pressure was applied to the walls of the inner space of the actuator, it was directly transmitted in the form of force exerted by the bellows actuator, resulting in a linear response. The measured force had an average error of 12.2% when compared with the numerically calculated force. It seemed that the heat conduction occurred between the end of the bellows and the thermal conductive surface of the scale, thus resulting in a discrepancy between them. The exerted force of 90.2 mN demonstrated that a single actuator with a thermopneumatic actuation mechanism has sufficient force to push an object [34].

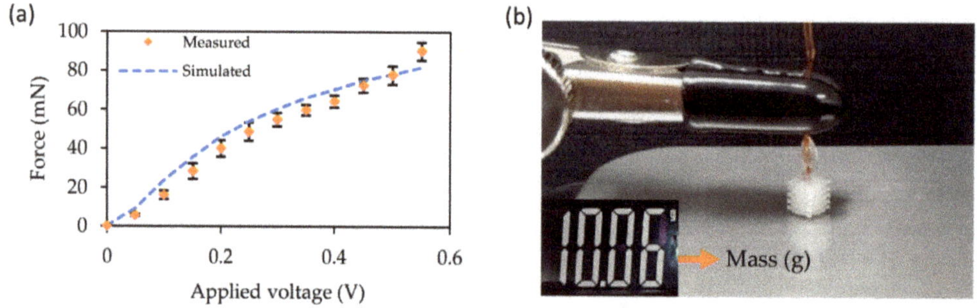

Figure 6. Steady-state force characteristics. (**a**) Experimental data (dot) and model prediction (solid line) of the thermopneumatic actuator, voltage was applied up to the maximum operating voltage of 0.55 V, and the force was measured at 0.05 V intervals. (**b**) Measuring the vertical force applied during expansion when 0.55 V, the maximum operating voltage of the thermopneumatic actuator, is applied. Inset shows measured mass in gram.

3.4. Standalone Operation of Thermopneumatic Bellows Actuators

The power supply can provide stable voltage and current to drive the thermopneumatic bellows actuator. However, compared to the entire volume and weight of the micro actuator, the commercially available power supply is generally heavy. In addition, the external power units must be wired to an outlet. To relieve the bulky payload, a thermopneumatic bellows actuator system for standalone driving was built with batteries and a control circuit. The standalone system depicted in Figure 7 and Video S1 is composed of an MCU (ATTINY13A-SSU, Microchip Technology, Chandler, AZ, USA), a switching MOSFET (FDS6680A, Fairchild Semiconductor, Sunnyvale, CA, USA), button cell batteries (LR732), and an alkaline battery. To supply a voltage of 0.40 V to the heater, a 1.5 V AAAA battery for the heater drive was connected to the thermopneumatic bellows actuator through the switching MOSFET.

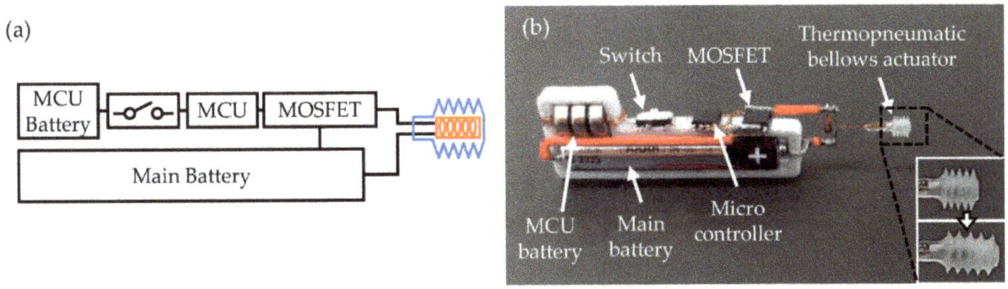

Figure 7. Independent drive system of thermopneumatic actuator. (**a**) Schematic diagram of the autonomous system for standalone operation of the thermopneumatic bellows actuator. (**b**) Photograph of an independent circuit (bird's eye view): button cell battery for MCU, MOSFET for switching, MCU for controlling the signal, a 1.5 V alkaline battery for supplying power to the heater.

4. Conclusions

Soft micro bellows actuators with thermopneumatic actuation for independent driving were designed, fabricated, and evaluated. The 3D thermopneumatic bellows actuator was manufactured using the 3D-printed soluble mold technique in a single body; therefore, no degradation of performance due to air leakage occurred until the fracture stress of the PDMS itself was reached. The performance of the actuator was evaluated by supplying a voltage of up to 0.55 V to prevent the thermal degradation of PDMS. It was observed that the thermopneumatic bellows actuator showed displacement up to 2184 μm and operated repeatedly. It was also confirmed that a standalone operation was possible by configuring the actuator with a control circuit and batteries instead of bulky external devices by changing the operation mechanism from pneumatic to thermopneumatic method. This implies that an autonomous micro soft robot can be constructed that achieves various motions by arranging several actuators using accurate adjustment of large displacements in an intended direction.

Supplementary Materials: The following are available online at https://www.mdpi.com/2072-666X/12/1/46/s1, Video S1: repeated operation of a standalone thermopneumatic actuator system.

Author Contributions: Conceptualization, S.A., W.J. and Y.H.; methodology, S.A., K.K. and Y.L.; investigation, K.K. and Y.L.; writing—original draft preparation, S.A. and W.J.; writing—review and editing, Y.H.; visualization, W.J. and C.L.; supervision, Y.H.; project administration, Y.H.; funding acquisition, Y.H. All authors have read and agreed to the published version of the manuscript.

Funding: This research was supported by the Basic Science Research Program through the National Research Foundation of Korea (NRF) funded by the Ministry of Education (Grants Nos.: NRF-2019R1F1A1059779).

Conflicts of Interest: The authors declare no conflict of interest.

References

1. Wehner, M.; Truby, R.L.; Fitzgerald, D.J.; Mosadegh, B.; Whitesides, G.M.; Lewis, J.A.; Wood, R.J. An integrated design and fabrication strategy for entirely soft, autonomous robots. *Nature* **2016**, *536*, 451–455. [CrossRef] [PubMed]
2. Mirvakili, S.M.; Hunter, I.W. Artificial Muscles: Mechanisms, Applications, and Challenges. *Adv. Mater.* **2018**, *30*, 1704407. [CrossRef] [PubMed]
3. Rus, D.; Tolley, M.T. Design, fabrication and control of soft robots. *Nature* **2015**, *521*, 467–475. [CrossRef] [PubMed]
4. Yu, Y.S.; Zhao, Y.P. Deformation of PDMS membrane and microcantilever by a water droplet: Comparison between Mooney-Rivlin and linear elastic constitutive models. *J. Colloid Interface Sci.* **2009**, *332*, 467–476. [CrossRef]
5. Bernacca, G.M.; O'Connor, B.; Williams, D.F.; Wheatley, D.J. Hydrodynamic function of polyurethane prosthetic heart valves: Influences of Young's modulus and leaflet thickness. *Biomaterials* **2002**, *23*, 45–50. [CrossRef]
6. Boubakri, A.; Guermazi, N.; Elleuch, K.; Ayedi, H.F. Study of UV-aging of thermoplastic polyurethane material. *Mater. Sci. Eng. A* **2010**, *527*, 1649–1654. [CrossRef]
7. Case, J.C.; White, E.L.; Kramer, R.K. Soft Material Characterization for Robotic Applications. *Soft Robot.* **2015**, *2*, 80–87. [CrossRef]

8. Calderón, A.A.; Ugalde, J.C.; Chang, L.; Cristóbal Zagal, J.; Pérez-Arancibia, N.O. An earthworm-inspired soft robot with perceptive artificial skin. *Bioinspir. Biomim.* **2019**, *14*, 56012. [CrossRef]
9. Mosadegh, B.; Polygerinos, P.; Keplinger, C.; Wennstedt, S.; Shepherd, R.F.; Gupta, U.; Shim, J.; Bertoldi, K.; Walsh, C.J.; Whitesides, G.M. Pneumatic networks for soft robotics that actuate rapidly. *Adv. Funct. Mater.* **2014**, *24*, 2163–2170. [CrossRef]
10. Kawun, P.; Leahy, S.; Lai, Y. A thin PDMS nozzle/diffuser micropump for biomedical applications. *Sens. Actuators A Phys.* **2016**, *249*, 149–154. [CrossRef]
11. Huh, D.; Matthews, B.D.; Mammoto, A.; Montoya-Zavala, M.; Yuan Hsin, H.; Ingber, D.E. Reconstituting organ-level lung functions on a chip. *Science* **2010**, *328*, 1662–1668. [CrossRef] [PubMed]
12. Kim, H.J.; Huh, D.; Hamilton, G.; Ingber, D.E. Human gut-on-a-chip inhabited by microbial flora that experiences intestinal peristalsis-like motions and flow. *Lab Chip* **2012**, *12*, 2165–2174. [CrossRef] [PubMed]
13. Wei, H.; Chueh, B.H.; Wu, H.; Hall, E.W.; Li, C.W.; Schirhagl, R.; Lin, J.M.; Zare, R.N. Particle sorting using a porous membrane in a microfluidic device. *Lab Chip* **2011**, *11*, 238–245. [CrossRef] [PubMed]
14. Hughes, J.; Culha, U.; Giardina, F.; Guenther, F.; Rosendo, A.; Iida, F. Soft manipulators and grippers: A review. *Front. Robot. AI* **2016**, *3*, 69. [CrossRef]
15. Nassar, J.M.; Khan, S.M.; Velling, S.J.; Diaz-Gaxiola, A.; Shaikh, S.F.; Geraldi, N.R.; Torres Sevilla, G.A.; Duarte, C.M.; Hussain, M.M. Compliant lightweight non-invasive standalone "Marine Skin" tagging system. *NPJ Flex. Electron.* **2018**, *2*, 13. [CrossRef]
16. Stevens, C.; Powell, D.E.; Mäkelä, P.; Karman, C. Fate and effects of polydimethylsiloxane (PDMS) in marine environments. *Mar. Pollut. Bull.* **2001**, *42*, 536–543. [CrossRef]
17. Slesarenko, V.; Engelkemier, S.; Galich, P.I.; Vladimirsky, D.; Klein, G.; Rudykh, S. Strategies to Control Performance of 3D-Printed, Cable-Driven Soft Polymer Actuators: From Simple Architectures to Gripper Prototype. *Polymers* **2018**, *10*, 846. [CrossRef]
18. Shintake, J.; Rosset, S.; Schubert, B.; Floreano, D.; Shea, H. Versatile Soft Grippers with Intrinsic Electroadhesion Based on Multifunctional Polymer Actuators. *Adv. Mater.* **2016**, *28*, 231–238. [CrossRef]
19. An, S.; Kang, D.J.; Yarin, A.L. A blister-like soft nano-textured thermo-pneumatic actuator as an artificial muscle. *Nanoscale* **2018**, *10*, 16591–16600. [CrossRef]
20. Schift, H. Nanoimprint lithography: 2D or not 2D? A review. *Appl. Phys. A* **2015**, *121*, 415–435. [CrossRef]
21. Xia, Y.; Whitesides, G.M. Soft lithography. *Annu. Rev. Mater. Sci.* **1998**, *28*, 153–184. [CrossRef]
22. Wu, H.; Odom, T.W.; Chiu, D.T.; Whitesides, G.M. Fabrication of complex three-dimensional microchannel systems in PDMS. *J. Am. Chem. Soc.* **2003**, *125*, 554–559. [CrossRef] [PubMed]
23. Jung, W.; Kang, Y.; Han, S.; Hwang, Y. Biocompatible micro, soft bellow actuator rapidly manufactured using 3D-printed soluble mold. *J. Micromech. Microeng.* **2019**, *29*, 125005. [CrossRef]
24. Duggan, T.; Horowitz, L.; Ulug, A.; Baker, E.; Petersen, K. Inchworm-inspired locomotion in untethered soft robots. In Proceedings of the 2019 2nd IEEE International Conference on Soft Robotics (RoboSoft), Seoul, Korea, 14–18 April 2019; pp. 200–205. [CrossRef]
25. Camino, G.; Lomakin, S.M.; Lazzari, M. Polydimethylsiloxane thermal degradation Part 1. Kinetic aspects. *Polymer* **2001**, *42*, 2395–2402. [CrossRef]
26. Kang, K.; Oh, S.; Yi, H.; Han, S.; Hwang, Y. Fabrication of truly 3D microfluidic channel using 3D-printed soluble mold. *Biomicrofluidics* **2018**, *12*, 14105. [CrossRef] [PubMed]
27. Michalski, D.; Strąk, K.; Piasecka, M. Comparison of two surface temperature measurement using thermocouples and infrared camera. *EPJ Web Conf.* **2017**, *143*, 2075. [CrossRef]
28. Sun, J.Y.; Xia, S.; Moon, M.W.; Oh, K.H.; Kim, K.S. Folding wrinkles of a thin stiff layer on a soft substrate. *Proc. R. Soc. A Math. Phys. Eng. Sci.* **2012**, *468*, 932–953. [CrossRef]
29. Kim, T.K.; Kim, J.K.; Jeong, O.C. Measurement of nonlinear mechanical properties of PDMS elastomer. *Microelectron. Eng.* **2011**, *88*, 1982–1985. [CrossRef]
30. Liu, M.; Sun, J.; Chen, Q. Influences of heating temperature on mechanical properties of polydimethylsiloxane. *Sens. Actuators A Phys.* **2009**, *151*, 42–45. [CrossRef]
31. Dong, F.; Sun, X.; Feng, S. Thermal degradation kinetics of functional polysiloxanes containing chloromethyl groups. *Thermochim. Acta* **2016**, *639*, 14–19. [CrossRef]
32. Xu, S.; Zhang, H.; Yu, F.; Zhao, X.; Wang, Y. Enhanced ethanol recovery of PDMS mixed matrix membranes with hydrophobically modified ZIF-90. *Sep. Purif. Technol.* **2018**, *206*, 80–89. [CrossRef]
33. Sawano, S.; Naka, K.; Werber, A.; Zappe, H.; Konishi, S. Sealing method of pdms as elastic material for mems. In Proceedings of the 2008 IEEE 21st International Conference on Micro Electro Mechanical Systems, Wuhan, China, 13–17 January 2008; pp. 419–422.
34. AbuZaiter, A.; Nafea, M.; Mohamed Ali, M.S. Development of a shape-memory-alloy micromanipulator based on integrated bimorph microactuators. *Mechatronics* **2016**, *38*, 16–28. [CrossRef]

micromachines

Article

High Permeability Photosintered Strontium Ferrite Flexible Thin Films

Abid Ahmad [1], Bhagyashree Mishra [2], Andrew Foley [3], Leslie Wood [3] and Maggie Yihong Chen [1,2,*]

1 Ingram School of Engineering, Texas State University, San Marcos, TX 78666-4684, USA; abidahmad.eee@gmail.com
2 Materials Science Engineering and Commercialization, Texas State University, San Marcos, TX 78666-4684, USA; b_m415@txstate.edu
3 Nanohmics, Inc., Austin, TX 78741, USA; afoley@nanohmics.com (A.F.); lwood@nanohmics.com (L.W.)
* Correspondence: maggie.chen@txstate.edu

Abstract: The paper is focused on the development and optimization of strontium ferrite nanomaterial and photosintered flexible thin films. These magnetic thin films are characterized with direct current (DC) and high frequency measurements. For photosintered strontium ferrite samples, we achieved relative complex permeability of about 29.5-j1.8 and relative complex permittivity of about 12.9-j0.3 at a frequency of 5.9 GHz.

Keywords: ferrite materials; magnetic thin films; strontium ferrite nanomaterial; relative permeability; relative permittivity

Citation: Ahmad, A.; Mishra, B.; Foley, A.; Wood, L.; Chen, M.Y. High Permeability Photosintered Strontium Ferrite Flexible Thin Films. *Micromachines* **2021**, *12*, 42. https://doi.org/10.3390/mi12010042

Received: 15 November 2020
Accepted: 27 December 2020
Published: 1 January 2021

1. Introduction

Ferrite materials, e.g., spinels and garnets, have attracted research interest in recent times because of their excellent microwave frequency characteristics such as low magnetic loss. Bulk ferrite materials have been widely used in the manufacture of microwave devices such as modulators, circulators, etc. [1]. However, thin films of ferrite materials have unique electrical, mechanical, and optical properties compared to their bulk counterparts. In many cases, these unique properties are a consequence of film thicknesses being below the length scale of magnetic domains in a material. As a result, desirable magnetic properties of ferrites can be engineered at the submicron level and are readily controlled. For example, the magnetic properties of zinc-ferrite thin film have been reported to show a sharp deviation from the bulk [2]. Furthermore, the permeability of a 0.6 μm a nickel-zinc ferrite thin film deposited on a 2 mm glass substrate by spin-spray exhibits a higher permeability than the Snoek's limit of bulk Ni-Zn [3]. Ferrite magneto dielectric materials are promising for antenna miniaturization and enhanced performance at microwave frequencies [4]. The measured properties of magnetic thin films can provide deeper insight into the fundamental structure and behavior of the materials. For example, soft magnetic thin films with high permeability and permittivity at MHz and GHz frequencies have received much recent research attention [5,6] because of the rapid growth of microwave technology sectors (e.g., WiFi and 5G). There are several techniques for determining the permeability and permittivity of magnetic thin films. Some are single frequency techniques while others are techniques that provide information over a broad frequency range [7]. In this work, we investigated thin ferrite materials properties such as permeability and permittivity at both direct current (DC) and high frequencies.

DC characterization of ferrite films was considered as fundamental research. In turn, it is also important to know the characteristics of ferrite materials at high frequency. Investigation at high frequency is a possible pathway for the future application of magnetic waveguide design in the radio frequency band of interest (i.e., C-band). Nicolson and Ross [8] developed a technique for determining permittivity and permeability from measurements of the reflection and transmission coefficients of a material sample placed in a transmission line. Weir [9] applied this technique to measurements made using a network analyzer. The basic concept is to measure the S-parameters of a sample placed in a transmission line and determine the intrinsic properties required to produce these measurements.

We developed ferrite materials and deposited them on flexible Kapton films. The ferrite thin films were examined with X-ray diffraction (XRD) to observe the material crystallinity. The XRD confirmed that the ferrite thin films had good crystallinity and indicated a high-quality film. We characterized properties of the ferrite thin films under DC and high frequency measurements. The results show high magnetic permeability and electrical permittivity, which supports the XRD assessment indicating high film quality. These high-quality ferrimagnetic thin films appear promising for flexible microwave devices due to their high permeability.

2. Materials and Methods

Hexaferrites such as Sr- and Ba-doped hexaferrites together with bismuth ferrite, have many properties, which make them suitable for GHz devices [10,11]. We will now describe our synthesis process for the materials we studied.

2.1. *Synthesis of Strontium Hexaferrite*

To synthesize nano particulate strontium hexaferrite, a Pechini sol–gel chemistry was used. Iron nitrate nonahydrate and strontium chloride were dissolved in water at a 12:1 molar ratio to produce strontium ferrite hexaferrite, $SrFe_{12}O_{19}$. Citric acid and ethylene glycol were added, which undergo a transesterification reaction forming a covalent polymer network to ensnare the dissolved metal ions. To this solution, 1 M sodium hydroxide was added dropwise initiating the reaction and causing red brown solids to precipitate out of solution as the reaction became progressively basic. Enough sodium hydroxide was added to achieve a pH between 13 and 14 to ensure conversion. The precipitated particulates were collected by centrifugation and rinsed several times with water to remove the NaCl byproduct. To achieve crystallinity, the dried powder was annealed in air at 500 °C for 1 h. It is during this annealing stage where the polymeric precursor previously formed with citric acid and ethylene glycol keeps the metal ions homogeneously dispersed throughout the network and slows particle growth or "ripening" under the high temperature conditions. In Figure 1, the SEM images show the particles to be 50–100 nm and have a quasispherical appearance.

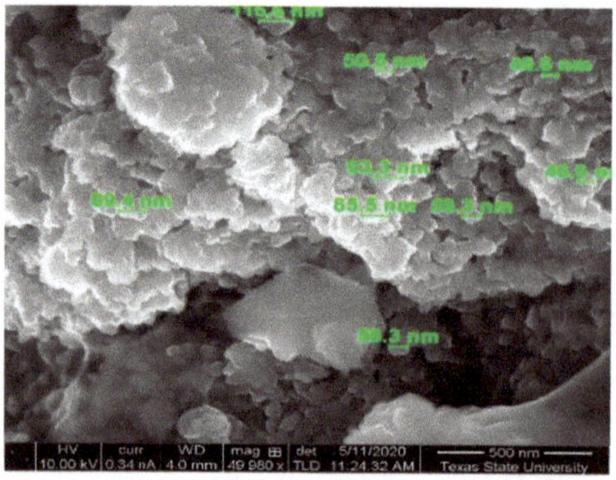

Figure 1. SEM of as synthesized strontium ferrite powders showing particle size between 50 and 100 nanometer. The scale bar in the lower right of the image is 500 nm.

2.2. Paste Formulation and Application of Strontium Ferrite Thin Films

The ferrite nanopowders were formulated into a paste consisting of terpineol (solvent), BYK 111 (dispersant), Pycal (plastisizer), polyvinyl pyrrolidone (PVP, binder) and the ferrite powder added to a mass fraction of between 60% and 80%. The mixture was then ball milled to achieve a homogeneous consistency. This paste was deposited onto Kapton and polyethylene terephthalate (PET) substrates by a simple doctor blade process. The applied films were dried in an oven at 100 °C to remove solvent. Figure 2 details the process.

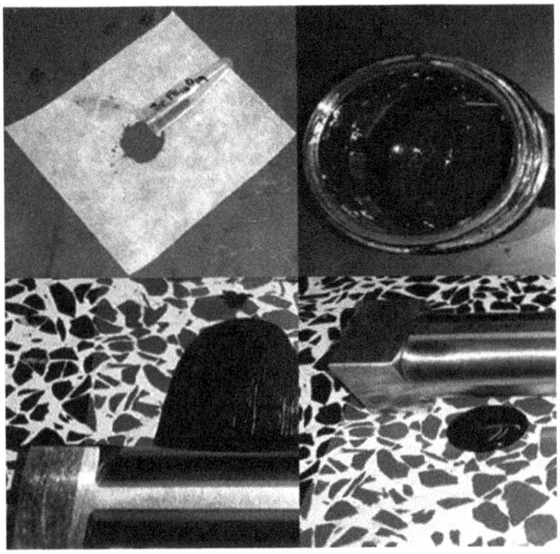

Figure 2. As synthesized strontium ferrite nanopowder, formulated ferrite paste and ferrite paste before and after spreading by doctor blade onto a plastic substrate.

2.3. Photosintering of Strontium Ferrite Thin Films

The deposited films were photosintered with a PulseForge® 1300 (NovaCentrix, Austin, TX, USA) using various voltages and pulse durations to determine the amount of incident photonic energy needed to reflow the particle grains and transform the dried particle film into a bulk material. A variety of single to multi pulse sequences with gradually increasing pulse durations up to 10 milliseconds were attempted at voltages ranging from 400 to 700 V. Longer pulse durations divided by 25–100 micropulses were found to be the most effective. An example sequence that delivers a photonic energy of 16.5 J/cm^2 at 575 V, 8 millisecond duration, divided in 100 micropulses and 50% duty cycle, which is the percentage of time the light is on during the entire sequence is shown in Figure 3. The expected temperature peak was 2200 °C and cools to a room temperature in around one second. The PulseForge® Simpulse® software (NovaCentrix, Austin, TX, USA) predicts a peak temperature of 1900–2530 °C for a voltage between 525 and 700 V keeping all other parameters the same as shown in Figure 3. This corresponds to an optical energy delivered per area between 13.8 and 24.9 J/cm^2 as the voltage increased. High temperature is required to effectively reflow the ferrite grains, which have a high melting point. Iron oxide powders are reported to have melting points between 1530 °C and 1570 °C.

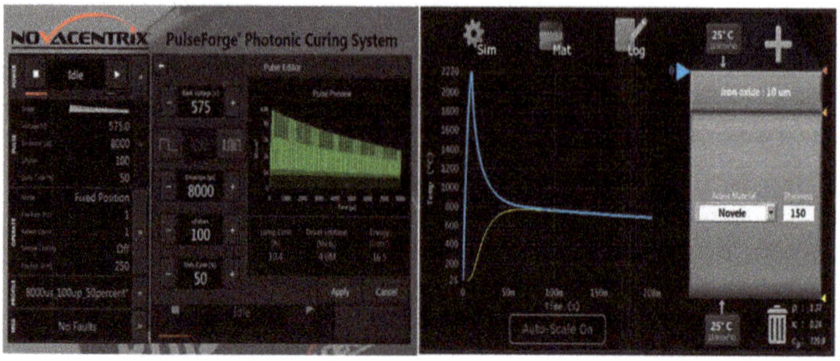

Figure 3. (**Left**) Screenshot of an example pulse sequence used on the PulseForge® 1300 to photosinter strontium ferrite thin films. (**Right**) SimPulse® predicted temperature profile for the pulse indicated.

For green films 10 micron thickness and below, it was found that a single sequence as shown in Figure 3 was sufficient to completely solidify, but thicker films 20 microns and above required lower voltages 450 V and for the sequence to be repeated multiple times with a few seconds between each sequence to prevent overheating of the substrate. This way the sinter can penetrate the thicker layers without melting the plastic substrates. SEM images found in Figure 4 show a deposited nanoparticle film before and after photosintering. The images on the left in Figure 4 are a top-down and cross section of the "green", as the deposited film shown in Figure 2. The film appeared granular and a crack formed upon drying can be seen traversing the film on the top-down image. The images to the right in Figure 4 were the corresponding photosintered films. The grains were clearly reflowed as evidenced by the smooth appearance to form a bulk material. The film appeared to lose 4-6 microns of film thickness due to ablation at the surface and densification. The granular layer just below the reflowed film in the photosintered cross section image was a fumed nano silica layer that is applied to PET printing media films (Novele™) to allow the reflowed material to bind strongly to the substrate.

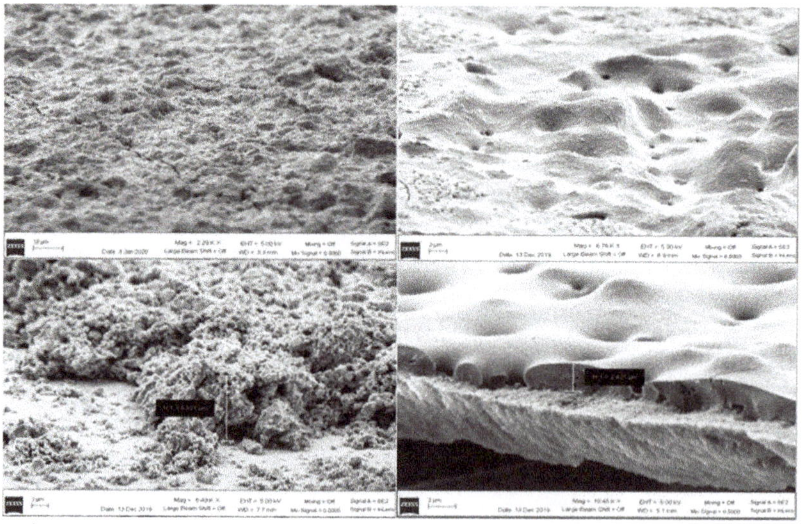

Figure 4. SEM images of the as synthesized strontium ferrite nanopowder film before and after photosintering. Top left and bottom left images in panel are the top-down and cross section views for the as made strontium ferrite film. Top right and bottom right images in the panel are the top-down and cross section views for the photosintered film.

3. Results

3.1. DC Characterization

We characterized the magnetic and electrical properties of strontium ferrite materials at DC. The DC measurement provides data for analysis of the static properties of the films. Magnetic properties of ferrite materials were evaluated using various characterization approaches. The X-ray diffraction (XRD) was used for assessing the crystallinity of the ferrite materials. The relative permeability and relative permittivity were analyzed using a vibrating sample magnetometer (VSM) and a mercury probe station respectively. Sheet resistance of the thin film was measured by the Van Der Pauw method at room temperature and environment using the Karl Suss four-point probe station and Keysight B1500A semiconductor device analyzer that uses tungsten probes.

3.1.1. XRD Analysis

X-ray diffraction (XRD) is a technique that gives us crystal structure. The intensity of the scattered X-rays was plotted as a function of the scattering angle, and the structure of the material was determined from the analysis of the location, in angle, and the intensities of scattered intensity peaks [12].

Both the raw synthesized strontium ferrite (Sr-ferrite) powder annealed at 500 °C and the fabricated thin film samples (as made and photosintered using a PulseForge 1300) were analyzed by a Rigaku SmartLab X-ray diffractometer (XRD) to determine the crystalline phases. Figure 5 displays the presence of multiple peaks between 30 and 40 and 2θ confirms the introduction of an additional structural phase of the ferrite material because the known undoped ferrite phases (i.e., nano magnetite (Fe_3O_4) and maghemite (Fe_2O_3)) are both known to only yield one strong peak in this region [13,14]. Further, the primary peak from the XRD of the synthesized strontium ferrite powder revealed that the coprecipitated strontium ferrite annealed at 500 °C did not match the XRD reported in the literature for strontium hexaferrite nanopowder; instead, matching the XRD for Fe_2O_3 as seen in Figure 5. Further, the larger magnitude of the hump in the 15° region compared to the literature values of the hexaferrite was indicative of substantial inclusion of amorphous structure in the synthesized powder. The lattice mismatch and additional amorphous structure of the synthesized compared to the literature XRD was reasonable due to the annealing temperature (500 °C) used for the synthesized powder being substantially lower than what is reported for the literature reference (850 °C).

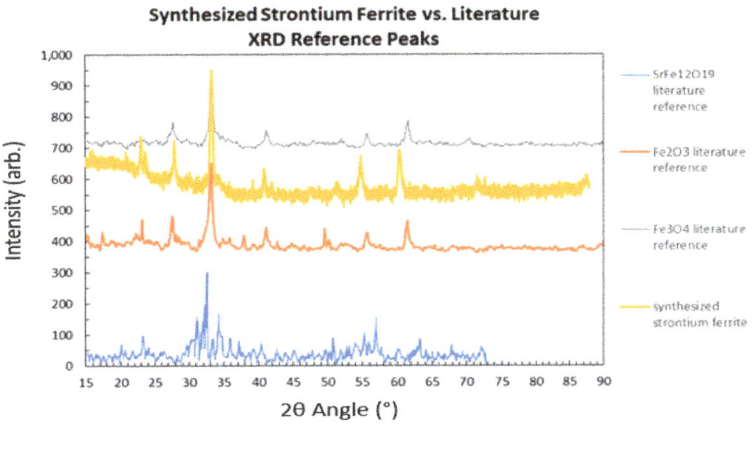

Figure 5. XRD analysis of synthesized strontium hexaferrite nanopowder compared to referenced XRD peaks.

Figure 6 shows the XRD for the as made strontium ferrite film followed by two samples independently photosintered at 525 V, two samples independently photosintered at 550 V and a final film photosintered at 575 V. The same sequence was used for each photosintered film (see Figure 3) by gradually increasing the voltage, the photonic energy and peak temperature was gradually increased. The predicted temperature for these sequences was 2000 °C. Going from the as made film to photosintered, the XRD did not change peak position, however, some peaks increased in intensity. There was no clear trend as the highest and lowest intensity for the peak just below 25° (2θ angle) was reached with the lowest voltage of 525 V.

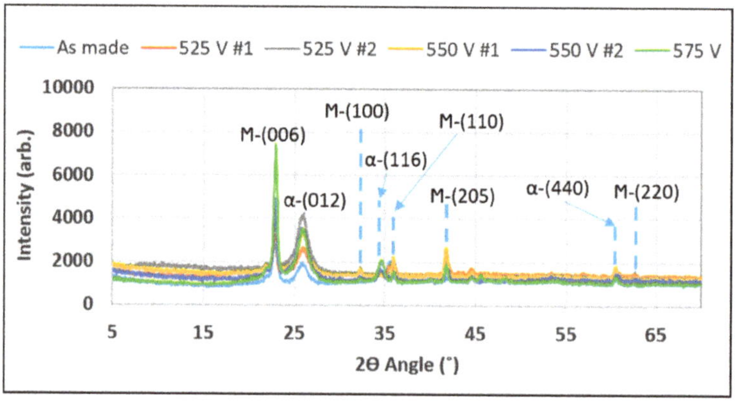

Figure 6. XRD analysis of as made and photosintered strontium ferrite films.

The XRD pattern in Figure 6 shows that the peaks became more intense and narrower as the voltage increased, which is an indication of an increase in crystal domain size. At the same, the diminishing height of the broad hump located at 2θ below 15° indicates a diminished volume of polycrystalline domains, which together with the increase in crystalline peak intensity indicates the conversion of polycrystalline volumes into crystalline ones. Given the stoichiometry of this Sr-Fe-O material system, these adjacent crystalline peaks were best identified as the 23.15° (006) peak of Sr-hexaferrite ($SrFe_{12}O_{19}$) and 24.20° (012) peak of hematite (α-Fe_2O_3), which would correspond to a combination of hexaferrite and hematite grains in the material. Similar inclusions of hematite impurities are observed elsewhere in the literature during Sr-hexaferrite synthesis [15].

3.1.2. VSM Analysis of Permeability

A microsense LLC or EZ9-HF VSM was used to measure the permeability behavior of magnetic ferrite materials. It operates on Faraday's law of induction, which states that a changing magnetic field will yield an electric field potential.

We used the system at low temperature range from −195 to 50 °C. The sample having 7 mm length, 6 mm width, and 128.5 micron thickness was mounted on the 8 mm transverse rod of the VSM. The applied field was varied from −21000 Oersted(Oe) to 21000 Oe in the sweep mode. The measured magnetic moment in emu vs. applied field is given in Figure 7.

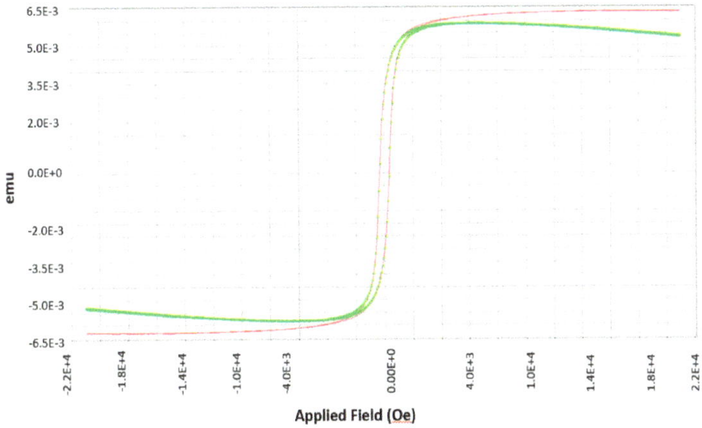

Figure 7. Magnetization (emu) vs. applied field (Oe).

In Figure 7, the green colored hysteresis is as measured results from the VSM using the EasyVSM Software (MicroSense, LLC, Lowell, MA, USA). This data shows a slope after it reaches saturation. As a result, the red hysteresis is calculated by subtracting the background signal and correcting the field lag and signal slope.

In Figure 8, the absolute permeability was the highest slope in hysteresis loop, which was 1.10×10^{-5} H/m. The relative permeability was calculated using the ratio of absolute permeability and permeability of air, where the permeability of air was $1.25663753 \times 10^{-6}$ H/m. Thus, the measured relative permeability was 8.75.

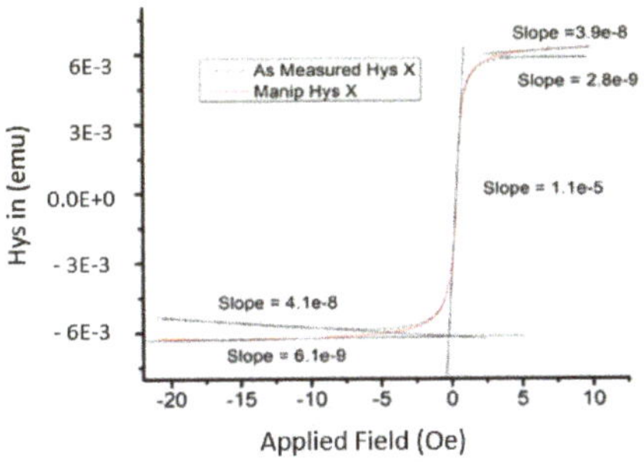

Figure 8. Slope calculation from the hysteresis graph.

3.1.3. Hermes Probe Station Analysis

The Hermes probe station is a specialized sample stage for probing thin film reactance (C, L, and R) without a top contact. Liquid mercury is applied via vacuum to a surface of the specimen and the back of the specimen was used as the other electrical contact. The thin films were deposited on silicon wafers to measures the permittivity at 0–10 MHz using this probe. The absolute permittivity and relative permittivity can be derived from $\epsilon = \frac{C \times d}{A}$ and $\epsilon_r = \frac{\epsilon}{\epsilon_0}$ where d is the thickness of the sample, area $A = 4.7 \times 10^{-7}$ m^2 and $\epsilon_0 = 8.85 \times 10^{-12}$ F/m. As the thin film was on Si wafer, it behaved as a MOS capacitor when connected to the Hermes probe. When DC bias voltage was applied the MOS capacitance will undergo accumulation, depletion and inversion mode. In accumulation and inversion mode charge carriers in Si substrate will be directly under the dielectric thin film, so it will give the exact capacitance (maximum capacitance in C–V curve) of the dielectric thin film. Whereas in the depletion mode, the charge carriers will be below the depletion layer; so here the total capacitance will be a series combination of both the thin film and depletion layer (minimum capacitance in the C–V curve). The relative permittivity of the Sr-ferrite film was measured using 802-150 MDC Mercury Probe station. Taking the maximum capacitance value in C–V curve with DC bias voltage from −5 to 5 V, the relative permittivity of Sr-ferrite film was found 6.48 at 1 kHz and 6.9 at 100 kHz.

3.2. *Microwave Frequency Characterization*

Since ferrimagnetic materials have promise in microwave frequency applications, we characterized the ferrimagnetic thin films within C-band (5.85–7.8 GHz) using the standard waveguide designated WR-137. The permeability was the most important parameter to define, because it governs the interaction between the electromagnetic wave and the material and was thus the origin of all magnetic phenomena. Therefore, we characterized permeability and permittivity at 5.85–7.8 GHz.

3.2.1. Waveguide Signal Perturbation: Basic Principle

Many methods are available to characterize thin films at microwave frequencies. An older and simpler approach is to insert the sheet of the film into a rectangular waveguide measure the S-parameters of the system over the frequency range of interest. A segment of a rectangular waveguide where a sample has been placed, filling the waveguide and leaving no air gaps is a typical measurement configuration.

Figure 9 shows such a segment whose axis is in the x-direction, as is the propagation direction. The electric fields at the three sections of the transmission line are E_{I}, E_{II}, and E_{III}, respectively. L is the length of the under-test unit, and the total length of the waveguide can be expressed as $L_{\mathrm{Total}} = L + L_1 + L_2$.

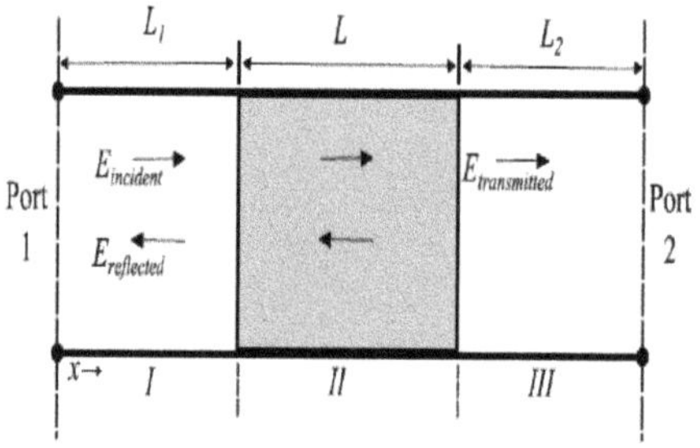

Figure 9. Incident, transmitted, and reflected electromagnetic waves in a filled waveguide.

S-parameters describe the input–output relationship between ports (or terminals) in an electrical system [16]. The S-parameters of the testing structure were measured directly using a vector network analyzer and served as the input for further calculation.

Nicolson–Ross–Weir (NRW) algorithm

The electric and magnetic behavior of a low-conductivity material was determined by two complex parameters, permeability, μ, and permittivity, ε. The Nicolson–Ross–Weir (NRW) algorithm [9] was used in order to calculate the permeability and permittivity of the ferrite film.

The Nicolson–Ross–Weir (NRW) algorithm combines and derives formulas for the calculation of permittivity and permeability.

$$X = \frac{\left(S_{11}{}^2 - S_{21}{}^2\right) + 1}{2S_{11}} \quad (1)$$

$$\gamma = X \pm \sqrt{X^2 - 1} \tag{2}$$

where γ is a complex number, which was used to describe the behavior of an electromagnetic wave along a transmission line. γ is also known as the reflection coefficient. The appropriate sign is chosen so that $\gamma \leq 1$ in order to express the passivity of the sample. The transmission coefficient is

$$T = \frac{(S_{11} + S_{21}) - \gamma}{1 - (S_{11} + S_{21}) * \gamma} \tag{3}$$

The complex permeability is calculated from

$$\mu_r = \frac{1 + \gamma}{(1 - \gamma) * \Delta * \sqrt{\frac{1}{\lambda_0{}^2} - \frac{1}{\lambda_C{}^2}}} \tag{4}$$

and the complex permittivity from

$$\varepsilon_r \mu_r = \lambda_0{}^2 (\frac{1}{\Delta^2} + \frac{1}{\lambda_C{}^2}) \tag{5}$$

where ε_r is the relative permittivity, μ_r is the relative permeability, λ_C is the cutoff wavelength of the transmission line section, and λ_0 the free space wavelength. Meanwhile, ε_0 and μ_0 are the permittivity and permeability of the free space.

Further,

$$\frac{1}{\Delta^2} = -[\frac{1}{2\pi D} \ln(\frac{1}{T})]^2 \tag{6}$$

where D is the thickness of sample, and $\frac{1}{\Delta} = \frac{1}{\lambda_g}$ where λ_g is the transmission line guide wavelength [17].

Equation (6) has an infinite number of roots since the imaginary part of the logarithm of a complex quantity T is equal to the angle of the complex value plus $2\pi n$, where n is equal to the integer of $\frac{L}{\lambda_g}$. The unwrapping method can be used to solve the problem of phase ambiguity [17].

This method with a network analyzer has three kinds of errors, systematic, random, and drift. Multiple measurements were taken to average out the random errors. The drift errors were minimized by the control of lab temperature and humidity. System errors were generated in the measurement setup due to coax cables, coax-to-waveguide adapters, etc. To de-embed these effects, mechanical calibration kit was used to calibrate the system to the point at the end of coax cables. The entire system without the sample was first measured and de-embed from the results with the sample inserted into the waveguides.

3.2.2. Experimental Setup

Measurements of the microwave properties of thin films were performed with the waveguide technique. Since the experimental setup involves several components such as cables and connectors, proper care has to be taken to ensure that the entire system remains stable over the measurement period. First, a bare commercial Kapton substrate was measured. The system errors were reflected and evaluated in the measured parameters of the commercial Kapton substrate. The samples examined were on a Kapton substrate one with an air-dried strontium ferrite thin film and one with a photosintered strontium ferrite thin film. The bare Kapton substrates had a thickness of 127 microns. Samples of these films were cut into a size suitable for insertion into a WR-137 waveguide shown in Figure 10 and covered the entire waveguide cavity with the film. The measurement was calibrated by normalizing to an air-filled cavity. Two independent scattering parameters S11 and S21 data were measured with a Keysight N9917A vector network analyzer (VNA) at 5.9 GHz.

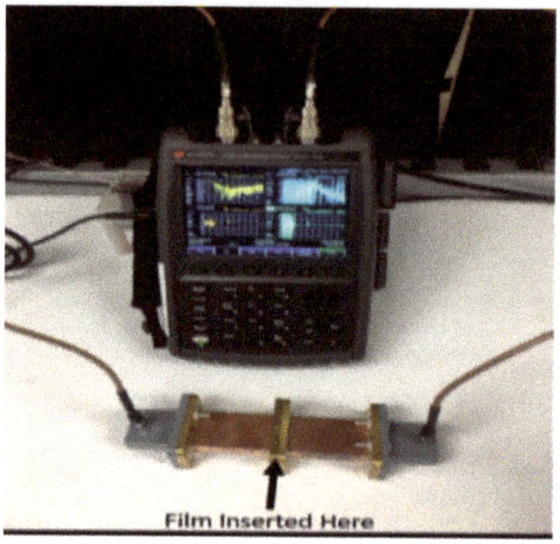

Figure 10. The waveguide setup for measuring thin film permeability and permittivity at 5.9 GHz. (**Top**) vector network analyzer (VNA) (**Bottom**) waveguide.

The parameters measured by the vector network analyzer were used to derive the real and imaginary parts of permittivity and permeability of the sample using the NRW algorithm. The accuracy of the constitutive material properties depends on the accuracy with which the parameters are measured.

3.2.3. Experimental Results

Since the ferrimagnetic films were deposited on the Kapton film, the Kapton film was characterized first. The Kapton film was inserted between two pieces of WR-137 waveguides, which were tightened up to avoid an air gap. The S parameters were measured with the Keysight N9917A vector network analyzer after a two-port calibration. Then the measured data were plugged into the NRW algorithm program to obtain the permeability and permittivity values. The relative permeability was calculated to be 1.15 at 5.9 GHz, which gave us an error below 15%. The error mainly came from the calibration of equipment and the air gap between the waveguides.

The strontium ferrite as-made thin film was deposited on top of Kapton. It was air dried with a thickness of 31 micron. The strontium ferrite thin film was then inserted between two pieces of WR-137 waveguides. Two samples were measured, and the values were averaged. The as-made air-dried strontium ferrite thin film had a measured relative complex permeability of 11.4-j3.6 and relative complex permittivity of 4.9-j0.9 at 5.9 GHz.

Another set of strontium ferrite thin film were photosintered right after deposition. The strontium ferrite thin film had a thickness of 22 microns. Two samples were measured, and the values were averaged. We found that the sintered strontium ferrite thin film had a relative complex permeability of 29.5-j1.8 and relative complex permittivity of 12.9-j0.3 at 5.9 GHz.

The measured data show that the photosintered strontium ferrite thin film had higher relative permeability and relative permittivity, which comes from the denser structure after photosintering.

4. Discussion

In summary, strontium ferrite nanomaterial was developed, and the material was deposited into thin films on flexible Kapton substrates for evaluation. DC and high frequency characterization of the strontium ferrite materials were carried out. The Sr-ferrite samples, which were photosintered, show a comparatively high relative complex permeability of 29.5-j1.8, and relative complex permittivity of 12.9-j0.3 at 5.9 GHz. In Ref. [10], powder strontium ferrite sample with an average grain size of about 50–100 μm and density 1.300 g/cm^3 was analyzed. The transmission/reflection technique and the cavity resonators technique were used to determine the permeability and permittivity. Measurements revealed that the real permittivity of strontium ferrite powder was in the range from 2.597 to 2.712 and it had an average real permeability of 1.063. The photosintered strontium ferrite thin film demonstrated in this work had much higher permittivity and permeability. This high relative permeability and relative permittivity would make this film a strong candidate to demonstrate a possible pathway for future microwave applications such as C-band communication waveguides. These higher magnetic properties of ferrimagnetic strontium ferrite will also be suitable for antenna design and application.

Author Contributions: Conceptualization, A.F. and M.Y.C.; methodology, A.A., L.W., A.F. and M.Y.C.; investigation, A.A., B.M. and L.W.; writing—original draft preparation, A.A., L.W. and M.Y.C.; writing—review and editing, M.Y.C.; supervision, A.F. and M.Y.C.; funding acquisition, A.F. and M.Y.C. All authors have read and agreed to the published version of the manuscript.

Funding: This research was funded by the U.S. Department of Transportation under award number 6913G619P800003.

Institutional Review Board Statement: Ethical review and approval were not applicable for studies not involving humans or animals.

Informed Consent Statement: Patient consent was not applicable for studies not involving humans.

Data Availability Statement: The data presented in this study are available on request from the corresponding author.

Acknowledgments: The authors would acknowledge Karl Stephan and Cecil Rich Compeau for the helpful discussion in microwave characterization.

Conflicts of Interest: The authors declare no conflict of interest.

References

1. Frans, T. New Developments in Ferromagnetic Materials. *Nature* **1948**, *161*, 666. [CrossRef]
2. Chen, J.; Srinivasan, G.; Hunter, S.; Babu, V.S.; Seehra, M.S. Observation of superparamagnetism in rf-sputtered films of zinc ferrite. *J. Magn. Magn. Mater.* **1995**, *146*, 291–297. [CrossRef]
3. Chao, L.; Sharma, A.; Afsar, M.N.; Obi, O.; Zhou, Z.; Sun, N. Permittivity and permeability measurement of spin-spray deposited Ni-Zn-ferrite thin film sample. *IEEE Trans. Magn.* **2012**, *48*, 4085–4088. [CrossRef]
4. Youngs, I.; Bowler, N.; Lymer, K.P.; Hussain, S. Dielectric relaxation in metal-coated particles: The dramatic role of nano-scale coatings. *J. Phys. D Appl. Phys.* **2005**, *38*, 188. [CrossRef]
5. Stamps, R.L. Dynamic magnetic properties of ferroic films, multilayers, and patterned elements. *Adv. Funct. Mater.* **2010**, *20*, 2380–2394. [CrossRef]
6. Seemann, K.; Leiste, H.; Krüger, K. Ferromagnetic resonance frequency increase and resonance line broadening of a ferromagnetic Fe–Co–Hf–N film with in-plane uniaxial anisotropy by high-frequency field perturbation. *J. Magn. Magn. Mater.* **2013**, *345*, 36–40. [CrossRef]
7. Clarke, R.N. (Ed.) *A Guide to the Characterisation of dielectric materials at RF and microwave frequencies*; The Institute of Measurement & Control (UK) & NPL: London, UK, 2003; ISBN 0904457389. Available online: http://eprintspublications.npl.co.uk/id/eprint/2905 (accessed on 31 December 2020).
8. Nicolson, A.M.; Ross, G.F. Measurement of the intrinsic properties of materials by time domain techniques. *IEEE Trans. Instrum. Meas.* **1970**, *19*, 377–382. [CrossRef]
9. Weir, W.B. Automatic measurement of complex dielectric constant and permeability at microiwave frequencies. *Proc. IEEE* **1974**, *62*, 33–36. [CrossRef]
10. Bahadoor, A.Y.; Afsar, M.N. Complex permittivity and permeability of barium and strontium ferrite powders in X, KU, and K-band frequency ranges. *J. Appl. Phys.* **2005**, *97*, 10F105. [CrossRef]
11. Kumar, A.; Rai, R.C.; Podraza, N.J.; Denev, S.; Ramirez, M.; Chu, Y.H.; Martin, L.W.; Ihlefeld, J.; Heeg, T.; Schubert, J.; et al. Linear and nonlinear optical properties of Bi FeO_3. *Appl. Phys. Lett.* **2008**, *92*, 121915. [CrossRef]
12. Inaba, K. High Resolution X Ray Diffraction Analyses of (La,Sr)MnO3/ZnO/Sapphire(0001) Double Heteroepitaxial Films. *Adv. Mater. Phys. Chem.* **2013**, *3*, 72–89. [CrossRef]
13. Ahmadi, A.; Ghanbari, D.; Nabiyouni, G. Facile synthesis of hexagonal strontium ferrite nanostructures and hard magnetic poly carbonate nanocomposite. *Main Group Met. Chem.* **2017**, *40*, 9–18. [CrossRef]
14. Esparza, R.; Rosas, G.; Perez, R. Effect of the surfactant on the growth and oxidation of iron nanoparticles. *J. Nanomater.* **2015**, *12*, 989–995. [CrossRef]

15. Kiani, E.; Rozatian, A.S.; Yousefi, M.H. Synthesis and characterization of $SrFe_{12}O_{19}$ nanoparticles produced by a low-temperature solid-state reaction method. *J. Mater. Sci. Mater. Electron.* **2013**, *24*, 2485–2492. [CrossRef]
16. Chen, L.F.; Ong, C.K.; Neo, C.P.; Varadan, V.V.; Varadan, V.K. *Microwave Electronics Measurement and Materials Characterization*; John Wiley & Sons Ltd.: Hoboken, NJ, USA, 2004. [CrossRef]
17. Vicente, A. The Step by Step Development of NRW Method. In Proceedings of the SBMO/IEEE MTT-S International Microwave and Optoelectronics Conference, Natal, Brazil, 29 October–1 November 2011; pp. 738–742. [CrossRef]

micromachines

Article

DNA Printing Integrated Multiplexer Driver Microelectronic Mechanical System Head (IDMH) and Microfluidic Flow Estimation

Jian-Chiun Liou [1,*], Chih-Wei Peng [1], Philippe Basset [2] and Zhen-Xi Chen [1]

[1] School of Biomedical Engineering, Taipei Medical University, Taipei 11031, Taiwan; cwpeng@tmu.edu.tw (C.-W.P.); hoppyshi@gmail.com (Z.-X.C.)
[2] ESYCOM, Université Gustave Eiffel, CNRS, CNAM, ESIEE Paris, F-77454 Marne-la-Vallée, France; philippe.basset@esiee.fr
* Correspondence: jcliou@tmu.edu.tw

Abstract: The system designed in this study involves a three-dimensional (3D) microelectronic mechanical system chip structure using DNA printing technology. We employed diverse diameters and cavity thickness for the heater. DNA beads were placed in this rapid array, and the spray flow rate was assessed. Because DNA cannot be obtained easily, rapidly deploying DNA while estimating the total amount of DNA being sprayed is imperative. DNA printings were collected in a multiplexer driver microelectronic mechanical system head, and microflow estimation was conducted. Flow-3D was used to simulate the internal flow field and flow distribution of the 3D spray room. The simulation was used to calculate the time and pressure required to generate heat bubbles as well as the corresponding mean outlet speed of the fluid. The "outlet speed status" function in Flow-3D was used as a power source for simulating the ejection of fluid by the chip nozzle. The actual chip generation process was measured, and the starting voltage curve was analyzed. Finally, experiments on flow rate were conducted, and the results were discussed. The density of the injection nozzle was 50, the size of the heater was 105 μm × 105 μm, and the size of the injection nozzle hole was 80 μm. The maximum flow rate was limited to approximately 3.5 cc. The maximum flow rate per minute required a power between 3.5 W and 4.5 W. The number of injection nozzles was multiplied by 100. On chips with enlarged injection nozzle density, experiments were conducted under a fixed driving voltage of 25 V. The flow curve obtained from various pulse widths and operating frequencies was observed. The operating frequency was 2 KHz, and the pulse width was 4 μs. At a pulse width of 5 μs and within the power range of 4.3–5.7 W, the monomer was injected at a flow rate of 5.5 cc/min. The results of this study may be applied to estimate the flow rate and the total amount of the ejection liquid of a DNA liquid.

Keywords: DNA printing; flow estimation; MEMS

Citation: Liou, J.-C.; Peng, C.-W.; Basset, P.; Chen, Z.-X. DNA Printing Integrated Multiplexer Driver Microelectronic Mechanical System Head (IDMH) and Microfluidic Flow Estimation. *Micromachines* **2021**, *12*, 25. https://doi.org/10.3390/mi12010025

Received: 29 November 2020
Accepted: 24 December 2020
Published: 29 December 2020

1. Introduction

Inkjet print head technology is very important, and the huge development of inkjet technology mainly started with the development of the principle of inkjet print head technology. Contains large-scale droplet generator for studies of inkjet printing [1–8]. The continuous inkjet system has the advantages of high frequency response and high-speed printing. However, the structure of the inkjet print head of this method is more complicated, and it requires a pressurizing device, a charging electrode, and a deflection electric field, which is difficult to mass produce. The inkjet print head of the on-demand inkjet system is simple in structure, easy to realize the multi-nozzle of the inkjet head, easy to digitize and color, and the image quality is relatively fine, but the general ink droplet ejection speed is low [9–11].

The total number of nozzles of the hot bubble inkjet head can reach hundreds or even thousands. The nozzles are quite fine, which can produce rich harmony colors and smoother mesh tone. The ink cartridge and the nozzle form an integrated structure, and the inkjet head is updated at the same time when the ink cartridge is replaced, so that there is no need to worry about the nozzle clogging, but it causes waste of consumables and relatively high cost. The on-demand inkjet technology ejects ink droplets only in the graphics and text parts that need to be ejected, and no ink droplets are ejected in the blank areas. This jetting method does not need to charge the ink droplets, and also does not need to charge electrodes and deflection electric fields. The nozzle structure is simple, and it is easy to realize the multi-nozzle of the nozzle, and the output quality is more refined; through pulse control, digitization is easy. However, the ejection speed of ink droplets is generally low. There are three common types: thermal bubble inkjet, piezoelectric inkjet, and electrostatic inkjet. Of course, there are other types.

The realization principle of piezoelectric inkjet technology is: placing many small piezoelectric ceramics near the nozzle of the print head, the piezoelectric crystal will deform under the action of an electric field. Extruded from the ink cavity, ejected from the nozzle, the pattern data signal controls the deformation of the piezoelectric crystal, and then controls the amount of ink ejection. Drop-on-demand hybrid printing using piezoelectric MEMS printhead [12]. The realization principle of thermal bubble inkjet technology is: under the action of heating pulse (recording signal), the temperature of the heating element on the nozzle rises, causing the ink solvent nearby to vaporize to generate a large number of nucleated small bubbles. The volume of the inner bubble continues to increase. When it reaches a certain level, the pressure generated will cause the ink to be ejected from the nozzle, and finally reach the surface of the substrate, reproducing the pattern information [13–18].

The meanings of "3D product printing" and "incremental rapid manufacturing" have evolved and refer to all incremental product manufacturing technologies. Although this has a different meaning from the previous production, it still reflects the common feature of the 3D work production process, which is stacking materials under automatic control [19–24].

This development system is thermal bubble jet technology. It is to design different heater diameter and cavity thickness to deploy DNA beads in this rapid array, and to evaluate the spray flow rate. The boost circuit system in the DNA jet chip is the signal source for driving large flow. The purpose is to adjust the amount of DNA solution sprayed and the output power. When the input voltage needs to be converted to a higher output voltage, a boost converter is the only choice. The boost converter achieves the purpose of boost output by charging the voltage through the internal metal oxide semiconductor field effect transistor (MOSFET), and when the MOSFET is turned off, the inductor is discharged through load rectification.

The conversion process between charging and discharging the inductor will reverse the direction of the voltage through the inductor, and then gradually increase the voltage higher than the input operating voltage. The switching duty cycle of the MOSFET definitely determines the boost ratio. The rated current of the MOSFET and the boost ratio of the boost converter determine the upper limit of the load current of the boost converter; the rated voltage of the MOSFET determines the upper limit of its output voltage. Some boost converters integrate the rectifier and MOSFET to provide synchronous rectification. The integrated MOSFET achieves accurate zero current turn-off, making the boost transformer more efficient. Real-time monitoring of the input power through the maximum power point tracking unit. When the input voltage reaches the maximum input power point, the boost converter starts to work, making the boost converter suitable for DNA printing on glass substrates with maximum power output point. Through the constant on-time generating circuit, the on-time is not affected by temperature and the corner angle of the chip, which enhances the stability of the system.

The technology employed in an inkjet print head is critical. The tremendous advances in inkjet technology primarily began with the development of theories behind inkjet print head technology, including the large-scale droplet ejector used in inkjet printing [1–8]. A continuous inkjet system has the advantages of high-frequency response and high-speed printing. The total number of the nozzles on the inkjet head can reach hundreds or even thousands, and these nozzles are exceedingly intricate. A nozzle can generate rich and harmonious color and smooth mesh tone [9–11]. Inkjets can be categorized into three major types: thermal bubble inkjet, piezoelectric inkjet, and electrostatic inkjet. Other types are also in use. Piezoelectric inkjets function as follows: Many small piezoelectric ceramics are placed near the nozzle of the inkjet head. Piezoelectric crystals deform under the electric field. Subsequently, ink is squeezed from the ink cavity and ejected from the nozzle. The data signal of the pattern controls the deformation of the piezoelectric crystals and then controls the amount of ink ejected. Piezoelectric microelectronic mechanical system (MEMS) inkjet heads are used for hybrid printing [12]. Thermal bubble inkjet technology functions as follows: Under a heating pulse (i.e., the recording signal), the temperature of the heating component on the nozzle increases, evaporating the ink solvent nearby, thereby generating a large amount of small nucleated bubbles. The volume of the internal bubbles continuously increases. When it reaches a certain level, the pressure generated leads to the ejection of the ink from the nozzle and the ink reaching the surface of the substrate, thereby presenting patterns and messages [13–18].

The evolution of three-dimensional (3D) product printing and rapid prototype techniques involves the production technology of all rapid prototypes. Although the rapid prototype technique differs from conventional production, it shares some characteristics of the production process of 3D product printing. Specifically, it stacks materials under automatic control [19–24].

The system developed in this study employed thermal bubble ejection technology. Different diameters for the heater and different cavity thickness were employed to deploy DNA beads in this rapid array. Subsequently, the spray flow rate was evaluated. The boosted circuit system in a DNA jet chip is the signal source for driving large flow. The goal is to adjust the amount of DNA liquid being sprayed and the output power. When the input voltage must be modified to a higher output voltage, a step-up converter is the only option. A step-up converter charges the internal metal oxide semiconductor field effect transistor (MOSFET) to enable an increased output voltage. When the MOSFET is turned off, inductance is discharged through load rectification. The process of changing an inductor between charging and discharging alters the direction of the voltage through the inductor. The voltage gradually increases to the point that it exceeds the input operating voltage. The duty cycle of the MOSFET's switch determines the boost ratio. The rated current and the boost ratio of the step-up converter of the MOSFET determines the upper limit of the load current of the step-up converter. The rated current of the MOSFET determines the upper limit of its output voltage. Some step-up converters integrate rectifiers and MOSFET to provide synchronous rectification. An integrated MOSFET can realize precise zero current shutdown, thereby making the set-up converter more efficient. The maximum power point tracking unit was used to monitor input power in real time. When the input voltage reaches the maximum input power point, the step-up converter starts operating. The step-up converter is used on the glass substrate with a maximum power output point for DNA printing.

2. MEMS Chip Design for Bubble Jet

This study designed five types of liquid ejection chamber structures by manipulating certain parameters, such as heater size, heater number, and loop resistance. Table 1 lists the measurement results. That system analyzed the loop resistance of different heaters. Because it passes through each single circuit series using two sets of heaters to complete the 100-heater design, when designing 100 heaters, the total loop resistance must be one heater larger than the total loop resistance of 50 heaters. In this study, during the process of bubble ejection from the MEMS chip, the sheet resistance of the resistance layer was 29 Ω/m^2. Therefore, the total loop resistance of Model A was the largest. It was twice that of the normal size models (Models B1, C, D, and E). The total loop resistance of Models B1, C, D, and E was approximately 29 Ω/m^2. According to Table 1, its error range was within the allowed design value. Thus, the single chips of each type designed in this study had the same production procedure results, and they were used for subsequent flow rate measurement.

Table 1. List of resistance measurement of single circuit resistance.

Type	A	B1	C	D	E
Heater Size (μm × μm)	105 × 105	105 × 105	78.75 × 105	105 × 78.75	132 × 132
Heater Number	100	50			
Loop Resistance (Ω)	60.1	32.1	25.6	42.2	32.3

Once the power in the DNA-sprayed chip was confirmed to be normal, the characteristics of the growth of the heater bubbles were tested and verified. The film thickness and film quality of the DNA-sprayed chip affect the operating conditions and spraying quality of the heater; therefore, understanding the bubble growth phenomenon and its growth characteristics helps clarify the operating conditions as well as the characteristics of the DNA-sprayed chip in this study.

The designed system adopted the open liquid supply method to observe the bubble growing conditions. For image observation, the synchronous flash method, employing light-emitting diodes (LED, Nichia NSPW500GS-K1, 3.1 V White LED 5 mm), was used to generate a synchronous delayed light source. The system also used a charged-coupled device (CCD, Flir Grasshopper3 GigE GS3-PGE-50S5C-C) to capture images. Figure 1 demonstrates the process of a bubble from nucleation, growth, bubble generation, to dissipation. The system confirmed the growth and dissipation process of bubbles, which can be used for observing starting voltage. Regarding the supply method of liquid in the microchannels, the time when the LED blinks was set as the time required to generate the largest bubble (15 μs). This design prevents incorrect judgments resulting from of an unsuitable blinking time as well as the inability to capture bubble images.

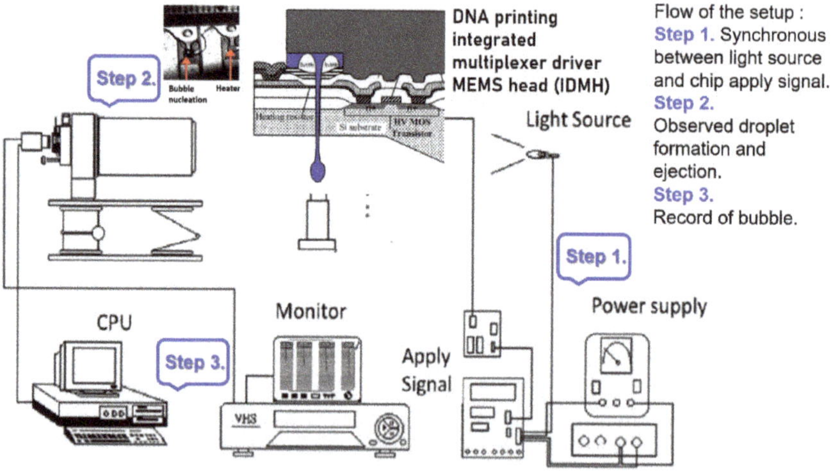

Figure 1. The system uses CCD to capture images. The process of bubble from nucleation, growth, bubble generation to dissipation (Flow of the setup: Step 1. Synchronous input power signal (for light source) and chip drive pulse (Apply signal), Step 2. Observe droplet formation and ejection trajectory through a microscope Observe droplet formation and ejection trajectory through a microscope, Step 3. Process record of bubble from nucleation, growth, bubble generation to dissipation).

In the open pool experiment, we set the operating frequency at 5 KHz. By adjusting the pulse width, we measured the starting voltage of various heaters. Table 2 presents the testing results.

Table 2. Open pool test starting voltage results.

Voltage (V)	Pulse Width (μs)							
	2 μs	2.5 μs	3 μs	4.5 μs	5 μs	5.5 μs	6 μs	6.5 μs
A Type	-	-	-	29.56 V	28.55 V	27.73 V	26.96 V	26.23 V
B Type	21.53 V	19.64 V	18.36 V	-	-	-	-	-
C Type	21.27 V	19.50 V	18.33 V	-	-	-	-	-
D Type	18.13 V	16.53 V	15.40 V	-	-	-	-	-
E Type	27.52 V	25.38 V	23.62 V	-	-	-	-	-

Application-specific integrated circuit (ASIC) chip was used in the DNA liquid ejection system. This system was developed with the aim to design a specific ASIC chip with different apertures and cavity thicknesses to place DNA beads in this rapid array.

This study aimed to develop a unique circuit system to manage the digital register for use in designing and producing ASIC chips for DNA liquid ejection. The system structure comprises serial input data and parallel output data. The "register digital circuit" function generates location "A" to search for the activity nozzle. The register serves as a bistable multivibrator model system.

The proposed technology was compared with the regular scan sequence, and the system with the novel technology only required approximately half the time of the conventional sequence. The serial input data and parallel output data are connected with a shift register (Figure 2). The input is a serial data bit, which is fed to the parallel-bit data output and connected to the boost circuit. Once the information has been input, the system transmits several serial data bits. The system can obtain the first-level parallel data output. Each trigger is a rising edge trigger model.

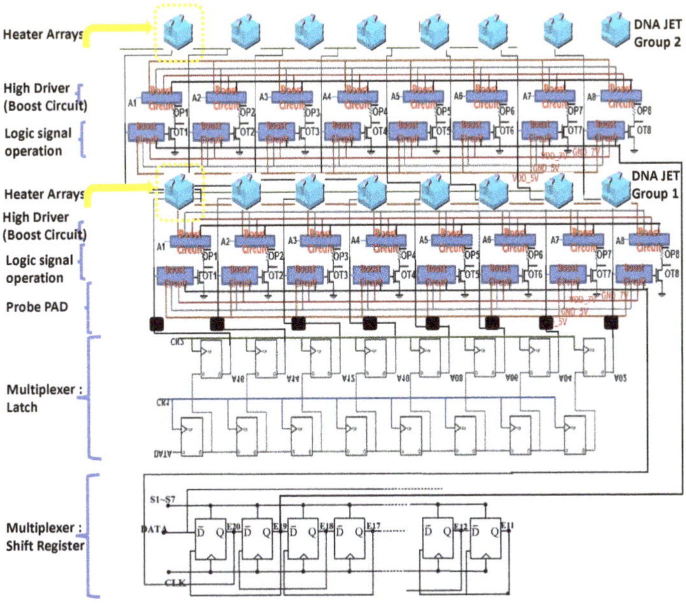

Figure 2. Serial input parallel output shift registers forms of connection.

3. Experiment and Results

We employed Flow-3D to simulate the internal flow field and flow rate distribution in a 3D spray room. The simulation procedure was used to calculate the time and pressure to generate thermal bubbles. Also, the corresponding mean outlet speed state of the fluid was calculated. The "outlet speed state" function in Flow-3D served as the power source for spraying liquid into a monomer. The actual chip generation process and the starting voltage curve were measured and analyzed. Finally, this study presents the flow rate experiment as well as the results and discussion.

3.1. Flow-3D Simulation

The software simulation was based on the actual DNA liquid cavity calculation to simulate the ejection cavity with single-channel supplying liquid. The geometric shape of the ejection cavity is illustrated in Figure 3.

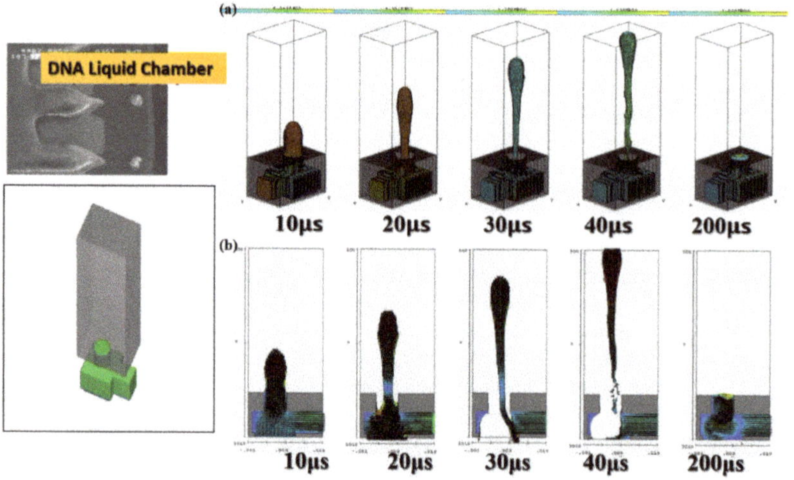

Figure 3. The geometry of the jet cavity. (**a**) The actual DNA liquid chamber, (**b**) the three-dimensional view of the microfluidic single channel. A single-channel jet cavity with 60 µm diameter and 50 µm thickness, with an operating frequency of 5 KHz, in (**a**) three-dimensional side view (**b**) X-Z two-dimensional cross-sectional view, at 10, 20, 30, 40 and 200 µs injection conditions.

The diameter of the ejection outlet was 60 µm. The thickness of the ejection orifice (volcano) was 50 µm. The process thickness of the dry film layer was 60 µm. The bottom design of the ejection cavity was 120 µm × 120 µm. The area of the heater was 105 µm × 105 µm. Under the operating frequency of 5 KHz, the 3D side view and the X-Z two-dimensional (2D) section view were analyzed. Simulation spraying was conducted for 10, 20, 30, 40, and 200 µs. This simulation calculated the area of the top of the thermal resistance. Initially, the mean droplet tip speed was estimated at 10 m/s, and the length of the tail of a droplet was approximately 300 µm. A 200-µm X-Z section view revealed the simulation results. The liquid clearly reached the fluid outlet. The liquid was in a stable state, indicating that the liquid had filled up the cavity as expected. The cavity had a uniquely designed shape. The simulation results revealed the behavior of the geometric shape of the 3D ejection cavity. The simulation results under a 5-kHz operating frequency revealed no shortage of DNA liquid. Therefore, the tail of the DNA droplet had the lowest quality and was easily affected by the asymmetry of the flow field force, resulting in an off-track flying trajectory.

Figure 3 reveals an obvious deflection in the droplet tail of the supplied liquid at 40 µs. As the flight time increased, the horizontal displacement deflection became more substantial. In inkjet printers or other industrial-use inkjet printing processed, the asymmetrical geometric shape of the relative ejection direction (Z-axis direction) must be redesigned to restore the ejection cavity. The direction-related problem did not affect the realization of the planned goals. Figure 4 reveals that, with an orifice diameter of 60 μm and a thickness of 50 μm, the 2D X-Y cross section was higher. The 2D cross-sections before and after the fluid ejection cavity was ejected under instant flow field could be clearly observed. In addition to the orifice, liquid droplets were ejected. The fluid ejection cavity pushes the liquid from the single channel. The results revealed that at 20 μs, the thermal air bubbles enter a state of dissemination. During that stage, fluid starts to refill, and the speed vector passes inside the cavity and points at the outlet channel. During the 30-μs and 40-μs stage, because liquid has been ejected from the orifice, a certain amount of liquid enters the nozzle, and the cavity is not completely filled with liquid. Under this circumstance, external air passes the orifice and enters the cavity while waiting for the fluid to pass the refilled object, and then wait to flow out through the outlet. If air is not completely ejected, it may block the bubbles. Bubbles may also occupy the injection cavity; hence, during the next injection, an adequate volume of liquid cannot be ejected. In addition, because fluid attenuates the tremendous amount of heat generated from the heater, it has a cooling effect. Therefore, the injection cavity inside the chip may accumulate heat. This harmful cycle reduces the ejection performance, finally resulting in the inability to eject liquid droplets; the excessive internal heat also damages the chip. Microfluid channels are not ideal for the injection cavity design. This may be because the inside retains bubbles, resulting in reduced performance. If the heater does not immediately cease operation, the excessive temperature can damage the chip. A major part of the overall design parameters is focused on addressing this concern.

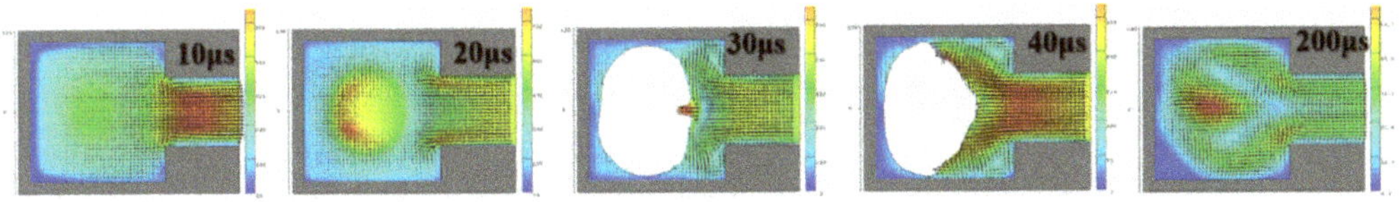

Figure 4. Calculate and simulate the injection of water in a single-channel injection chamber with a nozzle diameter of 60 μm and a thickness of 50 μm, at an operating frequency of 5 KHz, in the X-Y two-dimensional cross-sectional view, at 10, 20, 30, 40 and 200 μs.

Figure 5 depicts the calculation results of the 2D X-Z cross section. At 100 μs and 200 μs, the fluid injection orifice did not completely fill the chamber. This may be because the size of the single-channel injection cavity was unsuitable for the highest operating frequency of 10 KHz. Thus, subsequent calculation simulations employed 5 KHz as the reference operating frequency. The calculation simulation results were calculated according to the operating frequency of the impact. Figure 6 illustrates the injection cavity height as 60 μm and 30 μm and reveals the 2D X-Y cross section. At 100 μs and 200 μs, the fluid injection orifice did not completely fill the chamber. In those stages, the fluid was still filling the chamber, and the flow field was not yet stable.

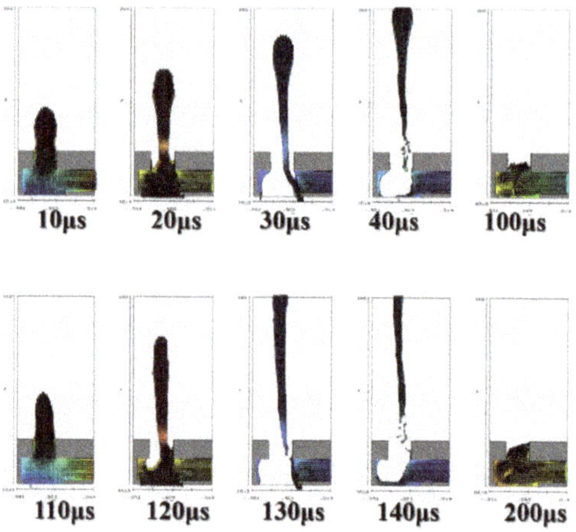

Figure 5. Calculate and simulate water in a single-channel jet cavity with a nozzle diameter of 60 μm and a thickness of 50 μm, with an operating frequency of 10 KHz, in the XZ two-dimensional cross-sectional view, at 10, 20, 30, 40, 100, 110, 120, 130, 140 and 200 μs injection situation.

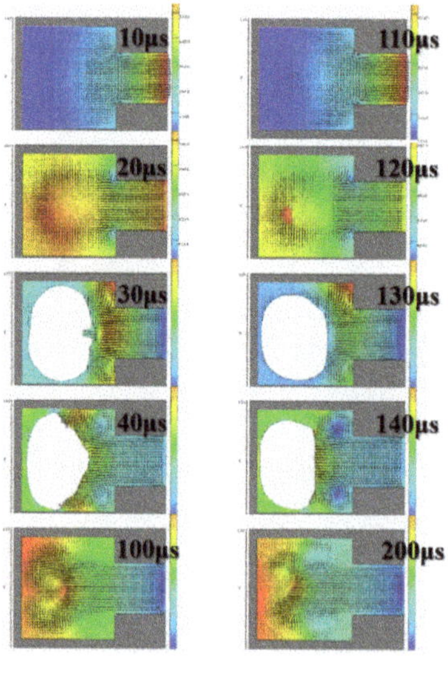

Figure 6. Calculate and simulate water in a single-channel spray chamber with a spray hole diameter of 60 μm and a thickness of 50 μm, with an operating frequency of 10 KHz, in an XY cross-sectional view, at 10, 20, 30, 40, 100, 110, 120, 130, 140 and 200 μs injection situation.

3.2. *Starting Voltage Curve*

The DNA printing integrated multiplexer driver MEMS head (IDMH) follows the 0.35-um 2P2M processing model standard complementary metal–oxide–semiconductor (CMOS) design rules to establish the model (Figure 7). This study involved high-pressure processing. For systems based on the 0.35-um DPDM 12 V/12 V or 12 V/5 V mixed model processing model design and allocating high voltage circuit, this study design may serve as a basic allocation guide. This study investigated the differences from the original 0.35-um CMOS mixed-mode (DPDM, 5 V) design principles. That is, during this process, we specified NDD, PDD, HV_OX rules, and high voltage equipment rules. Basically, the low voltage (LV) part adopts the 5-V logic, and its allocation rules are the same as those for the Taiwan Semiconductor Manufacturing Company common 0.6-um DPDM (including the PO1/PO2 capacitor options) ASIC design rules.

Figure 7. The DNA printing integrated multiplexer driver MEMS head (IDMH).

The measurement of the starting voltage curve was conducted on the ASIC on the actual manufactured MEMS chip. Figures 8 and 9 reveal the starting voltage and energy of the Type A chip, respectively. That chip was designed using dual heaters in series. The figures reveal that the starting voltage of the chip decreased as the pulse width increased. From the energy perspective, an increase in pulse width increased the required energy.

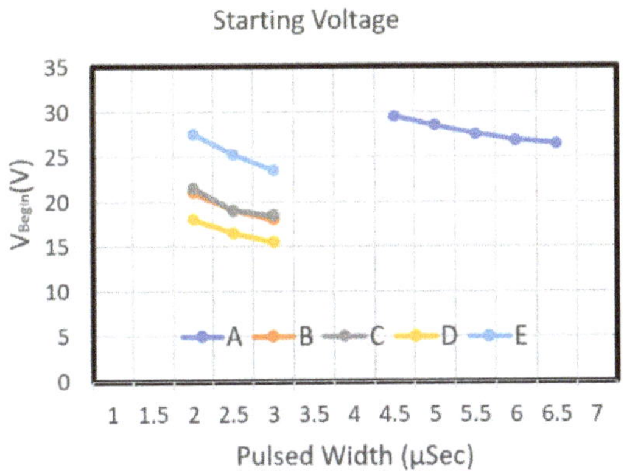

Figure 8. The initial voltage diagrams of chip number A,B,C,D,E type.

Figure 8 depicts the starting voltage graphs of Type B and Type E chips. The nozzle density of these two chips was 50, and their resistance was 35 Ω. As the pulse width increased, the starting voltage of Type B and Type E chip decreased. From the energy perspective, an increase in pulse width increased the required energy (Figure 9). The Type E design required more energy than did Type B. Although the two designs have the same circuit resistance and the same heater resistance (with an L/W value of 1), the area of Type E is greater than that of Type B. Therefore, regarding the power density of the work performed by the heater, when the same driving pulse voltage was input into these two heater designs, the power density of the work performed by the Type E heater was higher than that of Type B. Type B was designed very high; thus, the starting voltage curve of Type E is higher than that of Type B. Figure 8 illustrates the starting voltage of chip Type C and Type D. Both chips have a long heater design, and they have the same heater area. The L/W value of these two models was 0.75 and 1.33, respectively. The figure indicates that as the pulse width increased, the starting voltage of Type C and Type D declined. The starting voltage of Type D was higher than that of Type C. Concerning energy, an increase in pulse width increased the required energy (Figure 9). The energy required for the Type C design is higher than that of Type D.

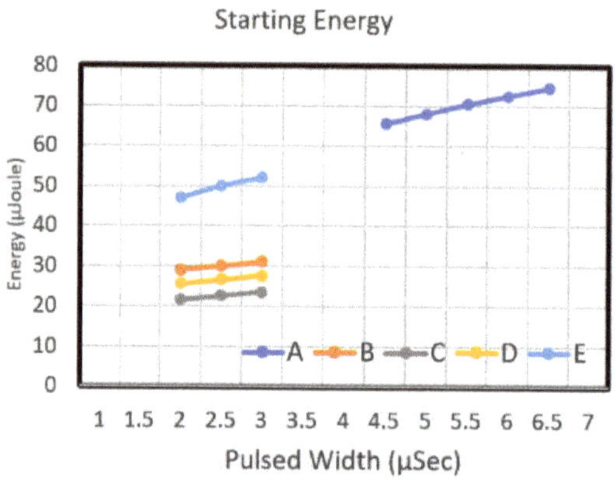

Figure 9. The initial energy diagrams of chip number A,B,C,D,E type.

3.3. *Flow Experiment and Results Discussion*

This study endeavored to estimate the amount of DNA being ejected and the amount of the DNA sample for testing. Therefore, we conducted tests on power conditions and DNA flow.

The spray DNA chip composed of 50 nozzles could be placed rapidly, and the heater area was increased. The maximum flow rate was 3.5 cc/min, and the DNA chip ejection speed was increased. In this study, we increased the density of the unit area nozzles without increasing the size of the chip, and the microchannel design method was used to expand the original two-line nozzles into four-line nozzles. The number of nozzles used on the spray DNA chip was substantially increased from 50 to 100. The increase in nozzle number was for testing maximum flow. The Type A chip had an increased density and 100 nozzles. On a chip with increased nozzle density, we first used chip No. B2-01 to conduct a flow test. We fixed the driving voltage at 25 V for the experiment, and we observed the flow curve under various pulse widths and operating frequencies (Figure 10).

An experiment was conducted to analyze different pulse widths. When the pulse width was 4 μs, the flow rate of the injected monomer linearly increased at a frequency of 1–2 KHz. The linear slope was 2.22. When the frequency was 2.5 KHz, the flow rate no longer increased linearly. When the operating frequency was increased to 3 KHz, the flow rate began to decrease. When the pulse width was 5 μs, the linearly increased frequency was only 1.5 KHz, and the linear slope was 2.68. When the pulse width was increased to 6 μs, the linearly increased frequency was smaller than 1.5 KHz, and the linear slope was 3.

We observed the experimental results under an operating frequency of 15 KHz (Figure 11). The volume of the spray dots increased from 330 pl under a pulse width of 4 μs to 500 pl. At 6 μs, the volume increased with the pulse width. The ejection rate of the ejection point's volume, in contrast to the chamber capacity, increased from 40% to 58%. The experimental results indicated that an increase in energy increased the pressure of bubbles and more liquid was expelled.

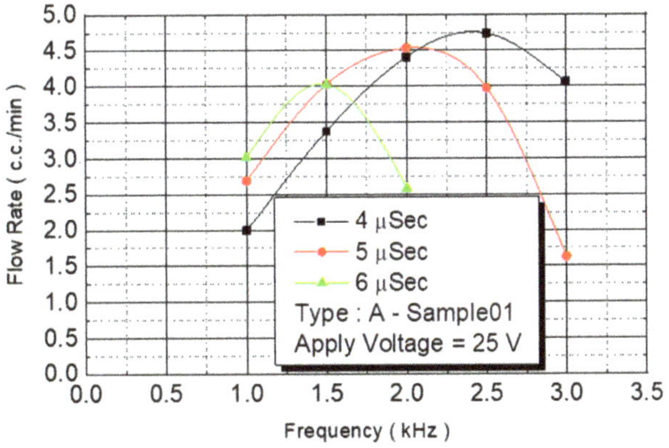

Figure 10. A Type-Sample01 flow test.

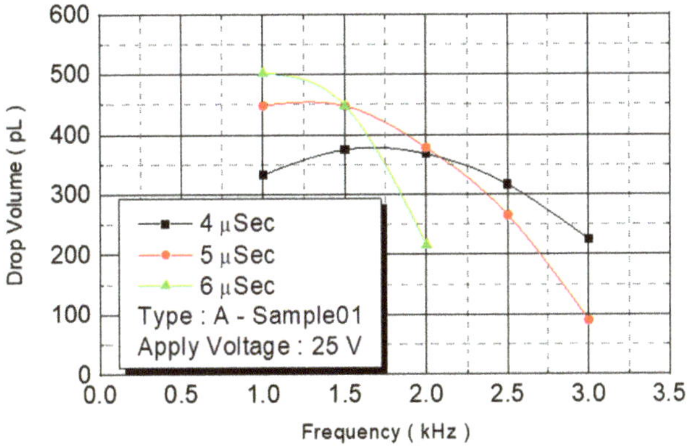

Figure 11. A Type-Sample01 drop volume.

We observed the flow rate changes under various power conditions, with an operating frequency of 2 KHz and a pulse width of 4 and 5 μs. With power ranging from 4.3 W to 5.7 W, the flow rate of the injected monomer was 5.5 cc/min (Figure 12). The experimental parameter was the maximum yield of the DNA liquid.

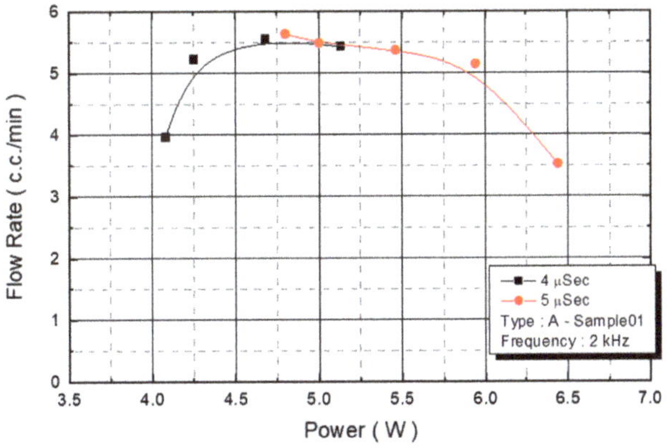

Figure 12. A Type-Sample01 flow rate.

This study tested the total flow rate of the ejected DNA chip. First, we injected a monomer in chips in the B1-oo series. Under an operating frequency of 2 KHz, various operating voltages were employed to test the flow rate. The results revealed that the amount of injected monomer was closely related to the provided power from the outside environment (Figure 13). Although the injected monomer had a different pulse width energy, under the same power condition, the energy obtained from injecting the monomer was the same. The bubbles generated on the heater had the same instant pressure. Therefore, the introduced flow rate was similar. The figure revealed that the flow rate increased with the added power. At a power condition of 4.5 W, the system had a maximum flow rate of 3.5 cc/min. When the power exceeded that level, the flow rate decreased.

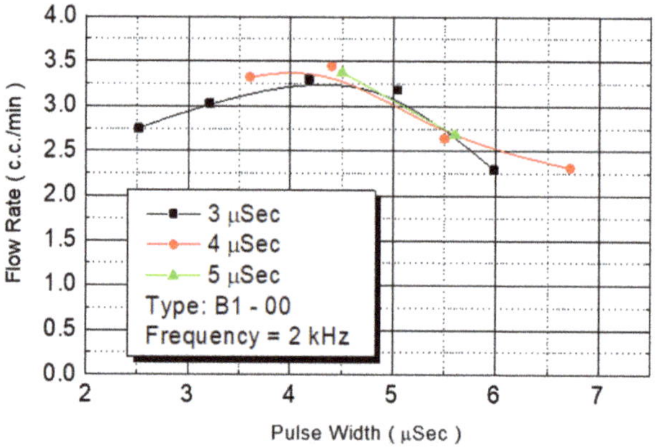

Figure 13. B1-00 flow test.

We employed the same experimental procedures for C Type-01 and D Type-02 flow rate test chips. The results are presented in Figures 14 and 15, respectively. Under the condition of a fixed operating frequency of 2 KHz, the amount of monomer injected changed according to the external power (Figure 14). The maximum flow rate (3.5 cc/min) of the injected monomer appeared when the power was 4.5 W. Figure 15 revealed the same result: The maximum flow rate (2.5 cc/min) of the injected monomer in D Type-02 was achieved under the power of 4.2 W. At 4-W added power, another D Type-02 chip achieved its maximum flow (3 cc/min).

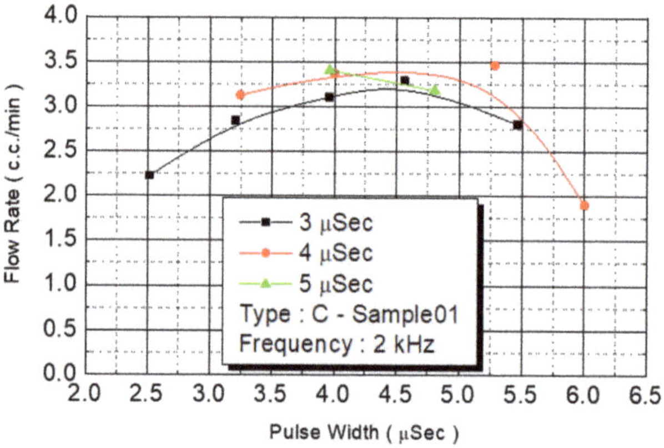

Figure 14. C Type-01 flow test.

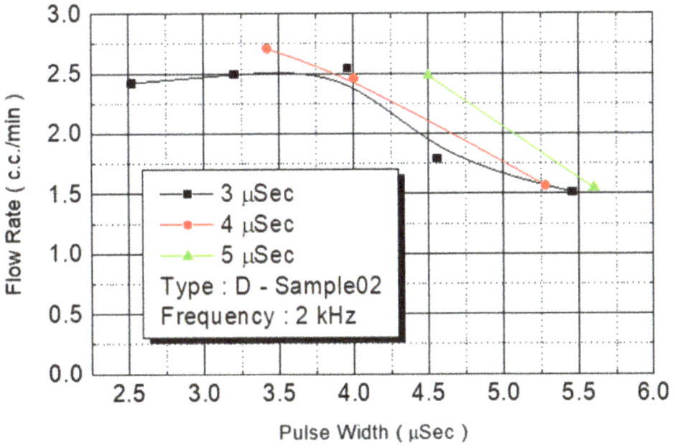

Figure 15. D Type-02 flow test.

In the aforementioned experiments, the chip types were heaters (size: 105 μm × 105 μm) with a nozzle density of 50. The nozzle outlet size was 80 μm for all chip designs. On the basis of the aforementioned results, the maximum flow rate for all B1 series chips was limited to 3.5 cc/min. The required power for generating the maximum flow rate was 3.5–4.5 W.

In an effort to increase the amount of monomer being injected, we first considered expanding the size of the heater. The unit heater size of chip No. E1-00 was increased from 105 μm × 105 μm to 132 μm × 132 μm. The size of the microfluid flow channel was increased from 120 μm × 120 μm to 150 μm × 150 μm. The injection cavity volume increased by 56%. The number of monomer injection nozzles was maintained at 50 for the experiment. Figure 16 reveals the flow rate test results. Under an operating frequency of 3 KHz and pulse width of 3 μs, the flow rate of the nozzle under 20–25 V of driving voltage range could be observed. The flow rate increased with the voltage. In excess of 25 V, any further voltage increase reduced the flow rate. When the pulse width was changed to 3.5 μs, the flow rate's increase range was reduced to 20–23 V. When the voltage exceeded 23 V, the flow rate began to decrease. The pulse width was contained to increase to 4 and 5 μs, and the increased voltage range was further reduced.

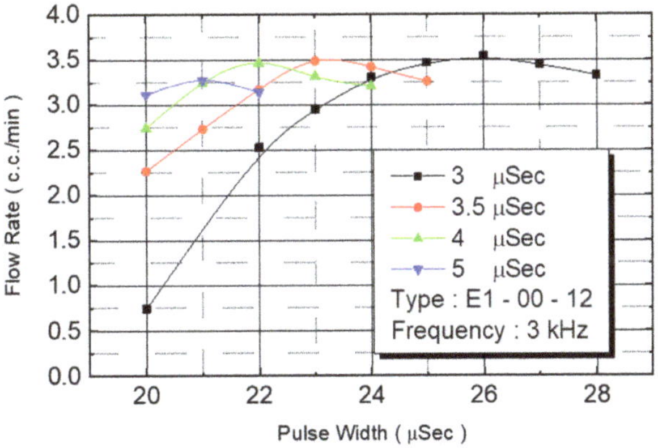

Figure 16. E1 type flow test.

We observed flow performance under various pulse widths under 20 V of driving voltage. As the pulse width increased from 3 μs to 3.5 μs, 4 μs, and to 5 μs, the corresponding flow rate increased from 0.75 cc/min to 23 cc/min, 2.7 cc/min, and to 3.1 cc/min. Figure 17 depicts the correlation between power and flow rate. Flow rate was still affected by power. The figure reveals that as the power increased, the flow rate increased from 0.75 cc/min to 3.5 cc/min until the power reached 6.5 W. When power increased further, the flow rate did not increase with the energy, revealing that the maximum flow rate of this design is 3.5 cc/min.

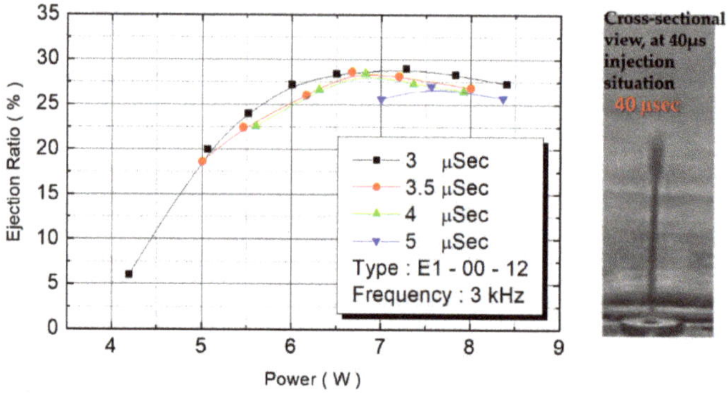

Figure 17. E1 type ejection rate relationship.

The summary of the experimental results is obtained from increasing the number of nozzles, increasing the input signal frequency, pulse width, the reduced heater size, and decreasing the input voltage (power) data to obtain the optimal DNA spray. DNA is not easy to obtain, it is estimated the amount of DNA sprayed outcomes from the data (increased number of nozzles, reduced heater size) while quickly deploying DNA according to the input parameters.

Spraying was conducted using a special 3D MEMS structure. DNA spraying technology can be employed to test the relationship between the flow rate of DNA on the glass slide and time. Because DNA cannot be obtained easily, rapidly deploying DNA while calculating the amount of DNA being sprayed is essential. Microarray biochip sample spots can probe clusters composed of DNA, RNA, peptide proteins, antibodies, cells, or human tissues. The sampling material was determined by the biological characteristics to be measured. For example, the DNA chips for testing the single nucleotide A, T, C, or G (single nucleotide polymorphism) are sample points of different single-stranded DNA sequence fragments being grouped; in this case, DNA sequences were used. The paring principle of AT and CG. When the designed single-stranded DNA sample spot can be paired with the sample and emit fluorescence, it indicates that the DNA sequence was changed by a single nucleotide, and the change in the DNA sequence can be used to explore the chromosome genome diversity of species (including humans). Another example is the protein biochip for detecting food-specific allergens. The conventional method is to place dozens to hundreds of different food allergen sampling point on a standard-size 1″ × 3″ glass. A probe design of 200–500 microns and its fluorescence strength can be used to screen and test dozens to hundreds of different food allergens.

Conventional experimental procedures require a long time to extract and purify DNA (DNA separation), removing polymerase chain reaction inhibitors to obtain DNA. Although DNA extraction is feasible, during the extraction process, precious samples may be lost. Human operation may lead to experimental errors and sample cross contamination. DNA nucleic acid is highly valuable. Through this experiment, we can precisely estimate the total amount of DNA required for thousands of spots on the microarray biochip sprayed with DNA.

4. Conclusions

This study provided a DNA printing IDMH and conducted microfluidic flow estimation. Under the designed DNA spray cavity and 20 V of driving voltage, the flow performance of different pulse widths was discovered to increase with the pulse width. The E1 type flow rate test revealed that as the corresponding flow rate increased to 3.1 cc/min, the flow rate was affected by changes in power. As the power increased, the flow rate increased from 0.75 cc/min to 3.5 cc/min, up to a power of 6.5 W. If the power was increased further, the flow rate did not increase with the energy. This reveals that this table design is the largest. The flow rate was 3.5 cc/min.

In a specifically designed DNA spray room structure with an operating frequency of 2 KHz and pulse widths of 4 μs and 5 μs, we observed changes in flow rate under various power conditions. Within the power range of 4.3–5.87 W, the flow rate of the injected monomer was 5.5 cc/min. This did not change as the power increased. DNA is precious and cannot be obtained easily. Through this experiment, we can precisely estimate the total amount of DNA required for thousands of spots on the microarray biochip sprayed with DNA

Author Contributions: Resources, Z.-X.C., J.-C.L., and C.-W.P.; data curation, J.-C.L., P.B., C.-W.P., and Z.-X.C.; original draft preparation, J.-C.L., P.B., C.-W.P., and Z.-X.C.; review and editing, J.-C.L. and C.-W.P. All authors have read and agreed to the published version of the manuscript.

Funding: This work was financially supported of the Higher Education Sprout Project by the Ministry of Education (MOE) in Taiwan (DP2-108-21121-01-O-05-04). DP2-109-21121-01-O-01-03, MOST-109-2918-I-038-002, TMU-NTUST-109-10, MOST- 109-2221-E-038 -013.

Conflicts of Interest: The authors declare no conflict of interest.

References

1. Pydar, O.; Paredes, C.; Hwang, Y.; Paz, J.; Shah, N.; Candler, R. Characterization of 3D-printed microfluidic chip interconnects with integrated O-rings. *Sens. Actuators Phys.* **2014**, *205*, 199–203. [CrossRef]
2. Ohtani, K.; Tsuchiya, M.; Sugiyama, H.; Katakura, T.; Hayakawa, M.; Kanai, T. Surface treatment of flow channels in microfluidic devices fabricated by stereolitography. *J. Oleo Sci.* **2014**, *63*, 93–96. [CrossRef]
3. Castrejn-Pita, J.R.; Martin, G.D.; Hoath, S.D.; Hutchings, I.M. A simple large-scale droplet generator for studies of inkjet printing. *Rev. Sci. Instrum.* **2008**, *79*, 075108. [CrossRef] [PubMed]
4. Asai, A. Application of the nucleation theory to the design of bubble jet printers. *Jpn. J. Appl. Phys. Regul. Rap. Short Notes* **1989**, *28*, 909–915. [CrossRef]
5. Aoyama, R.; Seki, M.; Hong, J.W.; Fujii, T.; Endo, I. Novel Liquid Injection Method with Wedge-shaped Microchannel on a PDMS Microchip System for Diagnostic Analyses. In *Transducers' 01 Eurosensors XV*; Springer: Berlin, Germany, 2001; pp. 1204–1207.
6. Xu, B.; Zhang, Y.; Xia, H.; Dong, W.; Ding, H.; Sun, H. Fabrication and multifunction integration of microfluidic chips by femtosecond laser direct writing. *Lab Chip* **2013**, *13*, 1677–1690. [CrossRef] [PubMed]
7. Nayve, R.; Fujii, M.; Fukugawa, A.; Takeuchi, T.; Murata, M.; Yamada, Y. High-Resolution long-array thermal ink jet printhead fabricated by anisotropic wet etching and deep Si RIE. *J. Microelectromech. Syst.* **2004**, *13*, 814–821. [CrossRef]
8. O'Connor, J.; Punch, J.; Jeffers, N.; Stafford, J. A dimensional comparison between embedded 3D: Printed and silicon microchannesl. *J. Phys. Conf. Ser.* **2014**, *525*, 012009. [CrossRef]
9. Fang, Y.J.; Lee, J.I.; Wang, C.H.; Chung, C.K.; Ting, J. Modification of heater and bubble clamping behavior in off-shooting inkjet ejector. In Proceedings of the IEEE Sensors, Irvine, CA, USA, 30 October–3 November 2005; pp. 97–100.
10. Lee, W.; Kwon, D.; Choi, W.; Jung, G.; Jeon, S. 3D-Printed microfluidic device for the detection of pathogenic bacteria using size-based separation in helical channel with trapezoid cross-section. *Sci. Rep.* **2015**, *5*, 7717. [CrossRef] [PubMed]
11. Shin, D.Y.; Smith, P.J. Theoretical investigation of the influence of nozzle diameter variation on the fabrication of thin film transistor liquid crystal display color filters. *J. Appl. Phys.* **2008**, *103*, 114905-1–114905-11. [CrossRef]
12. Kim, Y.; Kim, S.; Hwang, J.; Kim, Y. Drop-on-Demand hybrid printing using piezoelectric MEMS printhead at various waveforms, high voltages and jetting frequencies. *J. Micromech. Microeng.* **2013**, *23*, 8. [CrossRef]
13. Shin, S.J.; Kuka, K.; Shin, J.W.; Lee, C.S.; Oha, Y.S.; Park, S.O. Thermal design modifications to improve firing frequency of back shooting inkjet printhead. *Sens. Actuators Phys.* **2004**, *114*, 387–391. [CrossRef]
14. Rose, D. Microfluidic Technologies and Instrumentation for Printing DNA Microarrays. In *Microarray Biochip Technology*; Eaton Publishing: Norwalk, CT, USA, 2000; p. 35.

15. Wu, D.; Wu, S.; Xu, J.; Niu, L.; Midorikawa, K.; Sugioka, K. Hybrid femtosecond laser microfabrication to achieve true 3D glass/polymer composite biochips with multiscale features and high performance: The concept of ship-in-abottle biochip. *Laser Photon. Rev.* **2014**, *8*, 458–467. [CrossRef]
16. McIlroy, C.; Harlen, O.; Morrison, N. Modelling the jetting of dilute polymer solutions in drop-on-demand inkjet printing. *J. Non Newton. Fluid Mech.* **2013**, *201*, 17–28. [CrossRef]
17. Anderson, K.; Lockwood, S.; Martin, R.; Spence, D. A 3D printed fluidic device that enables integrated features. *Anal. Chem.* **2013**, *85*, 5622–5626. [CrossRef] [PubMed]
18. Avedisian, C.T.; Osborne, W.S.; McLeod, F.D.; Curley, C.M. Measuring bubble nucleation temperature on the surface of a rapidly heated thermal ink-jet heater immersed in a pool of water. *Proc. R. Soc. A Lond. Math. Phys. Sci.* **1999**, *455*, 3875–3899. [CrossRef]
19. Lim, J.H.; Kuk, K.; Shin, S.J.; Baek, S.S.; Kim, Y.J.; Shin, J.W.; Oh, Y.S. Failure mechanisms in thermal inkjet printhead analyzed by experiments and numerical simulation. *Microelectron. Reliab.* **2005**, *45*, 473–478. [CrossRef]
20. Shallan, A.; Semjkal, P.; Corban, M.; Gujit, R.; Breadmore, M. Cost-Effective 3D printing of visibly transparent microchips within minutes. *Anal. Chem.* **2014**, *86*, 3124–3130. [CrossRef] [PubMed]
21. Cavicchi, R.E.; Avedisian, C.T. Bubble nucleation and growth anomaly for a hydrophilic microheater attributed to metastable nanobubbles. *Phys. Rev. Lett.* **2007**, *98*, 124501. [CrossRef] [PubMed]
22. Kamei, K.; Mashimo, Y.; Koyama, Y.; Fockenberg, C.; Nakashima, M.; Nakajima, M.; Li, J.; Chen, Y. 3D printing of soft lithography mold for rapid production of polydimethylsiloxane-based microfluidic devices for cell stimulation with concentration gradients. *Biomed. Microdevices* **2015**, *17*, 36. [CrossRef] [PubMed]
23. Shin, S.J.; Kuka, K.; Shin, J.W.; Lee, C.S.; Oha, Y.S.; Park, S.O. Firing frequency improvement of back shooting inkjet printhead by thermal management. In Proceedings of the TRANSDUCERS'03. 12th International Conference on Solid-State Sensors.Actuators and Microsystems. Digest of Technical Papers (Cat. No.03TH8664), Boston, MA, USA, 8–12 June 2003; Volume 1, pp. 380–383.
24. Laio, X.; Song, J.; Li, E.; Luo, Y.; Shen, Y.; Chen, D.; Chen, Y.; Xu, Z.; Sugoioka, K.; Midorikawa, K. Rapid prototyping of 3D microfluidic mixers in glass by femtosecond laser direct writing. *Lab Chip* **2012**, *12*, 746–749. [CrossRef] [PubMed]

micromachines

MDPI

Article

Wireless Battery-Free Harmonic Communication System for Pressure Sensing

Deepak Kumar *, Saikat Mondal, Yiming Deng and Premjeet Chahal

Department of Electrical and Computer Engineering, Michigan State University, East Lansing, MI 48824, USA; mondalsa@msu.edu (S.M.); dengyimi@egr.msu.edu (Y.D.); chahal@egr.msu.edu (P.C.)
* Correspondence: kumarde2@msu.edu

Received: 12 November 2020; Accepted: 24 November 2020; Published: 27 November 2020

Abstract: In this paper, an efficient passive wireless harmonic communication system is proposed for the real-time monitoring of the pressurized pipelines. A pressure sensor is fabricated using the additive manufacturing technique and a harmonic radio frequency (RF) tag is designed to operate at the fundamental frequency (f_o) of 2 GHz that shifts the phase of the back reflected RF signal according to the applied pressure ranging from 0 to 20 psi. A power efficient phase modulation with virtually no losses is achieved using a hybrid coupler-based phase shifter that efficiently reflect back the incoming signal using an end coupled reactive impedance element/sensor. The phase delay introduced by the reactive element gets doubled with the second harmonic communication, which increases the sensitivity by a factor of two. The concept of harmonic backscattering is exploited to reduce the effects of multi-path interference and self jamming, as well as improving the signal-to-noise ratio (SNR).

Keywords: passive; pipeline; pressure; reactive impedance sensor; structural health monitoring; wireless

1. Introduction

Pipelines are the safest and economically viable arterial networks for transporting natural gases and oils across the globe. The pipeline network expands at a faster rate due to the projected growth in population and increased rate of urbanization, creating a demand in supply of natural gas for domestic and industrial use [1]. The natural gases are gathered, transported, and distributed at various pressure levels via metals, composites, and plastic pipelines. Any material subjected to a high or a low cyclic pressure over a prolonged period of time induces a stress on the material. This stress over time compromises the safety and reliability of the pipeline and may lead to a catastrophic failure [2–4]. In order to prevent such failures, there is a growing need for the development of economical, real-time, scalable, structural integrity monitoring, and sensing system for the pipeline infrastructure implemented in refineries, chemical plants and manufacturing facilities.

Related Work

In literature, a number of structural health monitoring (SHM) techniques exist to evaluate the integrity of pipes aiding in improving safety and lifespan of the pipeline infrastructure [5,6]. The different indicators that are monitored for elucidating the pipeline health are cyclic pressure, surface temperature, humidity, and vibration [5,6]. Among these indicators, pressure is the most important and found to be the primary cause of damage to the pipelines [7].

A number of techniques have been proposed to monitor the pipeline pressure; such as piezo based resistive sensors, micro-electromechanical systems (MEMS)-based capacitive sensors, and fiber Bragg grating (FBG)-based negative pressure sensors [7–10]. The fabrication process of all the sensors is very complex, which either requires a hefty clean room process or very expensive tools for precision.

For example, the MEMS-based capacitive sensors require expensive substrate material, chemical etching, wire bonding, and separate housing [10]. Additive manufacturing is a viable alternative of this process, which has been previously proven for electronic system and component design [11]. Three-dimensional printing technique was previously used for designing multiple components in a pressure sensor [12–14]. However, a fully 3D-printed pressure sensor is still a feat to achieve. Most of the published research is focused on various other applications and gas pipeline is not the primary focus, which leads to over complexity, limited communication range, and scalability issues.

One of the major limitations of these sensor units is that a direct wired connection is required for data acquisition. The direct probing or wired connection limits the application of the sensors and poses a problem for inaccessible environments, like under ground buried pipelines or large infrastructure with a requirement of thousands of sensing nodes.

A number of wireless sensing approaches have been proposed as an alternate solution to overcome the disadvantages of the wired sensing methods [15,16]. The wireless methods monitor the pressure in the pipeline by deploying a large number of active sensor nodes. The integrated nodes on the surface of the pipes transmit the data wirelessly to the interrogator. The disadvantage of active sensors is the need for an on-board battery, charging and/or replacement requirements increase the implementation and maintenance cost of the system [15–17]. Battery-free, low-cost, real-time sensing is ideal to provide a robust and economical system with longer life span and lower maintenance [18,19].

Inductor-capacitor (LC) resonator-based passive sensor tags have been proposed in literature for monitoring the pressure based on capacitive loading, which in turn shifts the resonance frequency [20]. LC resonant sensors operate in near-field configuration and have a limited interrogation distance [21]. The conventional far-field passive pressure sensors operate at a single frequency, which are prone to clutter due to multi-path interference and have a lower signal-to-noise (SNR) due to self-jamming [18]. To overcome these limitations, the harmonic communication approach is desired, that eliminates self jamming, reduces multi-path interference, lowers down phase noise, and increases the SNR. Harmonic-based sensors have been reported in literature for sensing temperature, and pH [22,23]. In this work, a passive harmonic RF tag with integrated pressure sensor is presented for the first time for the real-time monitoring of the pipeline infrastructure.

In this paper, a harmonic RF system is proposed for gas pipeline infrastructure that consists of an RF tag coupled with a pressure sensor and an RF interrogator. The harmonic RF sensor tag has a receiver (2 GHz) and transmitter (4 GHz) antenna, a harmonic doubler, and an integrated pressure sensor with the phase shifter. The integrated pressure sensor is designed with a cavity based parallel plate reactive impedance element/sensor that changes its reactance with the change in applied pressure from 0–20 psi. The designed sensor is fabricated using the additive manufacturing technique, which uses thin liquid crystalline polymer (LCP) substrate as its diaphragm. The phase shifter translates the change in reactance to the change in received 2 GHz signal's phase, with the power efficient modulation technique, which is backscattered at harmonic frequency of 4 GHz. A schematic of the proposed sensing system over the pipeline infrastructure is shown in Figure 1.

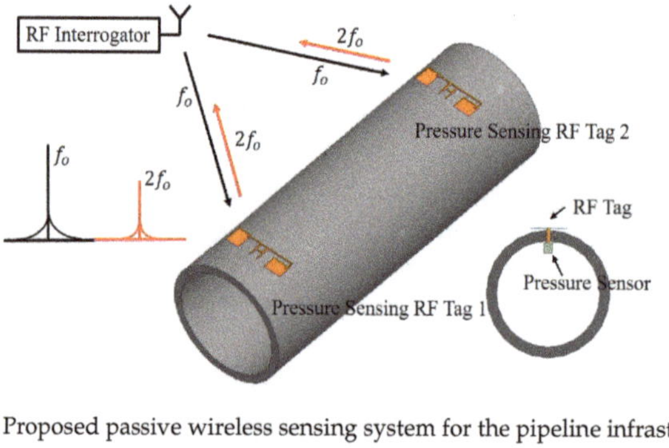

Figure 1. Proposed passive wireless sensing system for the pipeline infrastructure.

2. Design and Setup

The design of passive wireless harmonic system includes: (A) 3D printed pressure sensor (B) passive harmonic RF sensor tag, and (C) harmonic RF interrogator.

2.1. 3D Printed Pressure Sensor

The designed pressure sensor is an air cavity based parallel plate capacitor, where one electrode has a fixed position and second electrode is deformable due to the pressure as shown in Figure 2. The air cavity in between electrode acts as dielectric medium and holds 1 atmospheric pressure or reference pressure. The pressure is applied through a tunnel directly onto the diaphragm with copper electrode.

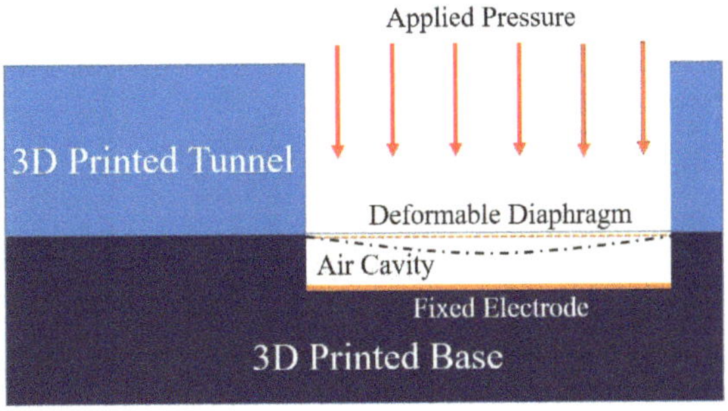

Figure 2. Air cavity-based fully 3D printed pressure sensor.

The pressure sensor is fabricated in three parts, as shown in Figure 3. The first component has the air cavity with the fixed electrode and electrical access holes for rigid coaxial cable. The copper/fixed electrode is plated into the cavity's bottom of the 3D printed part. The second component is the deformable electrode, which is realized using 18 μm-thick LCP film with copper plating. The LCP film is fixed on top of the cavity, such that it makes an electrical connection with the outer shell of the rigid coaxial cable. The third component is a pressure tunnel that directs the pressure onto the electrodes and hold the LCP film in its place. The top part has four alignment legs, which perfectly fits with the bottom component. An extra layer of polymer resin is coated on top of the fully assembled pressure sensor and cured under ultraviolet (UV) light, for eliminating any pressure leaks into or from the air cavity and acts as a seal of the pressure sensor. The outer resin layer also holds the three separately fabricated components together and replace any use of extra adhesive.

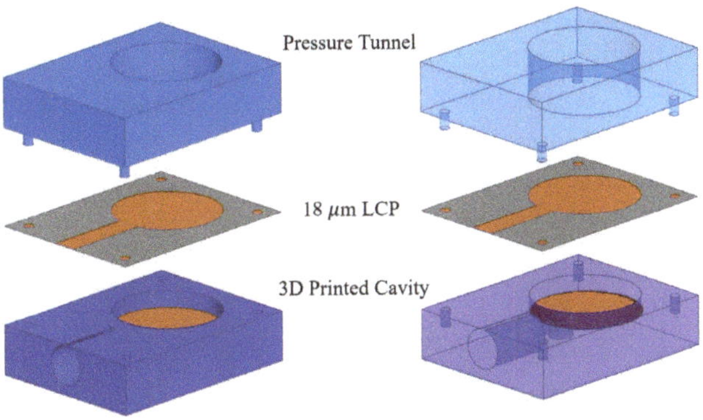

Figure 3. Exploded view of 3D printed pressure sensor with three main components.

Figure 4 shows a fully assembled view of the fabricated pressure sensor, where three components fit perfectly and allow a coaxial cable to make electrical connection with the electrodes for reactance measurement. The total volume of the fabricated sensor is 10 × 13 × 6 mm^3. The diameter of the pressure tunnel is 6.6 mm and the depth of the pressure cavity is 1 mm. The change in pressure deforms the LCP electrode and reduces the separation of two electrodes, therefore a change in electrical response, which can be measured using an impedance analyzing equipment.

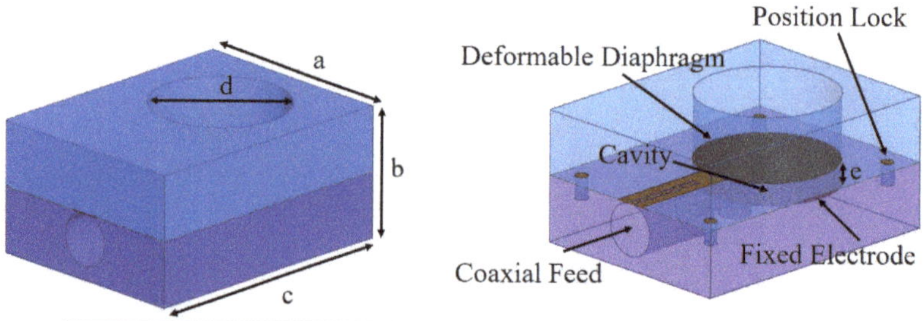

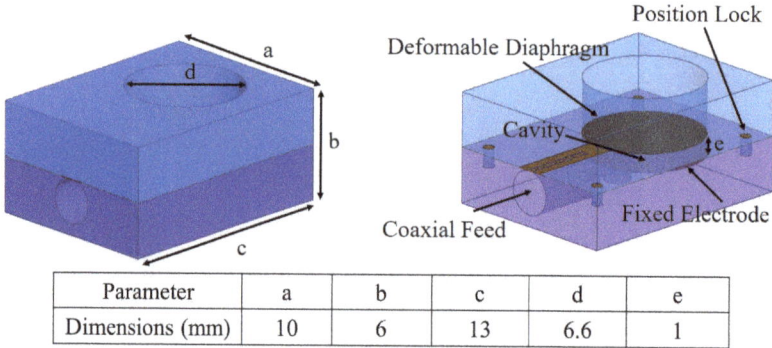

Parameter	a	b	c	d	e
Dimensions (mm)	10	6	13	6.6	1

Figure 4. Fully assembled 3D printed pressure sensor with component annotation and dimensions.

A sealed cavity-based sensor with deformable diaphragm in parallel plate configuration is used to measure the pressure, ranging from 0 to 20 psi. The pressure sensor is integrated inside a polypropylene plastic pipe and probed using an SMA connector from the outside, as shown in Figure 5A. A vector network analyzer (VNA) is used in a direct wired configuration to measure the impedance of the pressure sensor at 2 GHz, as shown in Figure 5B.

The measured impedance is plotted on a normalized smith chart for varying pressure, as shown in Figure 6. The reactive impedance changes from 10.23–17.80 Ω for the applied pressure. A small series resistance is observed in the acquired data, which is due to the parasitics associated with the high frequency and losses from the cable. The direct wired measurements support the hypothesis of significant change in reactive impedance due to pressure difference in between sealed cavity and pipeline pressure. Moreover, it provides an insight into the parasitic real impedance of the sensor and provide data for simulating ideal phase shift for a given range of impedance.

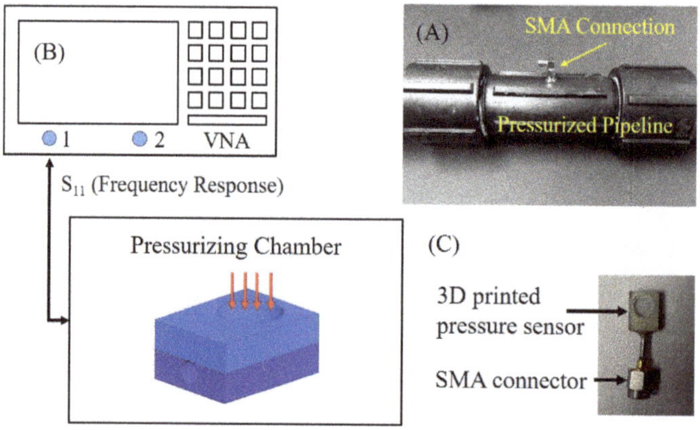

Figure 5. (**A**) Pressurizing chamber (pipe) with integrated pressure sensor, (**B**) Schematic of the measurement setup, and (**C**) Fabricated pressure sensor.

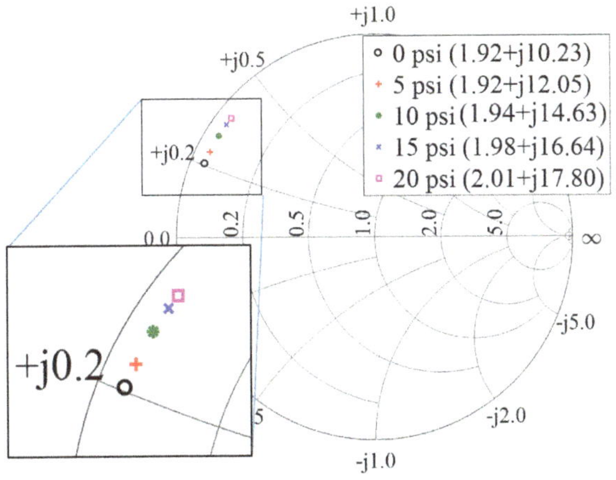

Figure 6. Measured impedance at 2 GHz for pressure range from 0 to 20 psi with a direct wired connection using vector network analyzer (VNA).

2.2. Passive Harmonic RF Sensor Tag

The passive harmonic RF tag uses a reactance-based phase shifting mechanism for wirelessly measuring the change in the applied pressure. The hybrid coupler is used as a phase shifter, which is a four port reciprocal device with an input, output, isolation, and coupled ports. The phase shifter is designed and simulated using ANSYS HFSS® (2019, ANSYS, Inc., Canonsburg, PA, USA) on an FR-4 substrate with dielectric constant (ϵ_r) of 4.4, thickness of 1.52 mm and a loss tangent (tan δ) of 0.02. The phase shifter operates at 2 GHz with a 3 dB coupling factor and a 20 dB isolation.

The phase shifter's input is connected to a 2 GHz patch antenna for receiving the interrogation signal. The received RF signal is fed to the output and the coupled ports of the coupler, to which a reactive impedance pressure sensor is connected. The RF signal at the output and the coupled ports experience a total internal reflection due to the termination by a reactive sensor element. The internally reflected signal is received from the isolated port without any internal power loss, but with a shifted phase according to the reactance. The isolated port is connected to a non-linear device (Schottky diode), which generate a second harmonic signal at 4 GHz. The Schottky diode (BAT 15-03) doubles the shifted phase along with a frequency. The phase modulated 4 GHz output of the frequency doubler is transmitted back using a patch antenna. The schematic of the designed harmonic RF tag is shown in Figure 7 along with the dimensions.

The fundamental and the harmonic antennae are designed in cross-polarization for increasing the SNR and minimizing the interference at the interrogator. The simulated and the measured frequency responses for both the fundamental and the harmonic patch antenna are shown in Figure 8. The measured response closely matches the simulated response, and the observed minimal differences are due to fabrication tolerances.

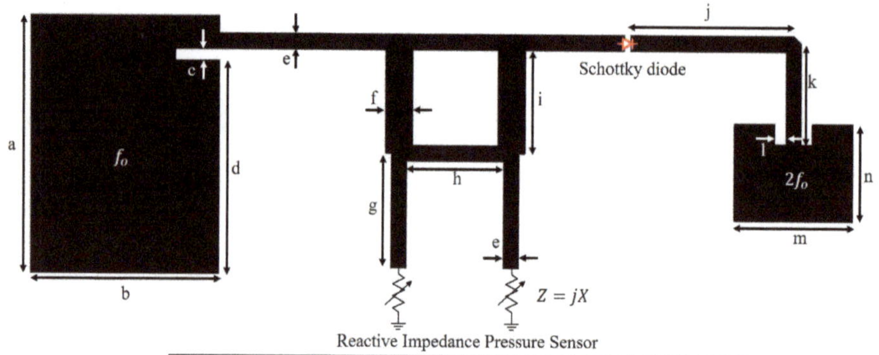

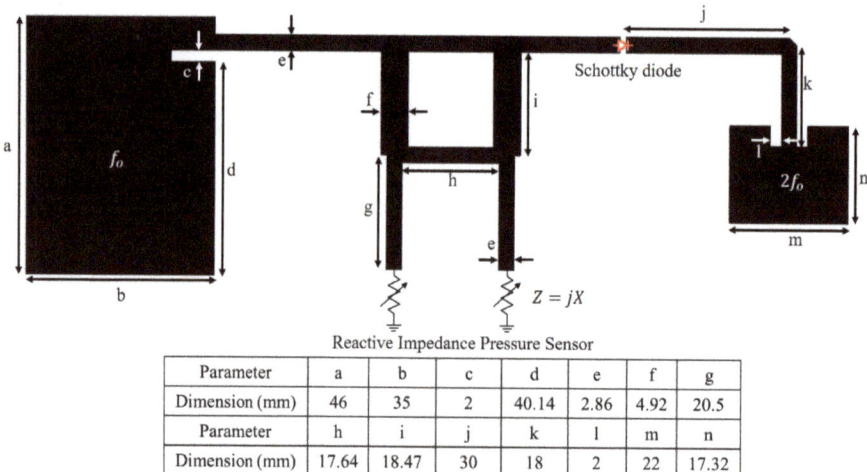

Parameter	a	b	c	d	e	f	g
Dimension (mm)	46	35	2	40.14	2.86	4.92	20.5
Parameter	h	i	j	k	l	m	n
Dimension (mm)	17.64	18.47	30	18	2	22	17.32

Figure 7. Schematic of the pressure sensing harmonic RF tag along with its dimensions. Signal received using 2 GHz antenna forwarded to the hybrid coupler for phase shift according to the integrated pressure sensor and the second harmonic signal with double shifted phase, generated using diode frequency doubler, is transmitted out using 4 GHz patch antenna.

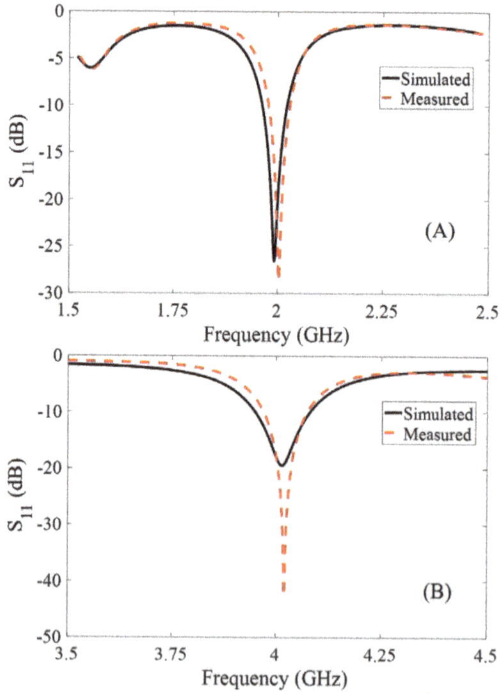

Figure 8. Simulated and measured frequency response; (**A**) Fundamental (2 GHz) and (**B**) Harmonic antenna (4 GHz).

2.3. *Harmonic RF Interrogator*

The harmonic RF interrogator is designed to communicate with the passive phase shifting-based RF tag. The interrogator generates an RF signal using a VNA with a signal strength of 13 dBm at 2 GHz and it is transmitted towards the harmonic RF tag using a Vivaldi antenna. The harmonic RF tag receives the fundamental frequency and backscatters the pressure sensor information modulated onto the 4 GHz harmonic signal. The harmonic interrogator receives the modulated signal using a Vivaldi antenna in a cross-polarized configuration. The received harmonic signal is amplified using LNA's (ZX60-53LNB-S+ and ZX60-43-S+) with a total amplification of 34 dB. The amplified 4 GHz signal is fed into the RF port of the mixer for down converting it to a 2 GHz signal. A reference signal is extracted from the 2 GHz transmission signal using an RF splitter and fed into the LO port of the mixer. The down converted output signal at 2 GHz with the modulated phase information is acquired using VNA at the IF port of the mixer. The schematic of the harmonic RF interrogator communicating with the harmonic RF sensor tag is shown in Figure 9.

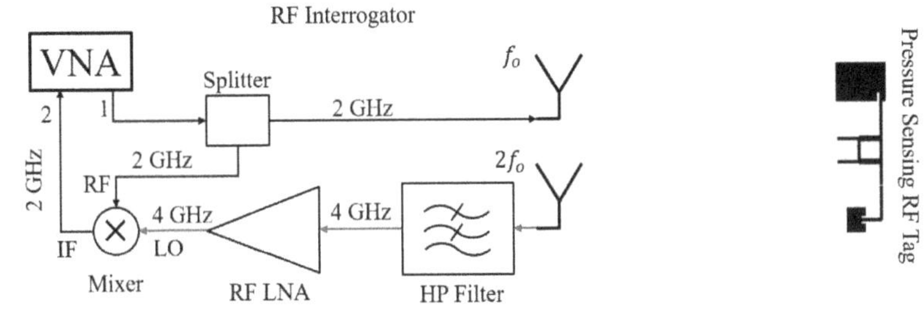

Figure 9. Schematic of the harmonic RF Interrogator consist of a VNA for RF source and phase measurement unit, transmitting and receiving antennas, and miscellaneous RF peripheral circuits for communicating with the harmonic RF sensor tag.

3. Results

Initially, a simulation-based validation study is performed in order to test the phase shifting mechanism using a reactive element. A four port phase shifter is designed in Keysight's ADS RF simulation software (2017, Keysight Technologies, Santa Rosa, CA, USA) with a coupled reactive impedance element. The range of reactive impedance is referenced from the previous direct wired measurements shown in Figure 6. The reactive elements are connected to the two of the four ports of hybrid coupler, while using two leftover ports as input and output of 2 GHz signal. The S-parameter study is used in ADS for validating the phase shifting mechanism. The simulated transmission coefficient (S_{21}) of the phase shifter is shown in Figure 10. The phase of the transmitted signal at 2 GHz is changed from −176.21° to 155.2° ($\Delta\Phi$ = 28.55°) with a change in coupled reactance from 6.82–36.73 Ω. The simulated phase shift due to reactive impedance is significant for practical implementation and can be easily detected using a standard phase measurement device. Moreover, the transmitted power at 2 GHz remains unchanged, shows no loss at the operating frequency for the entire range of the changing reactive impedance. Pressure information is modulated on the 2 GHz signal in the form of a phase with no power loss is a significant feat for passive RF tag technology, where a small amount of energy saving can lead to a huge improvement in range or signal-to-noise ratio. In this paper, the no power loss advantage is exploited for better harmonic generation using the Schottky diode, where the frequency conversion losses are inversely proportional to the input power of the signal [24].

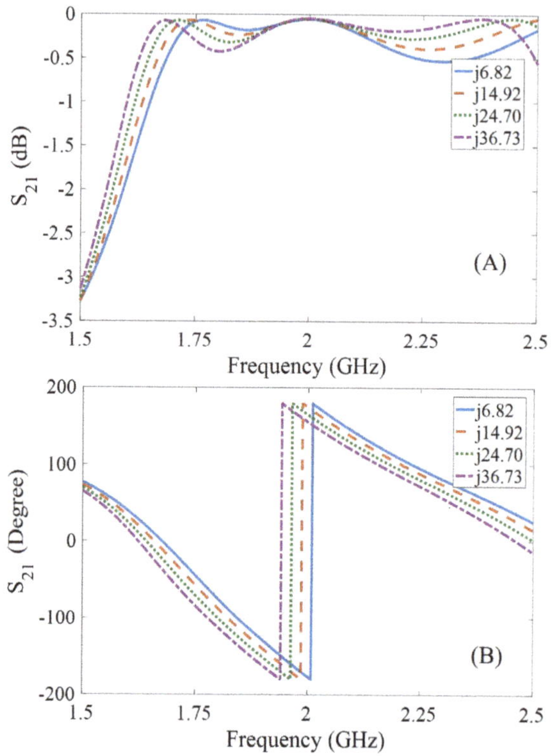

Figure 10. Simulated frequency response of the designed phase shifting hybrid coupler with a variable reactance from 6.82 Ω to 36.73 Ω: (**A**) No power loss is observed at 2 GHz due to changing reactance, (**B**) A total phase shift of 28.55° is observed at 2 GHz due to changing reactance.

The proposed technique is validated by performing two experiments. First, the phase shifter is designed in Ansys HFSS for 2 GHz and fabricated on a FR-4 board for validating the simulation results. The fabricated phase shifter's response, without a transmitter or a receiver antenna, is measured using the VNA at 2 GHz. A power combiner is used to connect a single reactive element (varactor diode) with the two ports of the phase shifter. The reactive impedance of the varactor diode is varied by applying a bias voltage from 0 V to 1.5 V. The change in phase, due to the change in reactance, of the transmission coefficient (S_{21}) at 2 GHz in direct wired configuration, is shown in Figure 11. The 1.5 V change in bias potential leads to a change of 18.28° in phase. The fabricated phase shifter performed in a similar way to the simulated results. Next, receiver and transmitter patch antennae are connected to the phase shifter with a harmonic doubler and the wireless phase measurement experiment is performed. The harmonic RF sensor tag is placed at a separation of 8 inches from the interrogator antenna for wireless communication. The wirelessly acquired phase change due to applied bias voltage is shown in Figure 11. A phase shift of 36.78° is observed for the wireless configuration, which is approximately double the single frequency wired configuration. In Figure 11, both responses are the average plots of three repetitions and error bars represent the standard deviation between all readings. A constant phase offset with reference to 0° is added in both the wired and the wirelessly acquired data.

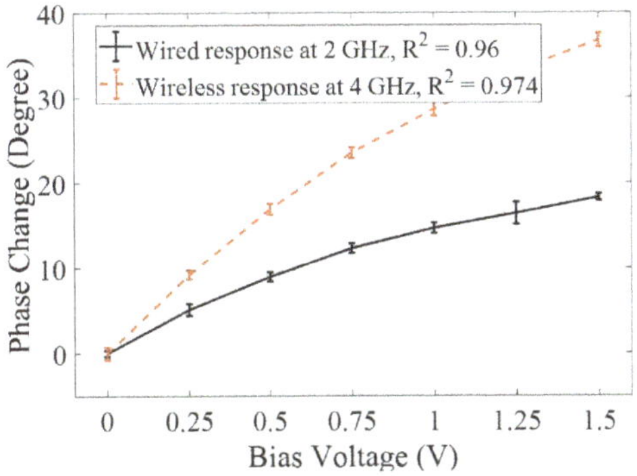

Figure 11. Measured phase change due to variable reactance (introduced with direct current (DC) biasing the varactor diode) in a wired configuration ($\Delta\Phi$ = 18.28°) using a single frequency approach and wireless configuration ($\Delta\Phi$ = 36.78°) using the harmonic frequency approach.

Second, a pressurized pipeline setup is designed to test the harmonic RF tag with integrated pressure sensor. A section of polypropylene pipe, 24 inches in length and 2 inches in diameter, is sealed using threaded polypropylene caps.

An 1/8 inch precision flow valve is connected to both the ends of the pipe for controlling the pressure. A high pressure nitrogen cylinder is used to pressurize the pipe through a compressed yor-lok fitting coupled to a precision flow valve. Due to the safety of research lab and personnel, natural gas pressure pipe application is demonstrated using nitrogen gas. A 20 psi safety valve is connected that sets the limit on the maximum applied pressure. The designed pressure sensor can also work at higher pressure ranges, but due to safety, the upper limit is set to 20 psi for this work. The working principle of the designed sensor is demonstrated for a limited pressure range from 0 to 20 psi. The integrated pressure sensor inside the pipeline is connected to the designed harmonic RF tag, as shown in Figure 12. The harmonic RF tag is capable of receiving RF signal at 2 GHz and modulate the phase of output signal according to the applied pressure.

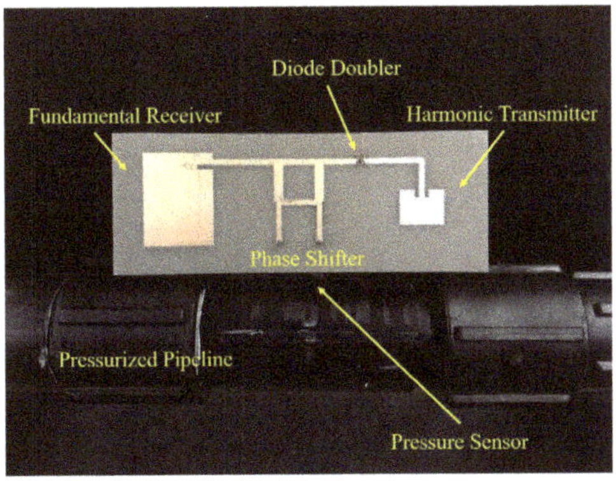

Figure 12. Pressurized polypropylene pipeline setup with the fabricated harmonic RF sensor tag for real-time wireless communication.

Discussion

In our experiment, the applied pressure range is from 0 to 20 psi. Due to applied pressure, the harmonic RF interrogator, placed at 18 inches' distance from RF tag, measured a phase shift of 16.6° in the received signal at 4 GHz. The wirelessly measured phase is shown in Figure 13, where the measurements are acquired at an interval of 5 psi. All measurements are performed thrice for checking the repeatability and standard deviation at each acquisition point, as shown in Figure 13. The plotted average data of three repetitions show a linearity of 0.9915 (R^2), which is comparable to existing commercial pressure sensors [25].

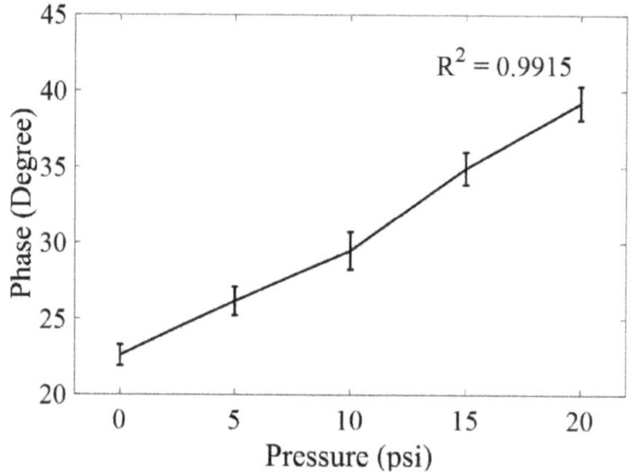

Figure 13. Phase change is wirelessly measured using harmonic RF interrogator due to applied pressure ranging from 0 to 20 psi. The linearity of change in phase due to pressure is represented using coefficient of determination (R^2 = 0.9915).

The designed harmonic RF sensor tag has a sensitivity of detecting a minimum change of 1 Ω in reactive impedance at 2 GHz and is compatible with other types of reactive impedance based sensors for measuring higher pressure or other physical parameters such as temperature, moisture, etc. The interrogation range of the system can be further enhanced by radiating the maximum allowable power of 4 W, whereas in this work, only 19 mW was used. A maximum communication range of 15 ft can be achieved with 36 dBm (4W) transmitted power at 2 GHz, 20 dB doubler diode conversion loss and −90 dBm receiver sensitivity at 4 GHz.

The 3D printed pressure sensor eliminated the complex fabrication requirement in clean room, while providing similar capability such as piezo-based resistive sensors, MEMS-based capacitive sensor, or FBG-based pressure sensors [7–10]. The passive wireless feature successfully communicated the pressure information from pipe to interrogator, while eliminating the battery requirement and lowering down the maintenance cost [18].

4. Conclusions

In this paper, a low cost 3D pressure sensor and passive wireless sensing mechanism is presented for monitoring the structural integrity of the gas pipelines. The developed harmonic pressure sensing system has a good sensitivity (1 Ω) towards reactive impedance, better linearity (R^2 = 0.9915), a higher SNR, lower modulation losses, and a longer communication range. The harmonic communication eliminates the self-jamming, reduces the multi-path interference, and doubles the measurement sensitivity (phase change) due to applied pressure. The coupler-based phase shifting mechanism efficiently modulated the pressure information on carrier harmonic frequency. The demonstrated detection range of applied pressure from 0 to 20 psi is practical for natural gas distribution and main pipeline infrastructure.

Furthermore, the logical next step of this research is to expand the application to higher pressure ranges and study the performance of sensor in liquid medium while focusing on the effects of non-dissolved particles. The developed long range wireless sensing platform allows a continuous, safe, reliable, and real-time monitoring of the pipeline and provides a cost-efficient solution for operations in refineries and chemical plants.

Author Contributions: Conceptualization, D.K. and S.M.; methodology, D.K. and S.M.; validation, D.K., S.M., Y.D. and P.C.; investigation, D.K. and S.M.; resources, Y.D. and P.C.; data curation, D.K.; writing—original draft preparation, D.K.; writing—review and editing, D.K.; visualization, D.K. and S.M.; supervision, Y.D. and P.C.; project administration, Y.D. and P.C. All authors have read and agreed to the published version of the manuscript.

Funding: This research was funded by USDOT grant number DTPH5615HCAP08L.

Acknowledgments: The authors would also like to acknowledge Brian Wright of MSU ECE shop for his assistance.

Conflicts of Interest: The authors declare no conflict of interest.

References

1. US-EIA. *Outlook for Future Emissons*; US-EIA: Washington, DC, USA, 2018.
2. Rajah, S.K.; Nasr, S. Longitudinal Stresses in Pressure Pipeline Design: A Critical Review. *Pipelines* **2008**. [CrossRef]
3. Wu, X.; Lu, H.; Wu, S. Stress analysis of parallel oil and gas steel pipelines in inclined tunnels. *Springer Plus* **2015**, *4*, 659. [CrossRef] [PubMed]
4. Makar, J.; Rogge, R.; McDonald, S. Circumferential failures in grey cast iron pipes. In Proceedings of the AWWA Infrastructure Conference, Chicago, IL, USA, 10–13 March 2002.
5. Liu, Z.; Kleiner, Y. State-of-the-Art Review of Technologies for Pipe Structural Health Monitoring. *IEEE Sens. J.* **2012**, *12*, 1987–1992. [CrossRef]
6. Obeid, A.M.; Karray, F.; Jmal, M.W.; Abid, M.; Qasim, S.M.; BenSaleh, M.S. Towards realisation of wireless sensor network-based water pipeline monitoring systems: A comprehensive review of techniques and platforms. *IET Sci. Meas. Technol.* **2016**, *10*, 420–426. [CrossRef]
7. Weber, J.; Shi, J.; Zhang, C. Embedded Capacitance Sensor Array for Structural Health Monitoring of Pipeline Systems. In Proceedings of the ASME Pressure Vessels and Piping Conference, Waikoloa, HI, USA, 9 March 2017; Volume 5. [CrossRef]
8. Zhou, G.; Zhao, Y.; Guo, F.; Xu, W. A Smart High Accuracy Silicon Piezoresistive Pressure Sensor Temperature Compensation System. *Sensors* **2014**, *14*, 12174–12190. [CrossRef] [PubMed]
9. Qingmin, H.; Liang, R.; Wenling, J.; Pinghua, Z.; Gangbing, S. An Improved Negative Pressure Wave Method for Natural Gas Pipeline Leak Location Using FBG Based Strain Sensor and Wavelet Transform. *Math. Probl. Eng.* **2013**, *2013*, 8. [CrossRef]
10. Jang, M.; Yun, K.S. MEMS capacitive pressure sensor monolithically integrated with CMOS readout circuit by using post CMOS processes. *Micro Nano Syst. Lett.* **2017**, *5*, 4. [CrossRef]
11. Ghazali, M.I.M. *Additive Manufacturing for Electronic Systems (AMES)*; Michigan State University, Electrical Engineering: East Lansing, MI, USA, 2019.
12. Hong, C.; Zhang, Y.; Borana, L. Design, Fabrication and Testing of a 3D Printed FBG Pressure Sensor. *IEEE Access* **2019**, *7*, 38577–38583. [CrossRef]
13. Park, J.; Kim, J.K.; Kim, D.S.; Shanmugasundaram, A.; Park, S.A.; Kang, S.; Kim, S.H.; Jeong, M.H.; Lee, D.W. Wireless pressure sensor integrated with a 3D printed polymer stent for smart health monitoring. *Sens. Actuators B Chem.* **2019**, *280*, 201–209. [CrossRef]
14. Yoon, J.; Choi, B.; Lee, Y.; Han, J.; Lee, J.; Park, J.; Kim, Y.; Kim, D.M.; Kim, D.H.; Kang, M.; et al. Monolithically 3D-printed pressure sensors for application in electronic skin and healthcare monitoring. In Proceedings of the 2017 IEEE International Electron Devices Meeting (IEDM), San Francisco, CA, USA, 2–6 December 2017; pp. 40.2.1–40.2.4.
15. Cattani, M.; Boano, C.A.; Steffelbauer, D.; Kaltenbacher, S.; Günther, M.; Römer, K.; Fuchs-Hanusch, D.; Horn, M. Adige: An Efficient Smart Water Network Based on Long-range Wireless Technology. In Proceedings of the 3rd International Workshop on Cyber-Physical Systems for Smart Water Networks, Pittsburgh, PA, USA, 18–21 April 2017; ACM: New York, NY, USA, 2017; pp. 3–6. [CrossRef]
16. Sadeghioon, A.M.; Metje, N.; Chapman, D.; Anthony, C. Water pipeline failure detection using distributed relative pressure and temperature measurements and anomaly detection algorithms. *Urban Water J.* **2018**, *15*, 287–295. [CrossRef]

17. Noel, A.B.; Abdaoui, A.; Elfouly, T.; Ahmed, M.H.; Badawy, A.; Shehata, M.S. Structural Health Monitoring Using Wireless Sensor Networks: A Comprehensive Survey. *IEEE Commun. Surv. Tutor.* **2017**, *19*, 1403–1423. [CrossRef]
18. Kim, J.H.; Sharma, G.; Boudriga, N.; Iyengar, S.; Prabakar, N. Autonomous pipeline monitoring and maintenance system: A RFID-based approach. *EURASIP J. Wirel. Commun. Netw.* **2015**, *2015*, 262. [CrossRef]
19. Mondal, S.; Wijewardena, K.; Karrapuswami, S.; Kumar, D.; Chahal, P.; Ghannam, M.; Cuddihy, M. A Wireless Battery-less Seat Sensor for Autonomous Vehicles. In Proceedings of the 2020 IEEE 70th Electronic Components and Technology Conference (ECTC), Orlando, FL, USA, 3–30 June 2020; pp. 2289–2294. [CrossRef]
20. Lin, L.; Ma, M.; Zhang, F.; Liu, F.; Liu, Z.; Li, Y. Fabrications and Performance of Wireless LC Pressure Sensors through LTCC Technology. *Sensors* **2018**, *18*, 340. [CrossRef] [PubMed]
21. Kumar, D.; Karuppuswami, S.; Deng, Y.; Chahal, P. A wireless shortwave near-field probe for monitoring structural integrity of dielectric composites and polymers. *NDT E Int.* **2018**, *96*, 9–17. [CrossRef]
22. Huang, H.; Chen, P.Y.; Hung, C.H.; Gharpurey, R.; Akinwande, D. A zero power harmonic transponder sensor for ubiquitous wireless μL liquid-volume monitoring. *Sci. Rep.* **2016**, *6*, 18795. [CrossRef] [PubMed]
23. Palazzi, V.; Alimenti, F.; Mezzanotte, P.; Orecchini, G.; Roselli, L. Zero-power, long-range, ultra low-cost harmonic wireless sensors for massively distributed monitoring of cracked walls. In Proceedings of the 2017 IEEE MTT-S International Microwave Symposium (IMS), Honololu, HI, USA, 4–9 June 2017; pp. 1335–1338. [CrossRef]
24. Mondal, S.; Kumar, D.; Chahal, P. A Wireless Passive pH Sensor With Clutter Rejection Scheme. *IEEE Sens. J.* **2019**, *19*, 3399–3407. [CrossRef]
25. NOVA Sensor. *NPI-19 Series Medium Pressure Sensor*; Amphenol NovaSensor: Fremont, CA, USA, 2015.

Publisher's Note: MDPI stays neutral with regard to jurisdictional claims in published maps and institutional affiliations.

micromachines

MDPI

Article

Fused Deposition Modeling of Microfluidic Chips in Polymethylmethacrylate

Frederik Kotz [1,2,*], Markus Mader [1], Nils Dellen [1], Patrick Risch [1], Andrea Kick [1], Dorothea Helmer [1,2,3] and Bastian E. Rapp [1,2,3]

1 Laboratory of Process Engineering, NeptunLab, Department of Microsystems Engineering (IMTEK), University of Freiburg, 79110 Freiburg, Germany; Markus.Mader@imtek.de (M.M.); Nils.Dellen@imtek.de (N.D.); Patrick.Risch@imtek.de (P.R.); Andrea.Kick@imtek.de (A.K.); Dorothea.Helmer@imtek.de (D.H.); Bastian.Rapp@imtek.de (B.E.R.)

2 Freiburg Materials Research Center (FMF), University of Freiburg, 79104 Freiburg im Breisgau, Germany

3 FIT Freiburg Center of Interactive Materials and Bioinspired Technologies, University of Freiburg, 79110 Freiburg im Breisgau, Germany

* Correspondence: Frederik.Kotz@imtek.de; Tel.: +49-761-203-7355

Received: 20 August 2020; Accepted: 18 September 2020; Published: 19 September 2020

Abstract: Polymethylmethacrylate (PMMA) is one of the most important thermoplastic materials and is a widely used material in microfluidics. However, PMMA is usually structured using industrial scale replication processes, such as hot embossing or injection molding, not compatible with rapid prototyping. In this work, we demonstrate that microfluidic chips made from PMMA can be 3D printed using fused deposition modeling (FDM). We demonstrate that using FDM microfluidic chips with a minimum channel cross-section of ~300 µm can be printed and a variety of different channel geometries and mixer structures are shown. The optical transparency of the chips is shown to be significantly enhanced by printing onto commercial PMMA substrates. The use of such commercial PMMA substrates also enables the integration of PMMA microstructures into the printed chips, by first generating a microstructure on the PMMA substrates, and subsequently printing the PMMA chip around the microstructure. We further demonstrate that protein patterns can be generated within previously printed microfluidic chips by employing a method of photobleaching. The FDM printing of microfluidic chips in PMMA allows the use of one of microfluidics' most used industrial materials on the laboratory scale and thus significantly simplifies the transfer from results gained in the lab to an industrial product.

Keywords: 3D printing; polymethylmethacrylate; additive manufacturing; microfluidics; fused deposition modeling

1. Introduction

Three-dimensional (3D) printing has gained great importance for rapid prototyping of microfluidic devices in the past decade, since it allows the fabrication of microfluidic chips on the laboratory scale, with the possibility to test a great variety of different chip designs early on in the development process [1–4]. Different printing technologies have been examined for 3D printing of microfluidic devices, such as stereolithography (SL), inkjet printing, multi-jet printing, two-photon polymerization, suspended liquid subtractive lithography and fused deposition modeling (FDM) [2,5–8]. On the laboratory scale these 3D printed microfluidic devices have been used in a great variety of applications: for mixers, droplet or gradient generators or active components like valves or pumps [9–12].

However, one major issue for most rapid prototyping methods in microfluidics, including 3D printing, is the conversion of a laboratory prototype to an industrial scale product [13]. 3D printing of microfluidic chips has so far been mainly realized using stereolithography printing, since it remains the method of choice for high-resolution 3D printing with affordable machinery. However, SL usually requires photocurable resins (mostly acrylic or epoxy based) which are strongly cross-linked thermosets, rendering them unsuitable for industrial replication processes like injection molding or hot embossing, which require thermoplastic materials. Due to the high surface-to-volume ratio in microfluidics, the chip material plays a major role. Therefore the transfer between lab-processes and industrial scale comes with a change in the final system behavior, which is a major issue [13]. One option to solve this problem is the use of 3D printing methods, which can process industrially relevant thermoplastic polymers already on the laboratory scale. FDM is such a 3D printing method in which a thermoplastic filament is melted, extruded through a nozzle and solidified by cooling. Furthermore, FDM is interesting since it is capable of multimaterial printing [14]. The complexity of FDM printed systems can be even extended by integration of components like membranes or electrodes by pausing the process, integrating the external components of choice before continuing the print (so called print-pause-print) [14]. Using FDM printing a variety of microfluidic concepts like mixers or chemical reactionware have already been realized [15,16]. Recently FDM printing has been used to fabricate simple channel geometries with sub-100 μm [17].

In theory, a wide range of thermoplastic materials can be processed using FDM, but only few materials have been studied. Most microfluidic chips are printed using poly (lactic acid) (PLA) or acrylonitrile-butadiene-styrene (ABS). FDM printing of cycloolefin copolymers (COC), thermoplastic urethane (PLU) or polypropylene being a notable exception [15,17]. Polymethylmethacrylate (PMMA) is one of the most important thermoplastic materials in microfluidics due to its high optical transparency and low autofluoresence combined with a high biocompatibility [18]. Furthermore, it is a rather hydrophilic material, making it interesting for capillary-driven microfluidic systems. However, PMMA is usually structured using industrial scale polymer structuring like injection molding or hot embossing [19,20]. Rapid prototyping of PMMA has been mainly conducted using subtractive processes like laser structuring or high-precision milling, as well as laboratory-scale imprinting technologies, such as solvent replication or room temperature imprinting [21–24]. PMMA prepolymers consisting of the monomer methylmethacrylate (MMA) and the polymer PMMA have been used for replication of microfluidic PMMA chips with high resolution from a variety of different master structures like polydimethlysiloxane (PDMS) or stainless steel [25,26]. Recently, we have introduced Liquid PMMA, a photocurable PMMA prepolymer which can be structured with tens of micron resolution using direct lithography [27]. However, all these rapid prototyping processes are only capable of fabricating open 2.5-dimensional microfluidic structures, which need to be closed during a subsequent bonding step. Furthermore, the integration of microstructures and biomolecular patterns is becoming increasingly important. Protein patterns with specific shapes and patterns are capable of inducing specific cellular responses and are important for the study of cellular behavior. Structured biochemical functionalization by means of photobleaching was established in the last decade and was already used for different substrate materials, like functionalized glass slides, polymer films and cellulose paper [28–30].

PMMA has so far not been studied as a material for FDM printing of microfluidic chips, which, however, could allow the direct printing of embedded microfluidic chip structures. In this work, we demonstrate that microfluidic PMMA chips can be directly printed using FDM with a minimum channel width of 300 μm. To ensure a high optical transparency in the region of interest, we evaluated two strategies: direct FDM printing on the print bed and printing on top of a commercial PMMA substrate. We further evaluate the influence of the nozzle distance on the transparency of the printed PMMA. Further, we show that high-resolution microstructures like microscale line-patterns can be integrated within the printed PMMA chip and that the microchannels can be easily functionalized with biomolecules. The FDM printing process allows the fabrication of a fully functional embedded microfluidic chip within 1 h, thus demonstrating that an industrially relevant thermoplastic polymer material can be structured using 3D printing. This effectively allows a direct transfer of results gained in the laboratory to an industrial scale.

2. Materials and Methods

2.1. Printing Materials

PMMA filament was purchased from materials4print, (Bad Oeynhausen, Germany). Methylmethacrylate (MMA) technical grade was purchased from VWR, (Darmstadt, Germany). Phenylbis(2,4,6-trimethylbenzoyl)phosphine oxide (BAPO), biotin (5-fluorescein) conjugate (F5B), streptavidin-Cy3 in buffered aqueous solution (STV-Cy3), phosphate buffered saline (PBS) and 2,2′-azobis(2-methylpropionitrile) (AIBN) were purchased from Sigma Aldrich, (Taufkirchen, Germany). PMMA substrates were purchased from Röhm GmbH (Darmstadt, Germany).

2.2. Fused Deposition Modeling

All designs were created using Autodesk Inventor Professional 2019 and exported as STL files. The STL files were imported into PrusaSlicer-2.1.0-rc for the slicing process. The FDM printer *Prusa i3 MK3S* (Prusa Research, Prag, Czech Republic) was used for printing. The printing parameters have been optimized to 3D print microfluidic channels with a minimum channel width of 300 µm. The parameters of the printing process were as follows: layer height: 50 µm (first layer: 100 µm), infill: 100%, printing speed: 30 mm/s and nozzle temperature: 230 °C. The bed temperature was increased significantly compared to other commercial filaments like PLA to 115 °C for the first layer to reduce warping effect, i.e., delamination from the print bed. During the print, the bed temperature was lowered to 110 °C for all following layers. A 45° angle was set for the infill orientation. Infill/contour overlap was set to 25% to reduce air gaps between the printed features. All parts were printed with a 0.4 mm nozzle. For high transparency of the printed microfluidic chips, the *z*-axis distance to the printing bed was reduced by doing a first layer calibration. The nozzle was moved 100 µm closer to the print bed compared to the standard printing distance. The specified standard *z*-axis distance of the nozzle is usually adjusted until the polymer sticks nicely to the print bed and is only slightly squished. Reducing the *z*-axis nozzle distance even more than specified results in broader strands, which make the FDM printed PMMA appear more homogenous and transparent (see Section 3.1). To reduce sagging during the 3D printing of embedded channels, the material extrusion was reduced by 60% while bridging the channel structures (bridge flow ratio: 0.4). By keeping the printing speed at 30 mm/s the bridging PMMA strands are stretched upon deposition which prevents excessive sagging. To evaluate different channel geometries, a series of embedded channels with different cross-section geometries were printed and compared to their original CAD geometry. The accuracy of the printable channel widths and heights were evaluated by printing open and embedded channels with varying channel widths from 0.2–1 mm and compared to the original CAD dimensions. Evaluation of the channel sizes and geometries was executed using a light microscope of type VHX 6000 (Keyence, Osaka, Japan).

To obtain microchannels with an improved transparency in the region of interest, the channels were printed directly on commercial PMMA substrates with a thickness of 2 mm. A similar strategy has been previously described for the fabrication of channels in PLA [31]. The channels were designed to be open at the bottom side, which is printed directly on the substrate. The printed PMMA structure bonds to the PMMA substrate to form sealed microfluidic channels. The *z*-axis printing height of the chip was adjusted during slicing to allow printing on the 2 mm PMMA substrate (z-offset: 2 mm). After heating and calibration, the printer was shortly paused to align and stick the substrates to the print bed.

2.3. PMMA Precursor Preparation

A thermally polymerized PMMA precursor was prepared by adapting a protocol described by Qu et al. [25]. The thermal initiator AIBN was dissolved in technical MMA (0.3 mg/mL) and the mixture was polymerized by heating the mixture to 93 °C within 20 min. After 15 min at this temperature the polymerization was stopped by cooling to room temperature with an ice bath. The mixture was blended with 1 wt% BAPO.

2.4. Casting Process

Thermally prepolymerized PMMA precursor was polymerized for 10 min between a commercial PMMA substrate and a metal microstructure using a UV-light source (Superlite 400, Lumatec, Oberhaching, Germany) at 415 nm (exposure intensity: 5.8 mW/cm^{-2}). A rectangular structure (2 × 12 mm^2) was cut out from a stripe of scotch tape (48 μm thickness), which was used as both mold and spacer between the metal microstructures and the PMMA substrate.

2.5. Microfluidic Experiments

In order to connect the inlets and outlet, tubing connectors were directly printed onto the microfluidic chips. Dispensing needles (1″, Vieweg, Kranzberg, Germany) were plugged into the printed connectors and fixed using epoxy glue. Colored water was pumped through the microfluidic channels using a syringe pump (Legato 210, KDScientific, Holliston, MA, USA) with a pump rate of up to 10 mL/min.

2.6. Contact Angle Measurements

Contact angles were measured with an OCA15 Pro (Data Physics, Filderstadt, Deutschland) using the sessile drop method. Static contact angles were measured using 5 μL water droplets at a temperature of 25 °C.

2.7. UV-Vis Measurements

Optical transmission of commercial PMMA and FDM printed PMMA with variable thickness was measured using a UV-visible spectrophotometer (Evolution 201, Thermo Fisher Scientific, Waltham, MA, USA).

2.8. Autofluorescence Measurements

Fluorescence intensity was characterized for commercial PMMA substrates and 3D printed PMMA with a thickness of 2 mm using an inverted fluorescence microscope (DMi8, Leica, Wetzlar, Germany) with Cy5, Cy3, FITC and DAPI filters and 130 ms exposure.

2.9. Biofunctionalization

The immobilization of proteins inside of FDM printed microfluidic channels was done similar to a previously described protocol [28]. F5B (80 μM in PBS) was injected into the channel and exposed for 3 min to 5 min at 490 nm (exposure intensity: 7.7 mW·cm^{-2}). After exposure with a custom build lithography system based on a digital mirror device the channels were rinsed with distilled water, PBS and, again, distilled water [28]. The patterns were visualized by incubating the microchannel with STV-Cy3 (5 μg/mL) for 30 min. Afterwards the channels were again rinsed with distilled water and PBS. The biomolecule patterns were visualized using an inverted fluorescence microscope (DMi8, Leica, Germany). Images were analyzed using ImageJ 1.53a. Regions of interest (ROI) were selected for background and signals (triangle shapes) and were analyzed in terms of brightness. Mean values of all ROI of background and signals respectively were averaged and average values were used to calculate signal-to-noise ratio (S/N).

3. Results and Discussion

We optimized the printing process to fabricate microfluidic chips in PMMA. Furthermore, several microfluidic devices were fabricated and assessed. The integration of high-resolution microstructures and biofunctionalization inside the channel was demonstrated.

3.1. Optimum Printing Parameters and Transparency Optimization

To print PMMA microfluidic chips with an optical transparency in the region of interest, we studied two different printing strategies: (1) standard printing procedure with direct printing on the print bed and (2) printing onto a commercial PMMA substrate. Both strategies and the respective printed microfluidic channels are shown in Figure 1a,b. First, we printed PMMA microfluidic chips using the standard printing setup, where the PMMA is deposited directly on the print bed. To ensure a completely sealed microfluidic structure, the printed channels are embedded in the PMMA chip, which is printed using 100% infill for high transparency. A minimum of two layers (0.15 mm) were deposited before printing the actual channel structures. This allows printing of embedded, leak-proof and transparent microfluidic chips, as shown in Figure 1b. However, the deposited PMMA strands from the two bottom layers are clearly visible and therefore reduce the transparency of the printed microfluidic chips. In the second strategy, we printed an open channel structure directly onto a commercial, highly transparent PMMA substrate. The heated PMMA structure bonds to the commercial PMMA substrate upon deposition and therefore allows printing of fully functional, embedded and transparent microfluidic channels. The absence of FDM printed PMMA layers between the printed channel structures and the commercial PMMA substrate allows 3D printing of PMMA microfluidics with a higher transparency than with the standard printing procedure.

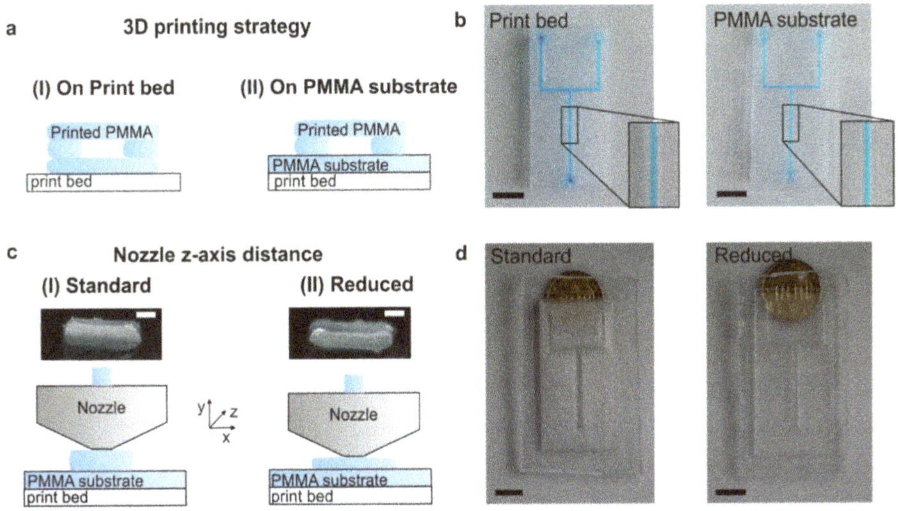

Figure 1. Fused deposition modeling of microfluidic chips in polymethylmethacrylate (PMMA). (**a**) Principle of printing microfluidic PMMA channels. PMMA was either directly printed on the print bed (**I**) or printed on top of a commercial PMMA slide (**II**). (**b**) Comparison of two identical microfluidic chips directly printed on the print bed and printed on a commercial PMMA substrate, respectively. As can be seen, the transparency in the region of interest is increased by printing on a commercial PMMA substrate (scale bars: 10 mm). (**c**) Increasing the optical transparency by reducing the nozzle distance: (**I**) standard configuration, (**II**) printing with a reduced nozzle distance which flattens the extruded filament. The nozzle is moving in the z-direction. The images show the 3D printed cross-section of the first layer calibration (scale bar: 100 μm). (**d**) Comparison of microfluidic channels printed with standard configuration and with reduced nozzle distance. As can be seen the transparency is increased for the printing setup with reduced nozzle distance (scale bar: 10 mm).

We further optimized the printing process to improve the overall transparency of the FDM printed PMMA. It was found that the z-axis distance of the nozzle for printing the first layer has a significant impact on the observed transparency of the FDM printed PMMA. Printing with the standard parameters for z-axis distance, specified by the Prusa manual, PMMA with poor optical transparency is obtained. Reducing the first layer z-axis distance of the nozzle by 100 μm compared to the standard parameters (Prusa manual) results in PMMA microfluidic devices with significantly higher transparency (see Figure 1c,d). The reduced nozzle distance results in slightly broader deposited PMMA strands that fuse together without having gaps in between thus increasing the transparency of the FDM printed component significantly.

We printed different channel cross-sections to evaluate the printing accuracy of these cross-sections and to determine the best cross-section for printing more complex embedded microfluidic channels. The CAD design and its respective printed PMMA chip can be seen in Figure 2a,b. FDM printing of a square cross-sections results in an excessive sagging of the top layer, which will clog the channel if small channel sizes are printed. The same effect was found for circular cross-sections. A major improvement was found when adding a triangular shaped roof on top of the square cross-section. The roof hereby compensates for the sagging effect and results in the square channel cross-sections. A similar optimization was found for elliptical and diamond shaped channels. Since the additional roof resulted in the most accurate microfluidic channels, all further structures were printed using this design.

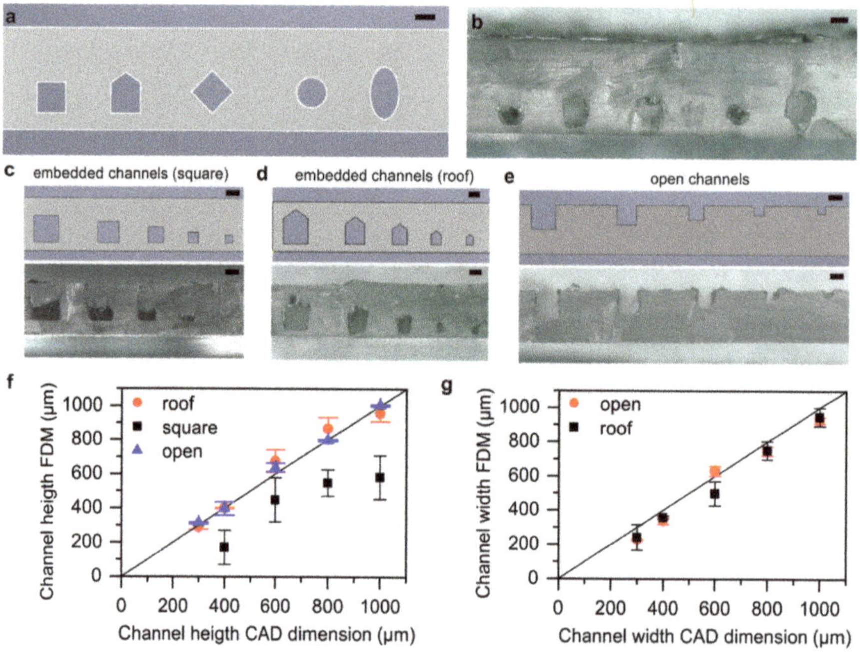

Figure 2. Characterization of the channel cross-section of a printed PMMA chip. (**a**) Image of the design of the printed channel cross-sections. (**b**) Different channel cross-sections with a channel width/diameter of 600 μm printed in PMMA. (**c**) Fused deposition modeling (FDM) printed embedded channels with a square cross-section for analysis of the printed channel height. Due to sagging the channel height deviates strongly from the original CAD design. (**d**) FDM printed embedded channels with a roof shaped cross-section for analysis of printed channel height and width. The addition of the roof compensates the sagging resulting in square channels. (**e**) Open square cross-section channels for analysis of the channel height and width. (**f**) Comparison of channel heights of embedded FDM printed channels using square and roof shaped cross-sections shown in (**c**,**d**) and open channels shown in (**e**) with the designed CAD heights. (**g**) Channel width of open and embedded roof shaped FDM printed channels from (**d**,**e**) compared to the designed CAD widths. (Scale bars (**a**–**e**): 500 μm).

The accuracy of the printed channel dimensions was evaluated by printing open and embedded (roof and square cross-section) test channels of 1 mm, 800 μm, 600 μm, 400 μm and 300 μm channel width and height, respectively (Figure 2c–e). The measured height of the resulting channel cross-sections compared to the CAD dimensions are shown in Figure 2f. Printing embedded square cross-sections resulted in excessive sagging which reduced the accuracy of the printed height compared to the CAD design significantly. Adding a roof on top of the square cross-section compensated for the sagging and allowed 3D printing of square channels with their channel heights corresponding to the square CAD design without the added roof. The open channels showed nearly no deviation from the CAD design. Both the open and embedded roof cross-section showed a high accuracy down to a channel height of 300 μm.

We further compared the width of the resulting channel cross-sections to the CAD dimensions (see Figure 2g). The actual printed channel widths of the open and roof-shaped cross-section channels are only slightly smaller than their respective CAD dimensions. This small deviation can be explained by the employed printing setup. Due to the reduced nozzle distance, used for printing of high transparency structures, the deposited PMMA strands are pressed to a slightly broader size than estimated by the slicing software. Printing of channels with a width and height below 300 μm resulted in partial clogging and was therefore not further investigated.

3.2. 3D Printed Microfluidic Devices

We printed several exemplary microfluidic devices to demonstrate the versatility of FDM 3D printing of PMMA (see Figure 3). An exemplary serpentine microfluidic mixer with a channel diameter of 600 μm was printed. It shows effective mixing along the channel cascade, resulting in a color gradient from yellow to blue (see Figure 3a). We further printed a simple microfluidic spiral with a channel width of 600 μm using the roof-shaped channel cross-section CAD design (see Figure 3b). We also show an enhanced mixer which consists of a 2 mm wide channel with 600 μm periodic geometries reaching into the channel (see Figure 3c). All three designs were printed using the roof-shaped channel cross-section CAD design. The mixers in Figure 3a,c show effective mixing of blue and yellow colored water. To demonstrate the feasibility of truly three-dimensional geometries, we also printed a 3D microfluidic spiral with 1.2 mm cross-section, which is intertwining around a straight channel (see Figure 3d). To test their performance, the printed channels were flushed with liquid at a flow rate of up to 10 mL/min for several minutes. No leakage occurred during these experiments and none of the channels showed clogging.

3.3. Integration of High-Resolution Microstructures

For a broader application scope, we investigated the potential to incorporate microstructures within 3D printed channels. In order to do so, a process was developed to allow for the integration of microstructures into the FDM printed microfluidic structures (see Figure 4a). To generate the desired microstructures, we used a photocurable PMMA prepolymer, which was structured on top of a commercial PMMA substrate. This PMMA substrate was then used as the PMMA substrate plate for FDM printing. This requires careful alignment of the PMMA substrate with the PMMA microstructure in the FDM printer to allow a precise print-on process. We first replicated a microscale lines-and-space structure using the PMMA prepolymer (Figure 4b), and then printed a channel around it, resulting in a channel with integrated microstructures (see Figure 4c). By doing so, we achieved the integration of high-resolution PMMA microstructures within the FDM printed PMMA channels.

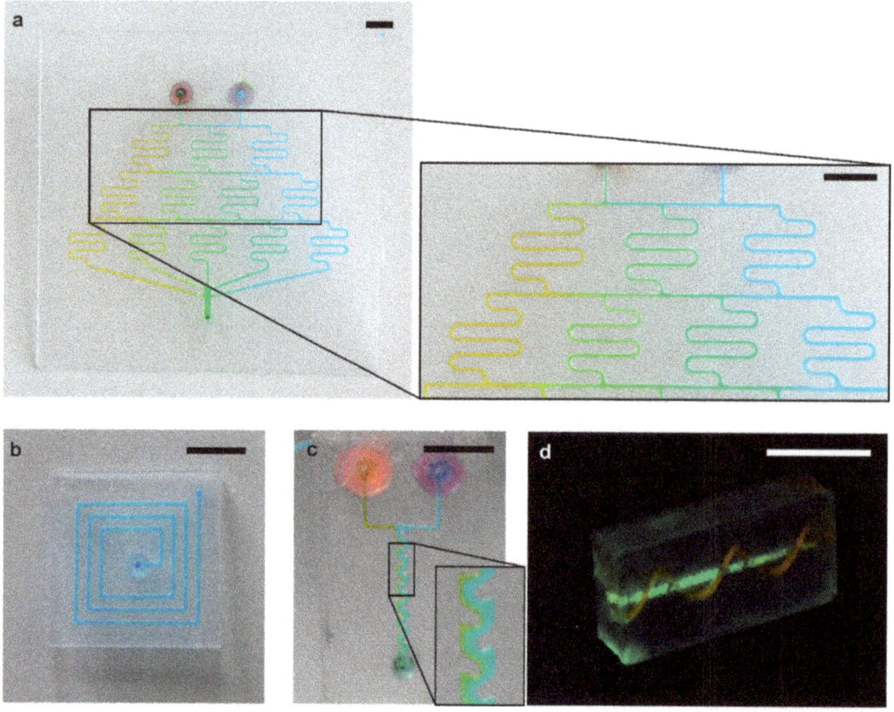

Figure 3. Microfluidic chips printed in PMMA. (**a**) Microfluidic mixer cascade for effective mixing of dyed water (scale bar: 10 mm). (**b**) 2.5-dimensional microfluidic spiral (scale bar: 10 mm). (**c**) Enhanced mixer structure of 600 μm periodic structures, effectively mixing dyed water (scale bar: 10 mm). (**d**) Three-dimensional microchannel printed in PMMA around a straight channel, visualized using an aqueous fluorescent dye (scale bar: 10 mm).

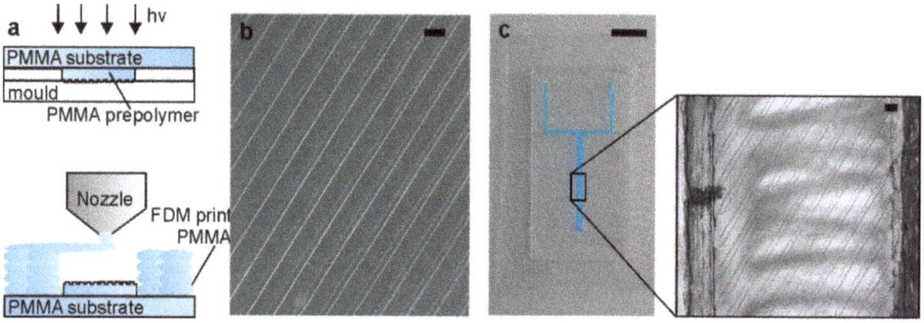

Figure 4. Integration of high-resolution PMMA structures inside a FDM-printed channel structure. (**a**) Principle of integration of high-resolution microstructures inside the FDM printed channel. The microstructure is fabricated by polymerizing a PMMA prepolymer directly on a commercial PMMA substrate using a lines-and-space structure. The PMMA substrate is then aligned and fixed within the FDM printer and the microfluidic channel is printed around the microstructure. (**b**) Polymerized PMMA microstructure fabricated using photocurable PMMA prepolymers (scale bar: 100 μm). (**c**) FDM printed PMMA chip including the microstructure from (**b**) (scale bar: 10 mm). The inset shows the integrated microstructure (scale bar: 100 μm).

3.4. Characterization of Printed PMMA

The wetting behavior of the chip material is important in microfluidics, to enable, e.g., filling of the channels and thus capillary-driven microfluidics in general. Therefore, the contact angle of water was measured on a commercial PMMA substrate and on a 3D printed substrate to characterize the wetting behavior. The printed PMMA and the commercial PMMA showed a water contact angle of 69°, respectively. These values are in good accordance with literature references which have been reported to be between 65° and 72° [32] and underline the hydrophilic character of the material.

The optical transmission of the 3D printed PMMA with the different nozzle distances described in Section 3.1, was characterized using UV/Vis measurements (see Figure 5a). Samples with a thickness of 100 μm, 400 μm and 1 mm were measured. The transparency was compared to the optical transmission of commercial PMMA with a thickness of 2 mm, which was used as the substrate for 3D printing of microfluidic channels directly on top of the substrate. The optical transparency of FDM printed PMMA parts was increased by using a printing strategy with reduced nozzle distance compared to the standard distance. The optical transmission, however, is still below that of commercial PMMA substrates. This is why for applications where high optical transmission is required in the region of interest, the demonstrated strategy of printing on top of a commercial PMMA substrate is preferred.

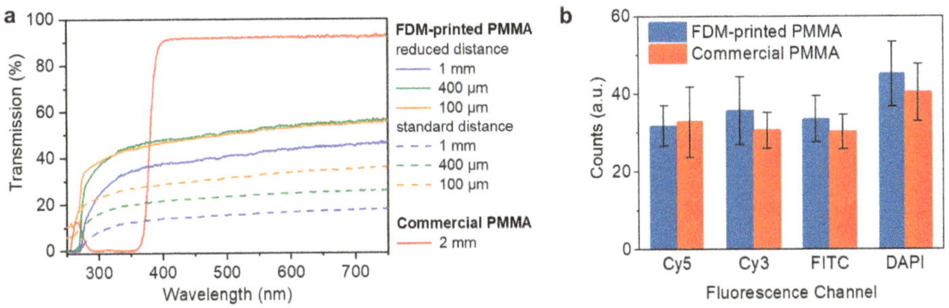

Figure 5. Characterization of FDM printed PMMA. (**a**) UV-Vis measurement of printed PMMA using different printing strategies (standard and reduced nozzle distance) and commercial PMMA substrates. (**b**) Fluorescence intensity of commercial PMMA substrate and FDM printed PMMA.

To ensure the applicability of 3D printed PMMA for fluorescence-based assays, the influence of autofluorescence of the material was investigated using an inverted fluorescence microscope. The printed and the commercial substrate show comparable low autofluorescence values (see Figure 5b).

3.5. Biofunctionalization Inside the Printed Microfluidic PMMA Chips

Protein patterns with specific shapes and patterns are capable of inducing specific cellular responses and are important for the study of cellular behavior. To demonstrate the possibility to locally create these patterns within the printed PMMA channels, we used a fast protein patterning method based on the immobilization of fluorescently labeled molecules via photobleaching which we have previously reported [28]. The photochemical immobilization of fluorescently labelled biomolecules is based on the transformation of fluorophores into reactive radicals as a consequence of photobleaching upon irradiation at the absorption wavelength. The radical species, which are only formed in the exposed areas, react with functional groups and thus bond covalently to the surface. The method is shown in Figure 6a. In this work, we created the pattern directly within the channel by printing a microfluidic PMMA channel on top of a transparent PMMA substrate (shown in Figure 6b), filling it with fluorescently labelled biotin (F5B) and illuminating it with a maskless projection lithography system. The patterns of photobleached F5B were visualized by staining with STV-Cy3. Biomolecule patterns with feature sizes of 125 μm were fabricated inside the microfluidic channel showing the facile biofunctionalization of 3D printed PMMA microfluidics (Figure 6c,d).

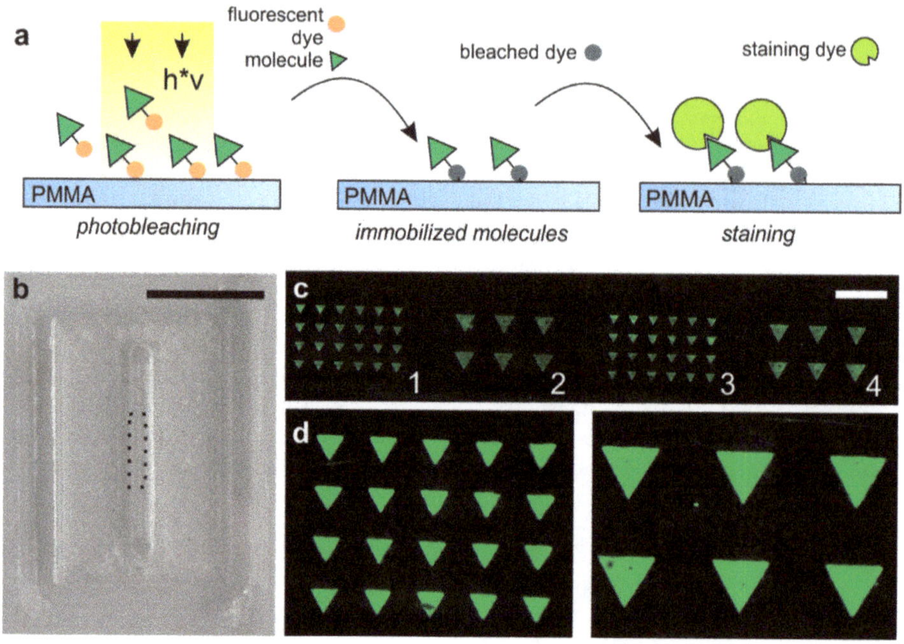

Figure 6. Biofunctionalization inside the 3D printed PMMA channel. (**a**) Fluorescently labelled biomolecules can be covalently bound to surfaces by photobleaching the fluorophore upon exposure with light patterns. (**b**) 2 mm wide embedded PMMA channel printed on a PMMA substrate with biofunctionalization inside the printed channel indicated by dotted lines (scale bar: 10 mm). (**c**) The bottom surface of the 3D printed channel on top of the commercial PMMA slide was functionalized with different biotin patterns by photobleaching of F5B. Patterns were illuminated for 3 min (patterns 1 and 2) and 5 min (patterns 3 and 4). The patterns were visualized via staining with STV-Cy3 (scale bar: 750 µm). Signal-to-noise ratios were calculated for all four images and were on average 29 for 3 min illumination and 11 for 5 min. (**d**) Close-up of patterns which were illuminated for 3 min.

4. Conclusions

In this work, we demonstrated that microfluidic PMMA chips can be fabricated down to channel diameters of 300 µm using a benchtop FDM machine. It was shown that the transparency of the printed chips can be significantly improved by using a commercial PMMA substrate as the bottom substrate for the printing process. The 3D printed PMMA was compared to commercial PMMA and shows the same low autofluorescence and hydrophilic wetting behavior, which is important for biocompatibility and channel wetting. We demonstrated that the integration of microstructures, as well as biofunctionalization in 3D printed PMMA channels, is possible. We showed that high-resolution microstructures can be produced using a simple prepolymer-casting process and that these structures can be easily integrated into the channel by directly printing onto the structured PMMA substrate. We have also shown that patterns of biomolecules can be attached to the commercial PMMA substrate using a photobleaching process. This effectively enables the formation of microfluidic channels with high-resolution biofunctionalization within the channel itself.

This work will allow the use of PMMA, one of the most important and industrially-relevant thermoplastic polymer materials in microfluidic mass-market manufacturing, in early stages of component development, where rapid prototyping and 3D printing facilitate fast design iterations and rapid experimental verifications. This closes an important gap in the academia-to-industry transition by allowing usage of the same material both during early-stage lab-scale development and industrial-scale mass-market production.

Author Contributions: F.K. concept idea, design of experiments and writing of the paper, N.D. and M.M. printing experiments, A.K. Biofunctionalization, D.H., M.M., P.R. and B.E.R. UV-Vis measurements and contribution to the writing of the paper. All authors have read and agreed to the published version of the manuscript.

Funding: This research was partly funded by the German Ministry of Education and Research (BMBF), funding code 03×5527 "Fluoropor". This research was partly funded by the German Research Foundation (Deutsche Forschungsgemeinschaft, DFG) for funding through the Centre for Excellence livMatS Exec 2193/1-390951807. This work was partly funded by the Research Cluster "Interactive and Programmable Materials (IPROM)" funded by the Carl Zeiss Foundation. This work was partly financed by the Baden-Württemberg Stiftung gGmbH within the program "Biofunctional Materials and Surfaces" (BiofMo-2, 3D Mosaic).

Conflicts of Interest: The authors declare no conflict of interest.

References

1. Waldbaur, A.; Rapp, H.; Lange, K.; Rapp, B.E. Let there be chip—Towards rapid prototyping of microfluidic devices: One-step manufacturing processes. *Anal. Methods* **2011**, *3*, 2681. [CrossRef]
2. Waheed, S.; Cabot, J.M.; Macdonald, N.P.; Lewis, T.; Guijt, R.M.; Paull, B.; Breadmore, M.C. 3D printed microfluidic devices: Enablers and barriers. *Lab Chip* **2016**, *16*, 1993–2013. [CrossRef] [PubMed]
3. Kotz, F.; Risch, P.; Helmer, D.; Rapp, B.E. High-Performance Materials for 3D Printing in Chemical Synthesis Applications. *Adv. Mater.* **2019**, *31*, 1805982. [CrossRef] [PubMed]
4. Kotz, F.; Helmer, D.; Rapp, B.E. Emerging Technologies and Materials for High-Resolution 3D Printing of Microfluidic Chips. In *Advances in Biochemical Engineering/Biotechnology*; Springer: Berlin/Heidelberg, Germany, 2020. [CrossRef]
5. Helmer, D.; Voigt, A.; Wagner, S.; Keller, N.; Sachsenheimer, K.; Kotz, F.; Nargang, T.M.; Rapp, B.E. Suspended Liquid Subtractive Lithography: One-step generation of 3D channel geometries in viscous curable polymer matrices. *Sci. Rep.* **2017**, *7*, 7387. [CrossRef] [PubMed]
6. Romanov, V.; Samuel, R.; Chaharlang, M.; Jafek, A.R.; Frost, A.; Gale, B.K. FDM 3D printing of high-pressure, heat-resistant, transparent microfluidic devices. *Anal. Chem.* **2018**, *90*, 10450–10456. [CrossRef]
7. Kotz, F.; Risch, P.; Helmer, D.; Rapp, B.E. Highly fluorinated methacrylates for optical 3D printing of microfluidic devices. *Micromachines* **2018**, *9*, 115. [CrossRef]
8. Sochol, R.D.; Sweet, E.; Glick, C.C.; Venkatesh, S.; Avetisyan, A.; Ekman, K.F.; Raulinaitis, A.; Tsai, A.; Wienkers, A.; Korner, K.; et al. 3D printed microfluidic circuitry via multijet-based additive manufacturing. *Lab Chip* **2016**, *16*, 668. [CrossRef]
9. Sachsenheimer, K.; Richter, C.; Helmer, D.; Kotz, F.; Rapp, B.E. A Nontoxic Battery with 3D-Printed Housing for On-Demand Operation of Microcontrollers in Microfluidic Sensors. *Micromachines* **2019**, *10*, 588. [CrossRef]
10. Shallan, A.I.; Smejkal, P.; Corban, M.; Guijt, R.M.; Breadmore, M.C. Cost-effective three-dimensional printing of visibly transparent microchips within minutes. *Anal. Chem.* **2014**, *86*, 3124. [CrossRef]
11. Rogers, C.I.; Qaderi, K.; Woolley, A.T.; Nordin, G.P. 3D printed microfluidic devices with integrated valves. *Biomicrofluidics* **2015**, *9*, 016501. [CrossRef]
12. Lee, Y.-S.; Bhattacharjee, N.; Folch, A. 3D-printed Quake-style microvalves and micropumps. *Lab Chip* **2018**, *18*, 1207. [CrossRef] [PubMed]
13. Becker, H. Mind the gap! *Lab Chip* **2010**, *10*, 271. [CrossRef] [PubMed]
14. Li, F.; Macdonald, N.P.; Guijt, R.M.; Breadmore, M.C. Multimaterial 3D printed fluidic device for measuring pharmaceuticals in biological fluids. *Lab Chip* **2019**, *19*, 35. [CrossRef] [PubMed]
15. Kitson, P.J.; Rosnes, M.H.; Sans, V.; Dragone, V.; Cronin, L. Configurable 3D-Printed millifluidic and microfluidic 'lab on a chip' reactionware devices. *Lab Chip* **2012**, *12*, 3267. [CrossRef] [PubMed]
16. Pranzo, D.; Larizza, P.; Filippini, D.; Percoco, G. Extrusion-based 3D printing of microfluidic devices for chemical and biomedical applications: A topical review. *Micromachines* **2018**, *9*, 374. [CrossRef]
17. Nelson, M.D.; Ramkumar, N.; Gale, B.K. Flexible, transparent, sub-100 µm microfluidic channels with fused deposition modeling 3D-printed thermoplastic polyurethane. *J. Micromech. Microeng.* **2019**, *29*, 095010. [CrossRef]

18. Chen, Y.; Zhang, L.; Chen, G. Fabrication, modification, and application of poly (methyl methacrylate) microfluidic chips. *Electrophoresis* **2008**, *29*, 1801. [CrossRef]
19. McCormick, R.M.; Nelson, R.J.; Alonso-Amigo, M.G.; Benvegnu, D.J.; Hooper, H.H. Microchannel electrophoretic separations of DNA in injection-molded plastic substrates. *Anal. Chem.* **1997**, *69*, 2626. [CrossRef]
20. Martynova, L.; Locascio, L.E.; Gaitan, M.; Kramer, G.W.; Christensen, R.G.; MacCrehan, W.A. Fabrication of plastic microfluid channels by imprinting methods. *Anal. Chem.* **1997**, *69*, 4783. [CrossRef]
21. Brister, P.C.; Weston, K.D. Patterned solvent delivery and etching for the fabrication of plastic microfluidic devices. *Anal. Chem.* **2005**, *77*, 7478. [CrossRef]
22. Sun, X.; Peeni, B.A.; Yang, W.; Becerril, H.A.; Woolley, A.T. Rapid prototyping of poly (methyl methacrylate) microfluidic systems using solvent imprinting and bonding. *J. Chromatogr. A* **2007**, *1162*, 162. [CrossRef] [PubMed]
23. Guckenberger, D.J.; de Groot, T.E.; Wan, A.M.; Beebe, D.J.; Young, E.W. Micromilling: A method for ultra-rapid prototyping of plastic microfluidic devices. *Lab Chip* **2015**, *15*, 2364. [CrossRef] [PubMed]
24. Suriano, R.; Kuznetsov, A.; Eaton, S.M.; Kiyan, R.; Cerullo, G.; Osellame, R.; Chichkov, B.N.; Levi, M.; Turri, S. Femtosecond laser ablation of polymeric substrates for the fabrication of microfluidic channels. *Appl. Surf. Sci.* **2011**, *257*, 6243. [CrossRef]
25. Qu, S.; Chen, X.; Chen, D.; Yang, P.; Chen, G. Poly (methyl methacrylate) CE microchips replicated from poly (dimethylsiloxane) templates for the determination of cations. *Electrophoresis* **2006**, *27*, 4910. [CrossRef] [PubMed]
26. Chen, Z.; Gao, Y.; Su, R.; Li, C.; Lin, J. Fabrication and characterization of poly (methyl methacrylate) microchannels by in situ polymerization with a novel metal template. *Electrophoresis* **2003**, *24*, 3246. [CrossRef] [PubMed]
27. Kotz, F.; Arnold, K.; Wagner, S.; Bauer, W.; Keller, N.; Nargang, T.M.; Helmer, D.; Rapp, B.E. Liquid PMMA: A high resolution polymethylmethacrylate negative photoresist as enabling material for direct printing of microfluidic chips. *Advanced Eng. Mater.* **2018**, *20*, 1700699. [CrossRef]
28. Waldbaur, A.; Waterkotte, B.; Schmitz, K.; Rapp, B.E. Maskless projection lithography for the fast and flexible generation of grayscale protein patterns. *Small* **2012**, *8*, 1570. [CrossRef]
29. Nargang, T.M.; Runck, M.; Helmer, D.; Rapp, B.E. Functionalization of paper using photobleaching: A fast and convenient method for creating paper-based assays with colorimetric and fluorescent readout. *Eng. Life Sci.* **2016**, *16*, 525. [CrossRef]
30. Waterkotte, B.; Bally, F.; Nikolov, P.M.; Waldbaur, A.; Rapp, B.E.; Truckenmüller, R.; Lahann, J.; Schmitz, K.; Giselbrecht, S. Biofunctional micropatterning of thermoformed 3D substrates. *Adv. Funct Mater.* **2014**, *24*, 442. [CrossRef]
31. Bressan, L.P.; Adamo, C.B.; Quero, R.F.; de Jesus, D.P.; da Silva, J.A.F. A simple procedure to produce FDM-based 3D-printed microfluidic devices with an integrated PMMA optical window. *Anal. Methods* **2019**, *11*, 1014. [CrossRef]
32. Briggs, D.; Chan, H.; Hearn, M.J.; McBriar, D.I.; Munro, H.S. The contact angle of poly (methyl methacrylate) cast against glass. *Langmuir* **1990**, *6*, 420. [CrossRef]

micromachines

Article

The Influence of Printing Orientation on Surface Texture Parameters in Powder Bed Fusion Technology with 316L Steel

Tomasz Kozior * and Jerzy Bochnia

Department of Manufacturing Technology and Metrology, Kielce University of Technology, 25-314 Kielc, Poland; jbochnia@tu.kielce.pl
* Correspondence: tkozior@tu.kielce.pl

Received: 24 May 2020; Accepted: 23 June 2020; Published: 29 June 2020

Abstract: Laser technologies for fast prototyping using metal powder-based materials allow for faster production of prototype constructions actually used in the tooling industry. This paper presents the results of measurements on the surface texture of flat samples and the surface texture of a prototype of a reduced-mass lathe chuck, made with the additive technology—powder bed fusion. The paper presents an analysis of the impact of samples' orientation on the building platform on the surface geometrical texture parameters (two-dimensional roughness profile parameters (*Ra*, *Rz*, *Rv*, and so on) and spatial parameters (*Sa*, *Sz*, and so on). The research results showed that the printing orientation has a very large impact on the quality of the surface texture and that it is possible to set digital models on the building platform (parallel—0° to the building platform plane), allowing for manufacturing models with low roughness parameters. This investigation is especially important for the design and 3D printing of microelectromechanical systems (MEMS) models, where surface texture quality and printable resolution are still a large problem.

Keywords: additive manufacturing technology; powder bed fusion; 3D printing; surface geometrical parameters (SGP); lathe chuck jaws

1. Introduction

1.1. 3D Printing

Additive manufacturing technologies were developed in the latter part of the 20th century, when the first additive technology was introduced—stereolithography [1]. In the subsequent years, we could notice the dynamic development of additive production technologies [2], manifested in many patent publications, such as FDM (fused deposition modeling) [3], SLS (selective laser sintering) [4], LOM (laminated object manufacturing) [5], SLM (selective laser melting) [6], and 3DP (3D printing) [7]. There are several ecological, environmental friendly additive manufacturing technologies such as FEF (freeze-form extrusion fabrication) [8,9].

Rapid prototyping of tools used in conventional production technologies such as machining, casting [10–12], or injection moulding allows for immediate subsequent printing necessary after testing. Moreover, additive technologies enhance optimization of production and enable fast adaptation of new solutions. Together with development of the technology, the increased precision of manufactured devices, as well as progress in laser technology, a dynamic evolution of additive technologies applications took place. In particular, this applies to powder-based technologies such as SLM, SLS, laser engineered net shaping (LENS), and so on. In many cases, studies confirm that the manufactured models can be characterized by significant density and ideal mechanical properties. Owing to the lack of uniform procedures concerning the studies that are available in the case of conventionally manufactured models, the development of additive technologies has partially decelerated. Owing to the layered nature of the manufacturing process, as presented above, with different methods of construction and connection of layers, the properties of the manufactured models depend on the correct selection of technological parameters [13] and further finishing processes [14]. There are numerous studies aimed at the determination of the impact of technological manufacturing processes parameters on properties and accuracy of manufactured models, as presented in the following publications [15–20]. Additive technologies can be classified based on many criteria [21], one of which is the type of input material used. In the case of this kind of classification, one differentiates the following technologies: powder-based (SLS, SLM, 3DP, and so on), liquid-based (SLA - stereolitography, PJ - PolyJet, PJM – PolyJet Matrix, MJM – Multi Jet Modeling, DODJET, FEF - freeze-form extrusion fabrication, and so on), and solid state material (FDM, FFF (fused filament fabrication), LOM, and so on) [22]. Each of the above-mentioned methods has its own advantages and drawbacks that will depend on the method of connecting layer and the delivery of the model and auxiliary material. Therefore, the application of the mentioned technologies differs as well. The following part of the paper explores additive technologies that use metal powder as a starting material. Within the last decade, a new type of additive production has been advanced, especially using FDM technology consisting of the construction of a physical object on other pre-existing models [23–26].

Additive technologies in the case of rapid tooling applications are broadly used for construction of casting moulds and models, 3D sand mold printing technology [27,28]; injection moulds, SLM and SLS (metal); and the printing of tools of complicated shapes (such as needed in the medical industry). Moreover, additive technologies using metal-based materials are commonly applied for the production of prototypes and fully functional elements in the automotive and aviation industries [29–31] as well as in robotics [32]. Current metal powder-based technologies are widely used in the odontology industry, where it is not necessary to construct large, rather but strong and corrosion-resistant dentures [33–35]. The medical industry pays even more attention to the application of additive technologies. This is highly related to the option to use three-dimensional models obtained directly from X-rays and computer tomography. Additive manufacturing technologies are also used in the food industry [36].

1.2. Literature Review

MEMS (microelectromechanical systems) is a potential application of 3D printing technology using metal-based materials [37–39]. In this application, models produced with 3D printing technologies, from both metal-based materials and plastics, must be characterized by appropriate surface quality and dimensional accuracy. The parameters of the surface layer obtained by additive technologies are largely influenced by technological parameters (e.g., grain size, layer thickness, printing direction, temperature, laser power, laser speed, cooling time) and the resolution of selected machines/3D printers. In the paper [37], the authors presented the possibilities of using 3D printing technology and selected machines for building MEMS models. The article mainly analyzes materials based on plastics and describes the disadvantages and advantages of currently used machines/printers in the context of the use of 3D printing to build MEMS models. It seems that increasing the accuracy of 3D printers is a natural consequence of their technological development, which will definitely increase the practical potential and uses 3D printing in MEMS production.

Studies related to the determination of geometrical texture quality of a model manufactured using both conventional and generative technologies utilizing strong, corrosion-resistant materials have been described in a few research papers.

In the papers of [40–42], different measurement systems have been used in order to analyse the measurement technology impact on the obtained results. The authors used both contact and optical measurement methods, considering X-ray computer tomography and multiscale 3D curvature analysis.

The authors of [43] carried out studies on models made using SLM technology and the powder AlSi10Mg. In the designed model, they determined basic parameters of geometrical texture on various surfaces oriented under variable angles in relation to the building platform. A measurement of the selected model dimensions was also performed.

Studies consisting of the analysis of geometrical texture quality of the models made using metal powders and SLM technology as well as the analysis of measurements methodology have been described in a few research works by [44–46].

The presented studies, based on the analysis of the impact of model orientation on the building platform on their SGPs (surface geometrical parameters), present actual application of the additive technologies in the construction of tools used in conventional methods of production, which must satisfy specified quality requirements of the top layer. The technological top layer plays a key role in the wearing process of mating machine parts. Additionally, it is responsible for a series of functional properties that condition the correct operation of the designed mechanisms. This is very important in the case of manufacturing prototypes, which are subjected to numerous tests that require the implementation of further design changes. Moreover, the quality of the top layer determines the further processes of model cleaning and type of finishing. It is also important that, in many cases, such as the production of elements of complicated internal shapes (e.g., turbine blades with complicated cooling channels), there is no option to perform finishing, which proves the presented studies concerning the impact of model orientation during construction to be considered as reasonable.

The influence of finishing on the surface roughness is presented in [47]. The authors of this work come to the fundamental conclusion that the finishing treatment significantly improves the surface roughness in relation to the surface obtained through additive technology and recommends finishing treatment after the additive shaping process. The results of roughness measurements for samples made with SLM technology and then machined mechanically (finish machining), dragging with a ceramic tool (drag finish), and vibration treatment (vibratory surface finish) are presented in this paper. The best results were obtained for drag finish. The roughness parameter *Ra* was used to evaluate the roughness. It should be emphasized here that, in conventional machining of machine elements, the roughness parameters *Ra* and *Rz* are used most often to assess surface roughness, because this is related to the path of the cutting edge of the tool on the surface of the workpiece. Usually, these are regular scratches, for example, after turning. The surface of the material created as a result of additive technology is more complex and more parameters characterizing its shape are needed to evaluate it. Surface roughness is just one of the elements of the assessment of the surface layer of elements produced by additive technologies. The quality of the laser-sintered powder is much more complex. Test methods and results for 316L steels are presented in paper [48]. Among other things, surface roughness was investigated, as well as two parameters *Ra* and *Rq*, for samples from 316L steels. Moreover, its corrosion resistance in biomedical applications is presented in [49]. The structure and tendency for corrosion of selectively laser-formed, additive fused (AM) 316 L stainless steel (AM 316L SS), and its wrought counterpart were analyzed. Increased corrosion resistance and improved biological response to 316 L stainless steel were obtained through additive technology.

Taking into account the above considerations, the authors of this article decided to present a wide range of surface roughness assessments, mainly geometric surface structure using modern measurement methods. The tests took into account the printing direction of the elements. The tests were carried out on samples made of 316L steel, which is a well-known and used material, so that the results could be applicable to other users.

The studies presented herein were performed, in part, within the scope of a grant of the Ministry of Science and Higher Education in Poland concerning construction of an innovative lathe chuck that allows for the compensation of centrifugal force acting on lathe chuck jaws, and thus machining of elements at increased rotational speed of the lathe spindles.

2. Materials and Methods

In order to build the sample models, a material based on 316L steel was used. The samples were designed in two variants. The first set of samples was produced in 30 pieces (10 per each direction of printing) according to the standard (ASTM E8/E8M-13 [50]) in order to perform further rheological studies. The second samples were produced in two pieces, and the design is a functional element of the innovative lathe chuck (self-centering chuck jaw). The jaws were manufactured using additive technology in order to produce the unique structure in which the jaws have empty internal cavities that reduce weight, as presented in Figures 1 and 2. 316L steel was also used to build the jaw prototype to avoid corrosion caused by operating fluids in future testing. This type of design (hollow chuck jaws) is unique and does not exist in any other produce lathe chuck. Because of additive technology, it was possible to reduce the lathe jaw weight, a fact that contributes to the reduction of the unfavourable impact of centrifugal force imposed by the jaws' weight. Sample model shapes were designed using CAD software - SolidWorks (Dassault Systèmes SolidWorks Corp., Waltham, Massachusetts) and saved as STL files (overall dimensions of the sample—40 mm × 78.3 mm × 57.45 mm^3). Sample models were arranged on the machine working platform in three characteristic orientation variants, as presented in Figure 1.

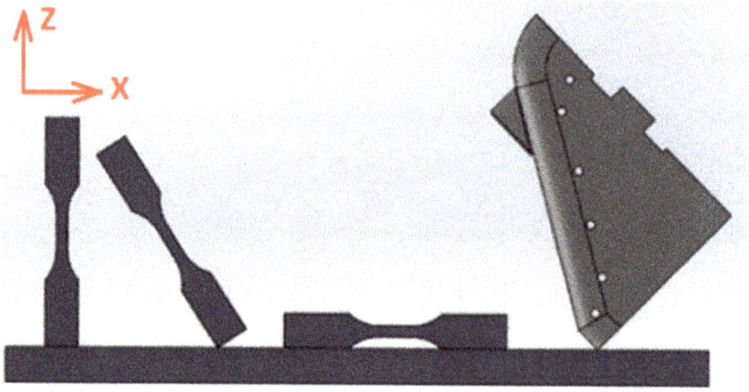

Figure 1. Orientation of samples on the building platform.

The samples were produced using the 3D printing machine, CONCEPT LASER M2 (www.concept-laser.de [51], Concept Laser GmbH, Lichtenfels, Germany). The machine has a working space of 250 × 250 × 350 mm^3 (x, y, z) and is equipped with a laser system having a power of 2 × 200 W. The machine is equipped to use a variety of materials, including 316L, CL 30/31AL, titanium, bronze, and nickel based powders that are used, especially for odontology applications. Upon completion of the manufacturing process, the samples were subjected to an annealing.

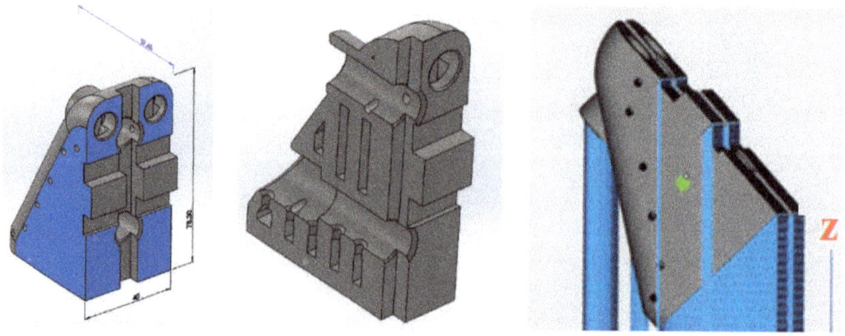

(**a**) (**b**) (**c**)

Figure 2. Jaws sample: (**a**) model dimension, (**b**) cross section, and (**c**) samples with support material.

The samples were heated in a furnace to 550 °C for 3 h and then soaked for 6 h to cool outside the furnace. The thickness of built layer (layer height) was 25 µm. Figure 3 presents sample models during metrology measurements using the optical profilometer Talysurf CCI Lite. The examination and determination of the surface texture parameters was performed according to applicable standards (ISO 4287:1997 [52], ISO 25178-2:2012 [53]).

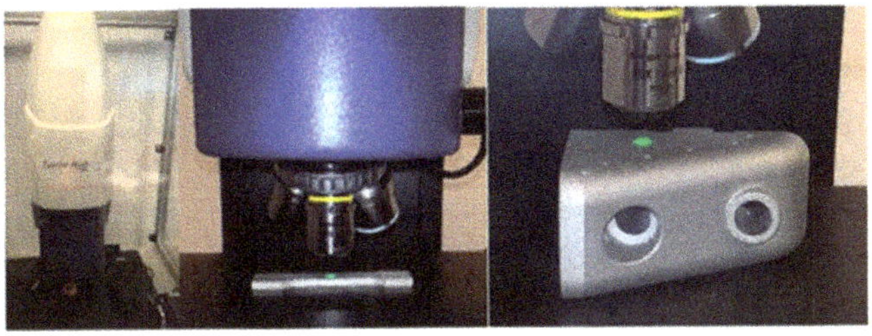

Figure 3. Talysurf CCI Lite—optical profilometer, samples during measurements.

Chemical composition and mechanical properties of steel 316L are presented in Tables 1 and 2.

Table 1. Chemical composition of steel 316L.

Component	Cr	Ni	Mo	Mn	Si	P	C	S	Fe
Indicative value, %	16.5–18.5	10.0–13.0	2–2.5	0–2.0	0–1.0	0–0.045	0–0.030	0–0.030	Balance

Table 2. Mechanical properties of steel 316L.

Properties	90°, Upright	45°, Polar Angle	0°, Horizontal
Yield strength, $R_{p0,2}$	374 ± 5 N/mm^2	385 ± 6 N/mm^2	330 ± 8 N/mm^2
Tensile strength, Rm	650 ± 5 N/mm^2	640 ± 7 N/mm^2	529 ± 8 N/mm^2
Elongation, A	(65 ± 4)%	(63 ± 5)%	(63 ± 5)%
Young's modulus		ca. 200 × 10^3 N/mm^2	
Hardness		20 HRC	

During measurements, two-dimensional parameters of surface roughness and spatial parameters were determined. Parameters of measurements were as follows: dimensions of measurement section 1.6 mm × 1.6 mm, number of measurement spots 1024, and applied magnifying lens 10×. In the case of 2D surface roughness parameters, attention was paid to amplitude parameters (*Rp, Rv, Rz, Rc, Rt, Ra, Rq, Rsk, Rku*), material ratio (*Rmr, Rdc*), and arrangement (*RSm, Rdq*). On the basis of the performed measurements, three-dimensional parameters were also determined, among other things, height (*Sq, Ssk, Sku, Sp, Sv, Sz,* and *Sa*), functional (*Smr, Smc,* and *Sxp*), and spatial (*Sal, Str,* and *Std*, as well as *Sdq* and *Sdr*) (ISO 25178-2:2012) [54].

Definitions of some roughness parameters are given below.

- *Sq*—mean square deviation of the surface from the reference surface (root mean square value of the ordinate values within a definition area (*A*)) calculated from the following equation:

$$Sq = \sqrt{\frac{1}{A}\iint_A z^2(x,y)dxdy} \tag{1}$$

- *Ssk*—surface asymmetry factor (slant) (quotient of the mean cube value of the ordinate values and the cube of *Sq* within a definition area (*A*)) calculated from the following equation:

$$Ssk = \frac{1}{Sq^3}\left[\frac{1}{A}\sqrt{\iint_A z^3(x,y)dxdy}\right] \tag{2}$$

- *Sku*—surface slope factor (kurtosis) (quotient of the mean quartic value of the ordinate values and the fourth power of *Sq* within a definition area (*A*)) calculated from the following equation:

$$Sku = \frac{1}{Sq^4}\left[\frac{1}{A}\sqrt{\iint_A z^4(x,y)dxdy}\right] \tag{3}$$

- *Sp*—largest peak height value within a definition area, *Sv*—minus the smallest pit height value within a definition area, and *Sz*—sum of the maximum peak height value and the maximum pit height value within a definition area calculated from the following equations:

$$\begin{aligned} Sp &= \sup[z(i,j)] \\ Sv &= \left|\inf[z(i,j)]\right| \\ Sz &= Sp + Sv \end{aligned} \tag{4}$$

- *Sa*—arithmetic mean of the absolute of the ordinate values within a definition area (A) calculated from the following equation:

$$Sa = \frac{1}{A}\iint_A \left|z(x,y)\right| dxdy \tag{5}$$

Many parameters were used to measure the geometrical surface texture, so that their significance in assessing the surface layer geometry of elements produced by the additive technology could be assessed. In machining, several parameters are usually used to assess the surface, for example, *Ra* and Rz. However, for elements manufactured with additive manufacturing technologies, a simple assessment using two or three parameters may not be sufficient.

3. Results

On the basis of the analysis of references and results of previous studies concerning the consumption and technological quality of top layer that depend on technological parameters such as location on the building platform, it was decided to perform complex 2D and 3D measurements together with consideration of statistical analysis. The parameters of surface texture are presented in Tables 3 and 4. The following elements were determined for each type of positioning on the building platform together with their symbols: *SD*—standard deviation of the test, $U_{A\beta}$—uncertainty of measurement, and X_β—average value of the test (β—angle of sample inclination against the building platform: 0°, 45°, and 90°). Moreover, for all 30 pieces, total statistical parameters identified with the below symbols were calculated: *X*, *SD*, and U_A. Standard deviation was calculated from Formula (6) and the standard uncertainty using method A from Formula (7). For the sample size $n = 10$, considering Student's t distribution, the uncertainty was evaluated based on Equation (8).

$$SD = \sqrt{\frac{1}{(n-1)}\sum_{i=1}^{n}(x_i - \overline{x})^2} \tag{6}$$

where n—sample size and $\overline{x}$—arithmetical mean of all measured values in a sample.

$$U_A = \sqrt{\frac{1}{n(n-1)}\sum_{i=1}^{n}(x_i - \overline{x})^2} \tag{7}$$

$$U_{A\beta} = k_\alpha \cdot U_A \tag{8}$$

where coefficient k_α is 2.

Table 3. Value of the 2D surface roughness parameters.

Sample Number	*Rp*, (µm)	*Rv*, (µm)	*Rz*, (µm)	*Rc*, (µm)	*Rt*, (µm)	*Ra*, (µm)	*Rq*, (µm)	*Rsk*	*Rku*	*Rmr*, (%)	*Rdc*, (µm)	*RSm*, (µm)	*Rdq*, $(^0)$
1	24.6	19.8	44.4	27.3	52.3	8.4	10.4	0.491	3.479	0.9	17.0	0.228	25.0
2	11.9	9.5	21.5	11.7	25.6	3.8	4.8	0.41	3.833	1.0	7.6	0.178	15.0
3	12.2	9.0	21.2	11.0	26.9	3.6	4.6	0.648	4.54	0.9	7.2	0.171	14.9
4	8.5	8.3	16.8	9.0	19.2	3.0	3.7	−0.007	2.943	1.1	6.4	0.16	13.1
5	9.4	8.5	17.9	9.4	20.5	3.1	3.9	0.167	3.129	1.1	6.6	0.15	14.4
6	7.9	8.1	16.0	9.0	18.2	2.9	3.6	−0.027	2.821	1.4	6.3	0.164	12.7
7	9.3	9.1	18.4	10.1	21.0	3.4	4.1	0.041	2.964	1.3	7.1	0.167	13.8
8	9.5	8.1	17.6	9.6	20.5	3.1	3.9	0.34	3.337	1.3	6.5	0.171	12.9
9	10.4	8.8	19.2	10.5	22.6	3.5	4.3	0.229	3.18	1.2	7.3	0.174	14.0
10	9.4	9.0	18.5	10.2	21.5	3.3	4.1	0.087	3.197	1.2	7.1	0.171	13.9
$\bar{x}_0$	11.3	9.8	21.2	11.8	24.8	3.8	4.8	0.238	3.342	1.1	7.9	0.173	15.0
SD	4.9	3.5	8.3	5.5	10.0	1.6	2.0	0.228	0.512	0.2	3.2	0.021	3.6
U_{A0}	1.5	1.1	2.6	1.7	3.2	0.5	0.6	0.072	0.162	0.1	1.0	0.007	1.1
11	9.2	9.6	18.8	11.4	21.7	3.6	4.4	−0.071	2.923	1.5	7.6	0.199	12.4
12	13.7	12.0	25.7	16.2	29.4	4.9	6.0	0.22	2.962	1.1	10.3	0.222	14.9
13	13.4	11.1	24.5	15.3	29.9	4.9	6.0	0.445	3.481	1.3	10.2	0.234	14.1
14	12.1	11.2	23.4	14.7	26.7	4.7	5.7	0.149	2.771	1.2	10.1	0.23	13.9
15	11.4	10.8	22.3	13.3	24.7	4.2	5.1	0.059	2.724	1.2	8.8	0.208	14.0
16	12.1	12.6	24.7	15.7	28.7	4.8	5.9	−0.028	2.735	1.1	10.4	0.232	14.3
17	11.8	11.4	23.1	15.0	26.2	4.5	5.5	0.081	2.727	1.2	9.6	0.223	14.0
18	10.4	10.1	20.5	11.8	23.4	3.8	4.7	0.048	3.08	1.2	7.8	0.183	14.3
19	11.3	10.5	21.8	12.4	25.8	4.1	5.0	0.099	3.238	1.2	8.5	0.205	13.4
20	12.0	11.6	23.5	14.4	27.1	4.6	5.6	0.127	2.871	1.2	9.7	0.212	14.6
$\bar{x}_{45}$	11.7	11.1	22.8	14.0	26.4	4.4	5.4	0.113	2.951	1.2	9.3	0.215	14.0
SD	1.3	0.9	2.1	1.7	2.6	0.5	0.6	0.143	0.251	0.1	1.0	0.016	0.7
U_{A45}	0.4	0.3	0.7	0.5	0.8	0.1	0.2	0.045	0.079	0.0	0.3	0.005	0.2
21	11.5	10.1	21.6	12.9	25.3	4.1	5.0	0.282	3.172	1.2	8.5	0.176	14.5
22	13.7	13.7	27.4	16.7	31.5	5.1	6.3	−0.01	2.97	1.0	10.6	0.203	15.8
23	11.9	12.5	24.5	15.4	27.3	4.7	5.7	−0.066	2.733	1.3	10.2	0.194	15.3
24	11.4	10.0	21.4	12.8	24.2	4.1	5.0	0.153	2.756	1.1	8.8	0.184	14.9
25	11.7	10.8	22.5	14.0	26.3	4.4	5.3	0.099	3.199	1.2	9.1	0.204	14.4
26	11.7	12.4	24.1	15.7	27.3	4.7	5.7	−0.105	2.747	1.2	10.2	0.195	14.6
27	12.1	10.5	22.6	12.4	25.5	4.0	5.0	0.247	3.213	1.0	8.2	0.178	14.5
28	11.0	10.1	21.2	12.6	23.8	3.9	4.8	0.162	2.966	1.3	8.1	0.182	13.1
29	11.0	10.5	21.5	12.6	24.8	4.0	5.0	0.046	2.903	1.2	8.5	0.183	13.4
30	12.1	13.1	25.2	16.8	29.8	5.2	6.2	−0.1	3.016	1.4	11.2	0.209	15.1
$\bar{x}_{90}$	11.8	11.4	23.2	14.2	26.6	4.4	5.4	0.071	2.968	1.2	9.3	0.191	14.6
SD	0.8	1.4	2.0	1.8	2.5	0.5	0.5	0.140	0.186	0.1	1.1	0.012	0.8
U_{A90}	0.2	0.4	0.6	0.6	0.8	0.1	0.2	0.044	0.059	0.0	0.4	0.004	0.3
$\bar{x}$	11.6	10.8	22.4	13.3	25.9	4.2	5.2	0.141	3.087	1.2	8.9	0.193	14.5
SD	2.8	2.3	5.0	3.5	6.0	1.0	1.3	0.1841	0.3814	0.1	2.1	0.0236	2.1
U_A	0.5	0.4	0.9	0.6	1.1	0.2	0.2	0.0336	0.0696	0.0	0.4	0.0043	0.4

Table 4. Value of the 3D surface roughness parameters.

Sample Number	*Sq*, (µm)	*Ssk*	*Sku*	*Sp*, (µm)	*Sv*, (µm)	*Sz*, (µm)	*Sa*, (µm)	*Smr*, (%)	*Smc*, (µm)	*Sxp*, (µm)	*Sal*	*Str*	*Std*,°	*Sdq*	*Sdr*, (%)
1	17.1	1.085	4.381	70.8	42.1	113.0	13.2	0.01	22.5	20.5	0.109	0.389	26.499	0.68	19.9
2	8.7	1.182	8.776	66.1	24.2	90.2	6.5	0.012	10.3	14.5	0.083	0.1	3.974	0.433	8.6
3	7.4	1.272	6.821	42.1	20.0	62.0	5.4	0.013	8.6	11.3	0.065	0.444	153.521	0.43	8.5
4	5.7	0.005	3.008	24.6	21.1	45.6	4.6	0.004	7.4	11.3	0.054	0.107	0.146	0.374	6.6
5	5.9	0.039	2.95	34.9	23.1	58.0	4.8	0.001	7.7	11.5	0.049	0.059	0.099	0.417	8.1
6	5.7	0.146	2.989	23.1	21.8	44.9	4.5	0.003	7.4	10.8	0.052	0.13	0.12	0.37	6.5
7	8.4	0.229	2.703	32.3	25.4	57.7	6.9	0.002	11.1	14.8	0.05	0.061	0.076	0.436	8.9
8	8.4	0.703	3.74	36.0	22.9	58.8	6.6	0.013	11.1	12.7	0.08	0.096	4.43	0.388	7.0
9	12.2	1.114	4.258	47.4	28.7	76.1	9.2	0.016	17.5	15.1	0.118	0.142	8.701	0.413	8.0
10	6.2	0.214	3.108	27.2	22.5	49.7	4.9	0.007	8.0	11.5	0.054	0.355	0.143	0.403	7.6
$\bar{x}_0$	11.6	0.650	3.745	49.0	32.3	81.3	9.1	0.009	15.3	16.0	0.082	0.372	13.321	0.542	13.8
SD	3.6	0.524	1.997	16.6	6.4	21.7	2.7	0.005	5.0	3.0	0.025	0.147	47.695	0.089	3.9
U_{A0}	2.3	0.332	1.263	10.5	4.1	13.7	1.7	0.003	3.2	1.9	0.016	0.093	30.165	0.057	2.5
11	5.4	−0.018	2.939	17.0	18.9	35.9	4.3	0.035	7.1	10.7	0.062	0.209	51.199	0.335	5.3
12	7.3	0.353	3.169	39.8	24.4	64.2	5.8	0.008	9.8	12.5	0.06	0.051	135.034	0.393	7.3
13	8.2	1.034	5.137	43.7	20.3	64.0	6.3	0.002	10.1	11.6	0.07	0.127	26.462	0.386	7.1
14	7.0	0.26	2.734	25.4	40.6	66.0	5.7	0.005	9.5	12.1	0.058	0.093	134.985	0.377	6.7
15	6.4	0.14	2.924	26.1	21.5	47.6	5.3	0.002	8.4	12.1	0.063	0.177	134.998	0.374	6.6
16	7.4	0.192	3.134	32.0	23.9	55.9	5.9	0.004	9.5	14.0	0.058	0.053	116.483	0.375	6.6
17	6.8	0.229	2.759	26.4	39.7	66.1	5.5	0.002	9.3	11.9	0.062	0.085	135.007	0.377	6.7
18	5.9	−0.039	3.209	22.3	26.9	49.2	4.7	0.005	7.5	11.7	0.052	0.256	47.899	0.385	7.0
19	6.5	0.378	3.407	29.2	20.1	49.2	5.1	0.002	8.3	11.7	0.065	0.255	141.207	0.363	6.2
20	7.4	0.278	2.799	26.0	24.9	50.9	5.9	0.001	9.9	12.6	0.07	0.06	134.966	0.39	7.1
$\bar{x}_{45}$	6.4	0.130	2.869	21.5	21.9	43.4	5.1	0.018	8.5	11.6	0.066	0.135	93.083	0.363	6.2
SD	1.4	0.209	0.099	6.4	4.3	10.6	1.1	0.024	2.0	1.3	0.006	0.105	59.232	0.039	1.3
U_{A45}	0.9	0.132	0.063	4.0	2.7	6.7	0.7	0.015	1.3	0.8	0.004	0.067	37.462	0.025	0.8
21	5.9	0.422	3.385	24.1	22.5	46.6	4.7	0.022	7.9	10.1	0.058	0.135	85.581	0.363	6.2
22	7.3	0.303	3.106	28.4	23.2	51.6	5.8	0.001	9.4	12.8	0.058	0.07	93.487	0.381	6.8
23	6.8	0.078	2.99	24.5	25.4	49.9	5.4	0.009	8.9	12.6	0.057	0.398	93.759	0.377	6.7
24	5.9	0.226	2.828	24.0	18.8	42.8	4.8	0.001	7.8	10.5	0.065	0.148	89.969	0.377	6.7
25	6.6	0.487	4.254	54.8	20.6	75.4	5.2	0	8.3	11.8	0.075	0.411	89.996	0.368	6.3
26	6.8	0.074	3.081	25.1	28.5	53.6	5.4	0.008	8.9	12.8	0.058	0.33	63.504	0.358	6.1
27	6.2	0.336	3.296	24.1	19.6	43.6	4.9	0.007	8.1	11.2	0.078	0.181	90.037	0.362	6.2
28	5.8	0.315	3.205	26.7	18.8	45.5	4.6	0.001	7.6	10.2	0.058	0.238	85.563	0.324	5.0
29	5.8	0.113	2.988	25.2	18.6	43.9	4.6	0.006	7.6	11.1	0.06	0.072	89.973	0.327	5.1
30	7.5	0.132	2.856	25.6	24.9	50.6	6.0	0.025	9.8	13.9	0.06	0.274	89.999	0.366	6.4
$\bar{x}_{90}$	6.7	0.277	3.121	24.9	23.7	48.6	5.3	0.024	8.8	12.0	0.059	0.205	87.790	0.365	6.3
SD	1.1	0.205	0.374	1.1	1.7	2.8	1.0	0.002	1.3	2.7	0.001	0.098	3.124	0.002	0.1
U_{A90}	0.7	0.130	0.237	0.7	1.1	1.8	0.6	0.001	0.8	1.7	0.001	0.062	1.976	0.001	0.1
$\bar{x}$	7.3	0.376	3.564	32.5	24.5	57.0	5.7	0.008	9.5	12.4	0.065	0.184	70.927	0.390	7.3
SD	2.3	0.381	1.306	12.9	6.2	15.7	1.7	0.008	3.1	2.0	0.016	0.123	52.658	0.061	2.6
U_A	0.4	0.070	0.238	2.3	1.1	2.9	0.3	0.001	0.6	0.4	0.003	0.022	9.614	0.011	0.5

With the expected value and the standard deviation, it is possible to present the results in the form of normal distribution based on the following formula:

$$f(x) = \frac{1}{SD\sqrt{2\pi}} e^{\frac{-(x-\overline{X})^2}{2SD^2}} \quad (9)$$

where $f(x)$—probability density function.

Figure 4 presents exemplary normal distributions of probability density in the function of the obtained results.

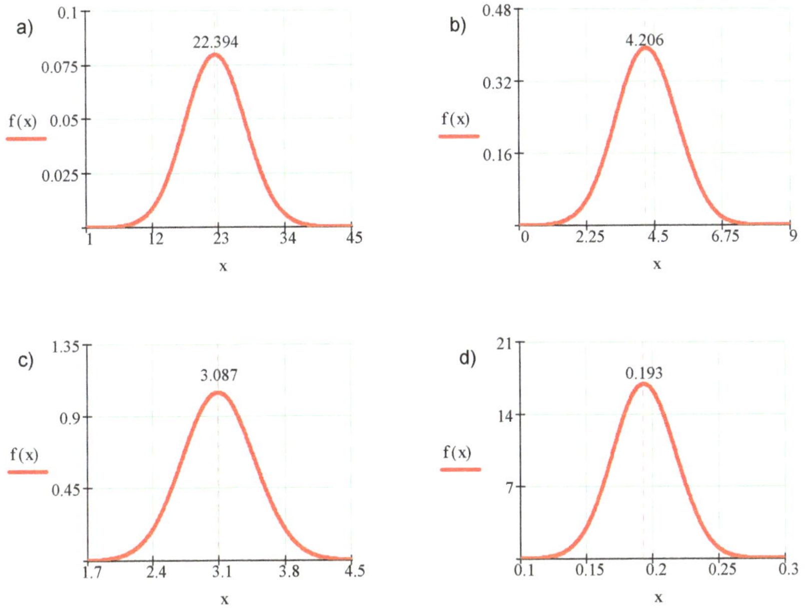

Figure 4. Normal distribution of probability density in the function of the obtained results: (**a**) maximum height of roughness profile—*Rz*, (**b**) mean arithmetic deviation of roughness profile—*Ra*, (**c**) relative material ratio of roughness profile—*Rmr*, and (**d**) mean element width of roughness profile—*Rsm*.

On the basis of the calculated statistical parameters and results presented in Graph 4, one may notice a scattering of roughness parameters values.

In order to assess the printing orientation with reference to individual values of surface texture parameters, relative values were introduced related to individual parameters and expressed in percentage. The mean value of a given parameter obtained based on sample surface measurements for all orientations of printing was adopted as the basis; in the considered case, this is the mean from the thirty samples. The mean value of a parameter from measurements concerning individual orientation of printing was referenced to the said basis. This is exemplified by the following Equations (10)–(12):

$$\Delta\overline{X}_0 = \frac{\left|\overline{X} - \overline{X}_0\right|}{\overline{X}} 100\% \tag{10}$$

$$\Delta\overline{X}_{45} = \frac{\left|\overline{X} - \overline{X}_{45}\right|}{\overline{X}} 100\% \tag{11}$$

$$\Delta\overline{X}_{90} = \frac{\left|\overline{X} - \overline{X}_{90}\right|}{\overline{X}} 100\% \tag{12}$$

where $\Delta\overline{X}_0$, $\Delta\overline{X}_{45}$, and $\Delta\overline{X}_{90}$—relative value of surface parameters for individual orientations of printing; $\overline{X}$—mean value of a parameter for all orientations of printing; $\overline{X}_0$—mean value of the measured parameters for printing orientation 0°; $\overline{X}_{45}$—mean value of the measured parameters for printing orientation 45°; and $\overline{X}_{90}$—mean value of the measured parameters for printing orientation 90°. For example, maximum height of the roughness profile *Rz* for printing orientation 0° calculated using Formula (10) based on data from Table 3 would be as follows:

$$\Delta\overline{Rz}_0 = \frac{\left|\overline{Rz} - \overline{Rz}_0\right|}{\overline{Rz}} 100\% \tag{13}$$

where $\Delta\overline{R}z_0$—relative maximum roughness profile height for printing orientation 0°, $\overline{R}z$—mean maximum roughness profile height for all printing orientations (22.4 based on Table 3), and $\overline{R}z_0$—mean maximum roughness profile height for printing orientation 0° (21.2 based on Table 3).

The value calculated based on Formula (13) is 5.5%. In this way, it is possible to determine all relative values of surface parameters with reference to printing orientation. The relative values of individual parameters are given in column graphs, Figures 5–7.

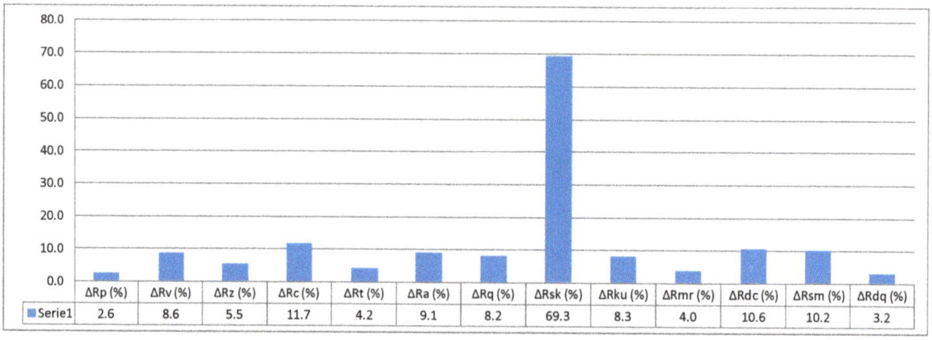

Figure 5. Relative values of surface parameters for printing orientation 0°.

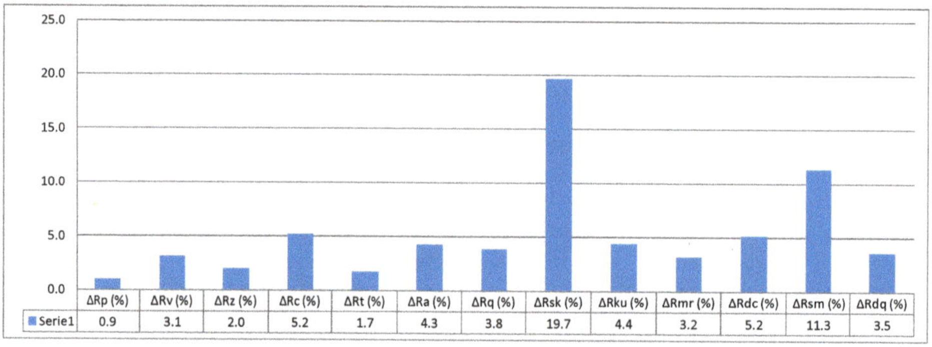

Figure 6. Relative values of surface parameters for printing orientation 45°.

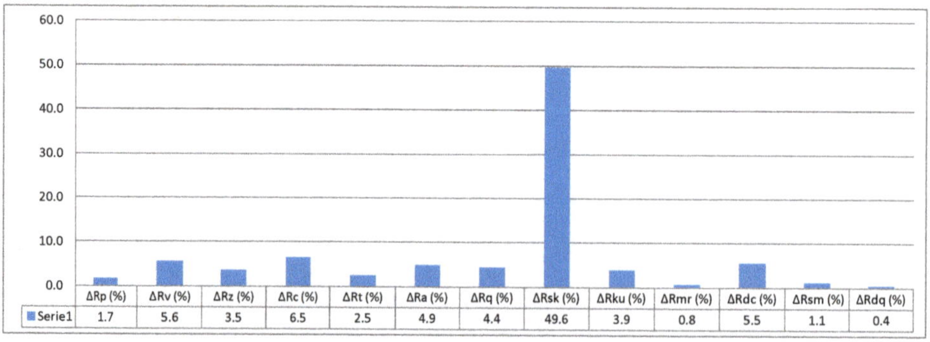

Figure 7. Relative values of surface parameters for printing orientation 90°.

When quantitatively analysing the results of the study, it can be said that there are distinct differences in the obtained surface texture parameter values depending on the models' orientation on the building platform. Concerning two-dimensional roughness parameters, the most advantageous orientation variant showing the least parameter values such as mean arithmetic deviation of roughness profile from mean line *Ra*; maximum profile height *Rz*; total profile height *Rt*, *Rv*; and other *Rq*, *Rdc*, and so on is the orientation at the angle of 0°, in other words, parallel to the base plane of the machine's working space. Ten out of thirteen measured parameters for the angle 0° show values less than mean values (measured for 30 samples) from 2.6% to 11.7%. Roughness profile asymmetry *Rsk* parameter shows the most significant differences from the printing orientation standpoint. This results from the fact that this parameter is inversely proportional to the cube of other roughness parameter *Rq* and is the third-order moment of the amplitude distribution curve determined along the measurement section. As the references show, these parameters directly affect the consumption of mating machine elements.

Figure 8 presents surface roughness profiles for three sample models' orientation variants.

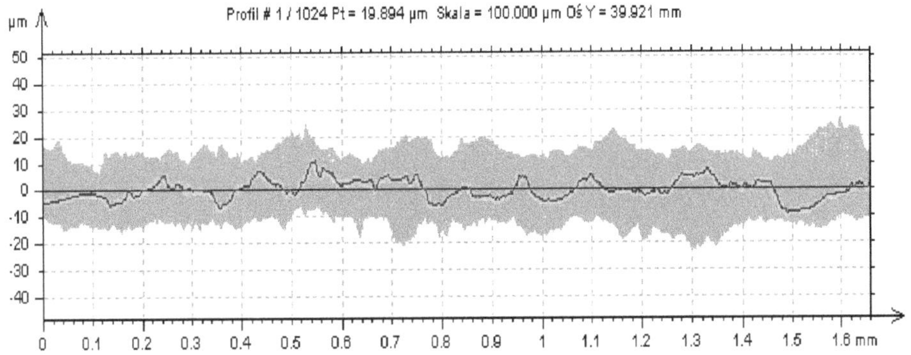

(**a**)

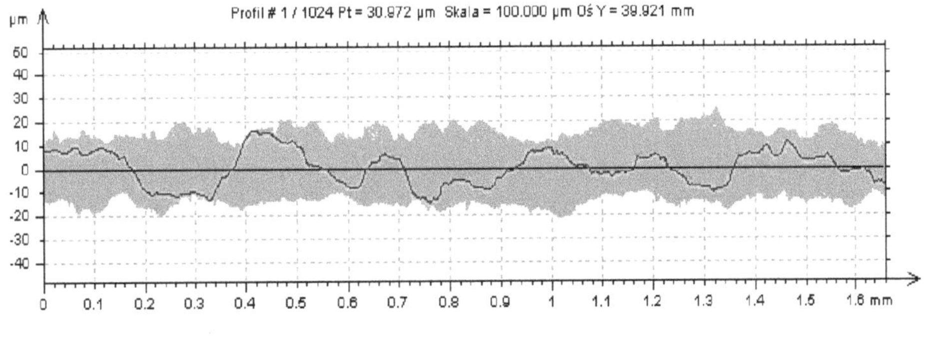

(**b**)

Figure 8. *Cont.*

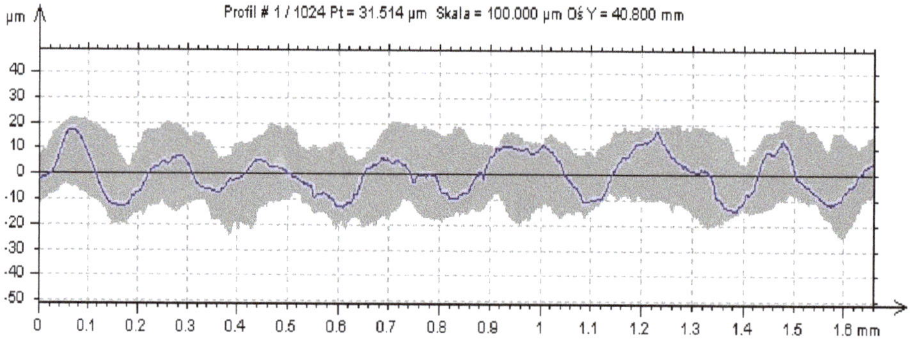

(**c**)

Figure 8. 2D roughness profile for samples: (**a**) 0°, (**b**) 45°, and (**c**) 90°.

Figure 9 presents the exemplary normal distribution of probability density in the function of the obtained results given in Table 4, calculated based on Formula (4).

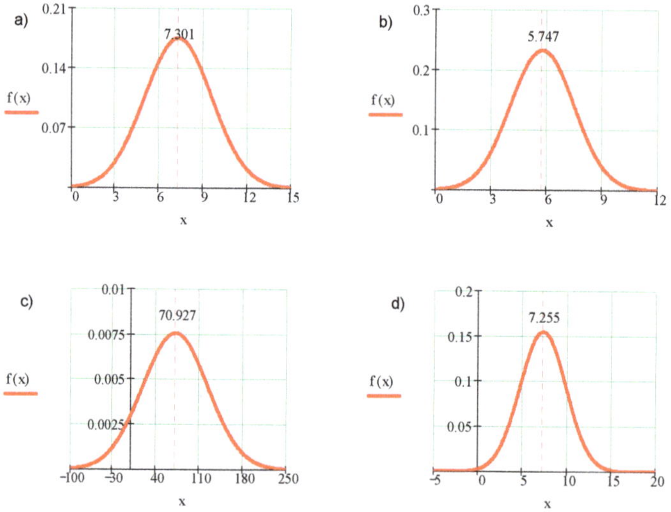

Figure 9. Normal distribution of probability density in the function of the obtained results: (**a**) height of Rms surface—*Sq*, (**b**) mean arithmetic surface height—*Sa*, (**c**) orientation of surface height—*Std*, and (**d**) developed ratio of surface interphasic area—*Sd*.

Relative values of individual surface geometrical parameters calculated based on formulae (5, 6, 7) and data given in Table 4 are presented in a column graph in Figures 10–12.

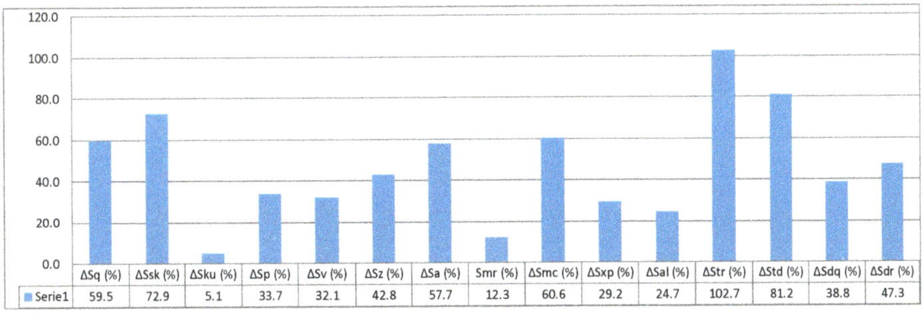

Figure 10. Relative values of surface parameters for printing orientation 0°.

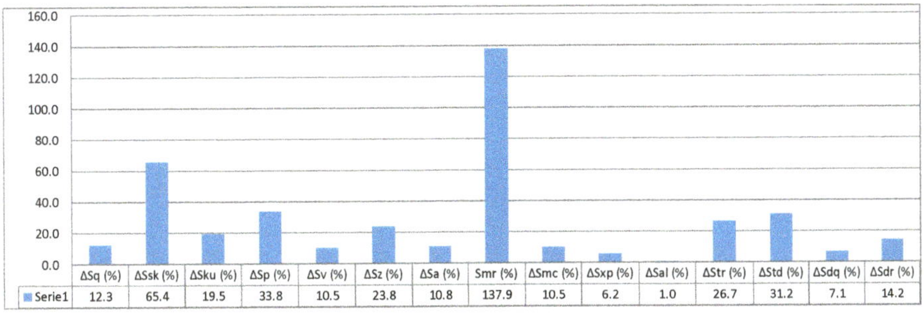

Figure 11. Relative values of surface parameters for printing orientation 45°.

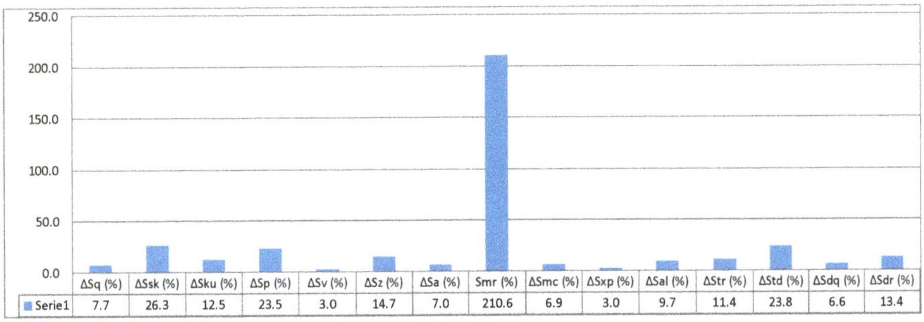

Figure 12. Relative values of surface parameters for printing orientation 90°.

Quantitative analysis of 3D spatial parameters measurement results, concerning height-related, functional, spatial, and hybrid, affirm what is similar to the case of 2D flat parameters of roughness: there are distinct differences in measurement results. In the case of samples measurements performed at a given angle 45° and 90°, mean values of parameters are in almost all cases less than mean values for the 30 samples. The only exception is the surface texture orientation parameter, wherein the values are less than mean values. For samples produced at the set angle 0°, orientation expressed with the *Std* parameter has values less than average. The value of the parameter is expressed in degrees, which means a different nature of surface texture, 0–13.3°, 45–93.1°, and 90–87.8°, respectively.

Figure 13 presents isometric views of geometrical texture of the examined surface. Attention must be paid to the functional parameter of surface material ratio — *Smr*. It is responsible for the surface roughness share expressed in percentage and shows the greatest relative deviations for the angles 45° and 90° at 137.9% and 210.6%, respectively, which can be particularly useful for assessing the impact of print orientation (print direction) on surface roughness quality. All values of spatial parameters for orientation at an angle of 0° show significant differences from the values of parameters measured for directions 45° and 90°.

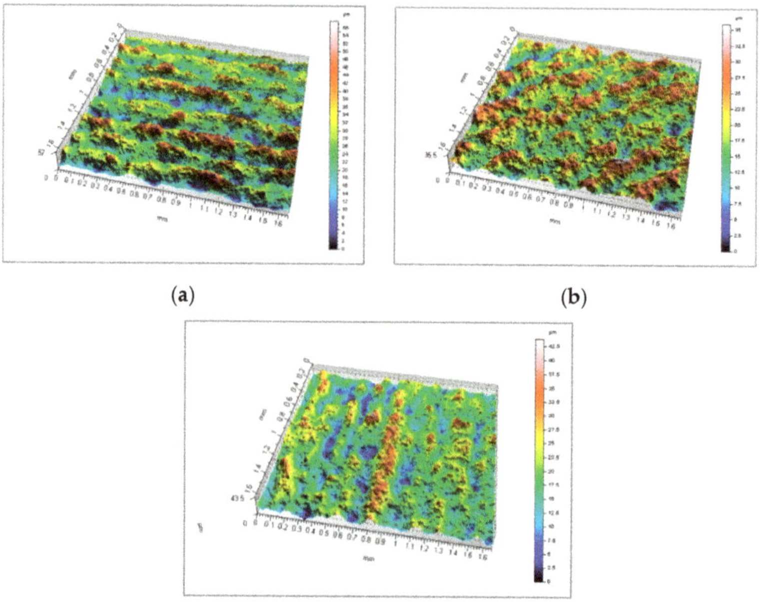

(**a**) (**b**)

(**c**)

Figure 13. 3D roughness profile for samples: (**a**) 0°, (**b**) 45°, and (**c**) 90°.

Owing to the layered nature of the additive manufacturing process, the images (Figure 13) show both deep valleys and high peaks, which is typical for the so-called step effect. This roughness is regular and corresponds to the angles of sample orientation on the working platform.

The test results presented above are an attempt to identify the geometrical texture of the surface of elements manufactured by additive technology—powder bed fusion. The surface texture of elements produced by methods known to date, for example, turning or milling, is quite well known and described in the literature. In mechanical engineering, several basic roughness parameters are usually used to describe it (*Ra, Rz, Rt*), usually only 2D parameters. In the case of a surface obtained by additive technology—as in the presented research results—one or two 2D parameters are not enough; it is necessary to use more complex tools and research methods (optical) enabling measurement of the geometrical surface texture—3D and identification of 3D parameters, giving much more information about the surface compared with the parameters of a single profile. In technological cases in which it is not possible to perform finishing treatment, for example, grinding, and the obtained surface can affect the functionality of the elements, a thorough knowledge of its geometric texture is of key importance. This may apply to different micro-mechanisms or, for example, micro-channels. Only the correct identification of the surface geometry structure, which has been presented above, enables further research, for example, in terms of its impact on various processes.

4. Conclusions

When analysing the above-mentioned study results and the state-of-the-art references, the following general conclusions can be derived.

The arrangement of sample models on the working platform of the machine affects the value of all tested parameters of surface texture.

The most advantageous variant of model orientation is the case where the surface is parallel to the building platform plane (samples orientation variant—1). The values of height parameters of surface roughness (2D) for individual orientations differ from the average value obtained for all orientations in the range from 0.4% to 69.3%. On the basis of normal distribution of probability density and values of included roughness (*Ra*, *Rz*), it can be said that they are close to the values of these parameters obtained during conventional machining (milling, turning).

On the basis of the spatial parameters value of the surface texture, providing a full image of the surface texture quality (3D image), it is possible to claim that the measurement results scattering presented in normal distributions are clearly greater for the *Std* parameter, which characterizes the orientation of the surface texture. At the same time, the graph shows the least value of the probability density function. The values of spatial surface roughness (3D) parameters for individual orientations differ from the average value obtained for all orientations in the range from 1% to 210.6%, while for the 0° orientation, these differences are greater than for other cases.

In summary, it is necessary to emphasize that, in the case of contact additive technologies, application of surface texture parameters measurements and their narrowing to the analysis of typical amplitude parameters, commonly applied in the industry, is not sufficient. The nature of the art allows for performing complete three-dimensional surface texture analysis, as suggested by the authors in the case of the studies of models manufactured using additive technologies. Understanding the impact of the orientation of the workpiece on the 3D printer's working platform on surface roughness parameters is important when planning allowances for further processing, for example, by grinding. Therefore, machining allowances should be planned depending on the orientation on the working platform, and this may especially apply to the processing of various micro parts. Identification of the geometric texture of the surface presented in this work can be particularly useful in cases where it is not possible to perform finishing technological operations, for example, on internal surfaces of elements (hollow). In these cases, 0° orientation can be recommended. Differences in the geometric texture of the surface may affect the fatigue strength of the elements, as well as their tribological wear, which may depend on the orientation of the models on the machine's working platform. The determination of specific values requires additional research, which the authors plan to do in the future.

Taking into account the results of surface texture research and the literature analysis, it can be concluded that 3D printing has very high potential applications and, in the near future, owing to the continuous development of technology, their precision, and resolution affecting the quality of the surface layer, it will be possible to produce full finished final models for MEMS technology.

Author Contributions: Conceptualization, T.K. and J.B.; methodology, T.K.; software, T.K.; validation, T.K. and J.B.; investigation, T.K.; writing—original draft preparation, T.K. and J.B.; writing—review and editing, T.K.; supervision, T.K.; project administration, T.K.; funding acquisition, T.K. All authors have read and agreed to the published version of the manuscript.

Funding: The project was financed by Ministry of Science and Higher Education within the project of Incubator of Innovation+, POIR.P.17.001.01 / 1.02.02.02.0001, task 1.1.4.2 (Grant No. 4/2017).

Conflicts of Interest: The authors declare no conflict of interest.

References

1. Hull, C. Apparatus for Production of Three-Dimensional Objects by Stereolithography. U.S. Patent 4575330A, 19 December 1986.
2. Campbell, I.; Bourell, D.; Gibson, I. Additive manufacturing: Rapid prototyping comes of age. *Rapid Prototyp. J.* **2012**, *18*, 255–258. [CrossRef]
3. Crump, S. Apparatus and Method for Creating Three-Dimensional Objects. U.S. Patent 5121329A, 9 June 1989.
4. Deckard, C.R.; Beaman, J.J.; Darrah, J.F. Method for Selective Laser Sintering with Layerwise Cross-Scanning. U.S. Patent 5155324A, 17 October 1986.
5. Feygin, M.; Shkolnik, A.; Diamond, M.N.; Dvorskiy, E. Laminated Object Manufacturing System. U.S. Patent 5730817A, 22 April 1996.

6. Meiners, W.; Wissenbach, K.; Gasser, A. Shaped Body Especially Prototype or Replacement Part Production. DE Patent Application No. 19649865, 2 December 1996.
7. Sachs, E.M.; Haggerty, J.S.; Cima, M.J.; Williams, P.A. Three-Dimensional Printing Techniques. U.S. Patent 5204055A, 20 April 1993.
8. Leu, M.C.; Guo, N. Additive manufacturing: Technology, applications and research needs. *Front. Mech. Eng.* **2013**, *8*, 215–243. [CrossRef]
9. Gibson, I.; Rosen, D.W.; Stucker, B. *Additive Manufacturing Technologies: 3D Printing, Rapid Prototyping, and Direct Digital Manufacturing*; Springer: New York, NY, USA, 2013.
10. Adamczak, S.; Zmarzły, P.; Kozior, T.; Gogolewski, D. Analysis of the dimensional accuracy of casting models manufactured by fused deposition modeling technology. *Eng. Mechan.* **2017**, 66–69.
11. Zmarzły, P.; Kozior, T.; Gogolewski, D. Dimensional and Shape Accuracy of Foundry Patterns Fabricated Through Photo-Curing. *Teh. Vjesn.-Tech. Gaz.* **2019**, *26*, 1576–1584. [CrossRef]
12. Zmarzły, P.; Kozior, T.; Gogolewski, D. Design guidelines for 3D printed molds for plastic casting. *J. Eng. Fibers Fabrics* **2020**, *15*, 1–10. [CrossRef]
13. Bochnia, J.; Blasiak, S. Anisotrophy of mechanical properties of a material which is shaped incrementally using polyjet technology. *Eng. Mechan.* **2016**, 74–77.
14. Kozior, T.; Mamun, A.; Trabelsi, M.; Sabantina, L.; Ehrmann, A. Quality of the surface texture and mechanical properties of FDM printed samples after thermal and chemical treatment. *Stroj. Vestn.* **2020**, *1*, 1–10. [CrossRef]
15. Dizon, J.R.C.; Espera, A.H.; Chen, Q.Y.; Advincula, R.C. Mechanical characterrization of 3D-printed polymers. *Addit. Manuf.* **2018**, *20*, 44–67. [CrossRef]
16. Damon, J.; Dietrich, S.; Vollert, F.; Gibmeier, J.; Schulze, V. Process dependent porosity and the influence of shot peening on porosity morphology regarding selective laser melted AlSi10Mg parts. *Addit. Manuf.* **2018**, *20*, 77–89. [CrossRef]
17. Kozior, T. Analysis of the Influence of Process Parameters of Selected Additive Technologies on the Geometry and Mechanical Properties of Products. Ph.D. Thesis, Kielce University of Technology, Kielce, Poland, 2018.
18. Puigoriol, J.M.; Alsina, A.; Salazar-Martin, A.G.; Gomez-Gras, G.; Perez, M.A. Flexural fatigue properties of polycarbonate fused-deposition modelling specimens. *Mater. Des.* **2018**, *155*, 414–421. [CrossRef]
19. Singh, S.; Prakash, C.; Antil, P.; Singh, R.; Krolczyk, G.; Pruncu, C.I. Analysis for Investigating the Quality Characteristics of Aluminium Matrix Composites Prepared through Fused Deposition Modelling Assisted Investment Casting. *Materials* **2019**, *12*, 1907. [CrossRef] [PubMed]
20. Gapinski, B.; Wieczorowski, M.; Bak, A.; Pereira Dominguez, A.; Mathia, T. The assessment of accuracy of inner shapes manufactured by FDM. In Proceedings of the 21st International ESAFORM Conference on Material Forming (ESAFORM 2018)—AIP Conference Proceedings, Palermo, Italy, 23–25 April 2018; Volume 1960. [CrossRef]
21. Niu, X.; Singh, S.; Garg, A.; Singh, H.; Panda, B.; Peng, X.; Zhang, Q. Review of materials used in laser-aided additive manufacturing processes to produce metallic products. *Front. Mechan. Eng.* **2019**, *14*, 282–298. [CrossRef]
22. Khaled, T. Available online: https://www.faa.gov/aircraft/air_cert/design_approvals/csta/publications/media/additive_manufacturing.pdf (accessed on 31 August 2015).
23. Unger, L.; Scheideler, M.; Meyer, P.; Harland, J.; Gorzen, A.; Wortmann, M.; Dreyer, A.; Ehrmann, A. Increasing Adhesion of 3D Printing on Textile Fabrics by Polymer Coating. *Tekstilec* **2018**, *61*, 265–271. [CrossRef]
24. Kozior, T.; Mamun, A.; Trabelsi, M.; Wortmann, M.; Sabantina, L.; Ehrmann, A. Electrospinning on 3D Printed Polymers for Mechanically Stabilized Filter Composites. *Polymers* **2019**, *11*, 2034. [CrossRef]
25. Kozior, T.; Mamun, A.; Trabelsi, M.; Sabantina, L.; Ehrmann, A. Stabilization of Electrospun Nanofiber Mats Used for Filters by 3D Printing. *Polymers* **2019**, *11*, 1618. [CrossRef]
26. Kozior, T.; Blachowicz, T.; Ehrmann, A. Adhesion of three-dimensional printing on textile fabrics: Inspiration from and for other research areas. *J. Eng. Fibers Fabrics* **2020**, *15*, 1–6. [CrossRef]
27. Coniglio, N.; Sivarupan, T.; Mansori, M.E. Investigation of process parameter effect on anisotropic properties of 3D printed sand molds. *J. Adv. Manuf. Technol.* **2018**, *94*, 2175–2185. [CrossRef]
28. Casting 3D Printer Manufacturer. Available online: https://www.exone.com (accessed on 12 July 2018).

29. Rokicki, P.; Budzik, G.; Kubiak, K.; Bernaczek, J.; Dziubek, T.; Magniszewski, M.; Nowotnik, A.; Sieniawski, J.; Matysiak, H.; Cygan, R.; et al. Rapid prototyping in manufacturing of core models of aircraft engine blades. *Aircr. Eng. Aerosp. Technol.* **2014**, *86*, 323–327. [CrossRef]
30. Adamczak, S.; Zmarzły, P. Research of the influence of the 2D and 3D surface roughness parameters of bearing raceways on the vibration level. *J. Phys. Conf. Ser.* **2019**, *1183*, 1–10. [CrossRef]
31. Segonds, F. Design by Additive Manufacturing: An application in aeronautics and defence. *Virtual Phys. Prototyp.* **2018**, *13*, 237–245. [CrossRef]
32. Blasiak, S.; Takosoglu, I.E.; Laski, P.A. Optimizing the Flow Rate in a Pneumatic Directional Control Valve. *Eng. Mechan.* **2014**, 96–99.
33. Borsuk-Nastaj, B.; Młynarski, M. Selective laser melting (SLM) technique in fixed prosthetic restorations. *Protet. Stomatol.* **2012**, *LXII*, 203–209.
34. Dikova, T.; Vasilev, T.; Dzhendov, D.; Ivanova, E. Investigation the fitting accuracy of cast and SLM Co-Cr dental bridges using cad software. *J. IMAB* **2017**, *23*, 1688–1696. [CrossRef]
35. Lipowicz, A. Development of Designing and Fabrication Methodology of biomedical Models with Generative Technologies. Ph.D. Thesis, Wrocław University of Science and Technology, Wrocław, Poland, 2008.
36. Voon, S.L.; An, J.; Wong, G.; Zhang, Y.; Chua, C.K. 3D food printing: A categorised review of inks and their development. *Virtual Phys. Prototyp.* **2019**, 1–16. [CrossRef]
37. Blachowicz, T.; Ehrmann, A. 3D Printed MEMS Technology-Recent Developments and Applications. *Micromachines* **2020**, *11*, 434. [CrossRef] [PubMed]
38. Joshi, N.; Kohler, E.; Enokssen, P. MEMS based micro aerial vehicles. *J. Phys. Conf. Ser.* **2016**, *757*, 012035. [CrossRef]
39. Wang, Y.J.; Gao, L.B.; Fan, S.F.; Zhou, W.Z.; Li, X.; Lu, Y. 3D printed micro-mechanical device (MMD) for in situ tensile testing of micro/nanowires. *Extreme Mechan. Lett.* **2019**, *33*, 100575. [CrossRef]
40. Thompson, A.; Senin, N.; Giusca, C.; Leach, R. Topography of selectively laser melted surfaces: A comparison of different measurement methods. *CIRP Ann. Manuf. Technol.* **2017**, *66*, 543–546. [CrossRef]
41. Thompson, A.; Körner, L.; Senin, N.; Lawes, S.; Maskery, I.; Leach, R. Measurement of internal surfaces of additively manufactured parts by X-ray computed tomography. In Proceedings of the 7th Conference on Industrial Computed Tomography 2017, Leuven, Belgium, 7–9 February 2017.
42. Bartkowiak, T.; Brown, C.A. Multiscale 3D Curvature Analysis of Processed Surface Textures of Aluminum Alloy 6061 T6. *Materials* **2019**, *12*, 257. [CrossRef]
43. Kamarudin, K.; Wahab, M.S.; Shayfull, Z.; Ahmed, A.; Raus, A.A. Dimensional Accuracy and Surface Roughness Analysis for AlSi10Mg Produced by Selective Laser Melting (SLM). *Matec Web Conf.* **2016**. [CrossRef]
44. Townsend, A.; Senin, N.; Blunt, L.; Leach, R.K.; Taylor, J.S. Surface texture metrology for metal additive manufacturing: A review. *Precis. Eng.* **2016**, *46*, 34–47. [CrossRef]
45. Chlebus, E.; Kurzac, J.; Szymczyk, P.; Ziółkowski, G. Application of X-ray CT method for discontinuity and porosity detection in 316L stainless steel parts produced with SLM technology. *Arch. Civ. Mechan. Eng.* **2014**, *14*, 608–614. [CrossRef]
46. Kurzynowski, T.; Gruber, K.; Stopyra, W.; Kuźnicka, B.; Chlebus, E. Correlation between process parameters, microstructure and properties of 316 L stainless steel processed by selective laser melting. *Mater. Sci. Eng.* **2018**, *718*, 64–73. [CrossRef]
47. Kaynaka, Y.; Kitayb, O. The effect of post-processing operations on surface characteristics of 316L stainless steel produced by selective laser melting. *Addit. Manuf.* **2019**, *26*, 84–93. [CrossRef]
48. Heiden, M.J.; Deibler, L.A.; Rodelas, J.M.; Koepke, J.R.; Tung, D.J.; Saiz, D.J.; Jared, B.H. Evolution of 316L stainless steel feedstock due to laser powder bed fusion process. *Addit. Manuf.* **2019**, *25*, 84–93. [CrossRef]
49. Lodhia, M.J.K.; Deenb, K.M.; Greenlee-Wackerc, M.C.; Haidera, W. Additively manufactured 316L stainless steel with improved corrosion resistance and biological response for biomedical applications. *Addit. Manuf.* **2019**, *27*, 1–19. [CrossRef]
50. Available online: https://www.astm.org/DATABASE.CART/HISTORICAL/E8E8M-13.htm (accessed on 28 June 2020).
51. Concept Laser M2 - 3D Printer Manufacturer. Available online: https://www.concept-laser.de (accessed on 12 July 2019).

52. *ISO 4287: Geometrical Product Specification (GPS)-Surface Texture: Profile Method-Terms, Definition and Surface Texture Parameters*; International Organization for Standardization (ISO): Geneva, Switzerland, 1997.
53. *ISO 25178-2: Geometrical Product Specifications (GPS)—Surface Texture: Areal—Part 2: Terms, Definitions and Surface Texture Parameters*; International Organization for Standardization (ISO): Geneva, Switzerland, 2012.
54. Whitehouse, D.J. *Handbook of Surface and Nanometrology*, 2nd ed.; CRC Press: Boca Raton, FL, USA, 2011.

micromachines

MDPI

Article

The Fabrication of Micro Beam from Photopolymer by Digital Light Processing 3D Printing Technology

Ishak Ertugrul

Department of Mechatronics, Mus Alparslan University, 49250 Mus, Turkey; i.ertugrul@alparslan.edu.tr

Received: 9 April 2020; Accepted: 9 May 2020; Published: 20 May 2020

Abstract: 3D printing has lately received considerable critical attention for the fast fabrication of 3D structures to be utilized in various industrial applications. This study aimed to fabricate a micro beam with digital light processing (DLP) based 3D printing technology. Compound technology and essential coefficients of the 3D printing operation were applied. To observe the success of the DLP method, it was compared with another fabrication method, called projection micro-stereolithography (PμSL). Evaluation experiments showed that the 3D printer could print materials with smaller than 86.7 μm dimension properties. The micro beam that moves in one direction (*y*-axis) was designed using the determined criteria. Though the same design was used for the DLP and PμSL methods, the supporting structures were not manufactured with PμSL. The micro beam was fabricated by removing the supports from the original design in PμSL. Though 3 μm diameter supports could be produced with the DLP, it was not possible to fabricate them with PμSL. Besides, DLP was found to be better than PμSL for the fabrication of complex, non-symmetric support structures. The presented results in this study demonstrate the efficiency of 3D printing technology and the simplicity of manufacturing a micro beam using the DLP method with speed and high sensitivity.

Keywords: digital light processing; projection micro-stereolithography; 3D printing; micro beam; fabrication

---

1. Introduction

The micro-electro-mechanical system (MEMS) is a process technology comprising of miniaturized mechanical and electronic parts, that includes the transformation of a measured mechanic signal into a readable signal. This signal may be force, pressure, heat, or chemical. MEMS has created serious innovations in micro- and nano-study fields since the early 1980s [1]. Primarily, MEMS has been improved for different implementations for example force, navigation, optical transmitting, radio frequency, biological and medical, microfluidics, and gyroscope applications [2–6]. Currently, MEMS has become an essential part of various study fields such as material, mechanical, and electrical engineering studies [7–9]. The MEMS device is usually formed of four components, micro-structures, micro-sensors, micro-actuators, and micro-electronics for data utilization [10,11].

MEMS technologies have been manufactured by traditional methods such as lithography, galvano forming, and photolithography etc. These methods are dependent on additive or subtractive procedures that operate minuscule capacities of materials in the shape of thin layers on the surface of silicon wafers [12–16]. These conventional techniques are extremely precise and appropriate for the production of planar geometries [17]. Although precise, these operations are associated with some drawbacks such as multiple processing steps, the requirement of a cleanroom, an advanced work environment, lengthy fabrication, inconsistency of flexible materials, and a costly fabrication process [18]. With the improvement of 3D printing-additive manufacturing, the fabrication costs and processing steps of MEMS devices have been gradually reduced. According to these improvements, it is possible to fabricate MEMS devices in atmospheric air without the need for multiple operations and cleanrooms [19,20]. Due to such advantages, it is now possible to use 3D printing technology for micro beam fabrication.

3D printing technology has become progressively popular in the fabrication of MEMS, since it can be employed to manufacture complicated structures clearly from digital files, for instance computer-assisted design drawings. This technology helps to improve the design and production and especially facilitates the production and repair of complex parts using printing through layer- by-layer deposition of the constituent materials [21–23]. With the progress of 3D printing techniques, interest in using 3D printing for building MEMS systems has grown remarkably in the areas of biomedical, electronics, wearable devices, soft robots, and automotive applications [24–29]. Unlike traditional manufacturing processes such as machining and punching, 3D printing does not entail on-site process control, cutting tools, coolers, or other additional resources. One of the important factors of 3D printing methods is its capability to make miniaturized complex structural geometries using easy steps that are not achievable by traditional manufacturing methods. Besides that, 3D printing methods offer many other characteristics, for example flexibility in geometrical designs, excellent feature size and shapes, and the ability to print functionally classified materials [30,31].

Micro beams are utilized as important components of different sensing and actuation systems such as sensors, gyroscopes, micro actuators, and resonators [32–34]. Their easy geometries make them very advantageous in terms of design, and microfabrication. In many applications, ranging from residual stress measurement mass flow sensors to biomedical or DNA analysis, the sensing mechanism is linked to the sensitivity of the micro beam to some applied stimulation [35–37]. Many studies have been conducted, especially in the field of DNA [38–40]. In a study, a multi-scale analytical model was created to define the relationship between the surface mechanical features of DNA self-assembled 2D films and the detection signals of DNA-micro beam systems [41]. Micro beams exactly predict the dynamic properties of the device, such as its natural frequencies and forced- vibration response.

Fabrication methods are very significant in designing and researching micro beams. Until now, conventional MEMS fabrication methods have been used, such as photolithography and surface micromachining, etc. [42,43]. These methods are usually time-consuming, with high fabrication costs, and multi-step fabrication processing. Alternatively, the 3D printing or additive manufacturing method solves these problems by manufacturing the structure directly. In recent years, various 3D printing methods have been employed, for instance frontal polymerization (FP), projection micro-stereolithography (PμSL), laser micro sintering (LMS), selective laser melting (SLM), etc. [44–47]. Among the different 3D printing techniques, the digital light processing (DLP) technique using photocurable resins is appealing, since it can be used to manufacture a single layer of the 3D object through spatially-controlled solidification using a projector light [48]. This light produces benefits such as fast fabrication, high sensitivity, and surface quality. Besides, it is feasible to adapt the final features of the printed object by only altering the photocurable resin formulations [49]. In this way, it is feasible to reach a large diversity of systems for the fabrication of structures with excellent features and functions [50].

In this study, we present our development on utilization of DLP technology for fast and highly sensitive production of a micro beam with sub-millimeter scale properties. The DLP and PμSL methods were compared based on the fabrication results. With this study, a micro beam was fabricated for the first time using the 3D printing method. It is expected that this paper will contribute to the current literature in terms of manufacturing a micro device through the use and comparison of different techniques.

This study is arranged as follows. Section 2 describes the design of the micro beam. Section 3 explains the fabrication process, results, and discussion. Section 4 displays the result of the studies.

2. Materials and Methods

2.1. Design Conditions

The possibility of coupling thermal, electrical, and structural characterization by fabrication of a micro beam is accomplished with a model. For characterization, the displacement of the micro beam is produced by passing a current through a beam; heat is produced by the current, and the rise in temperature causes a displacement through thermal expansion. The displacement of the micro beam is formed in these situations.

The MEMS-based micro beam is designed to move in one direction (y-axis). For the beam to move on the y-axis, DC voltage must be applied. The feet of the micro beam at both ends are rigidly bound to a substrate, and DC voltage is applied at both ends. The applied voltage induces an electric current in the micro beam; current passing through the structure causes some retardation to the free flow of electrons by which energy is dissipated in the form of heat. This generated heat induces thermal stress on the beam and displaces the beam. The dimensioning and geometric structure of the micro beam, designed as a 3D (3-dimensional plane) using CAD software, is shown in Figure 1.

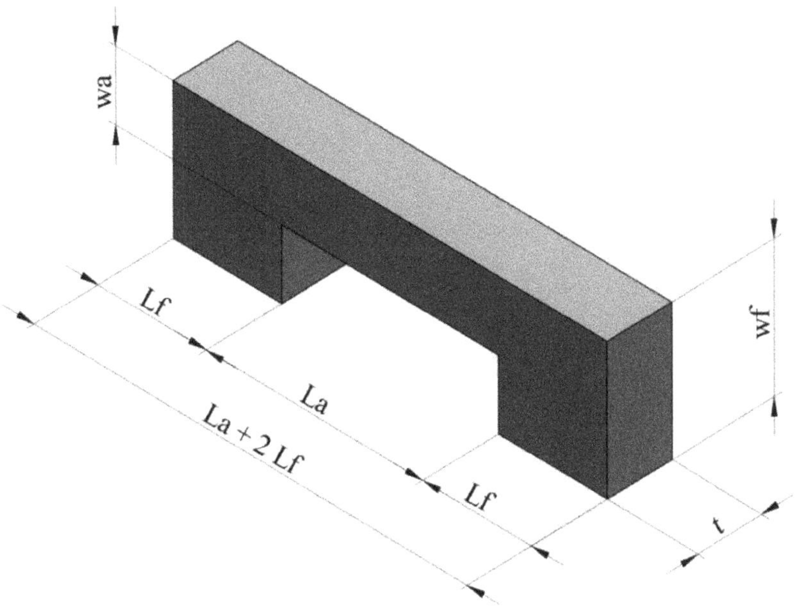

Figure 1. Dimensions of the micro beam.

All the dimensions of our micro beam are shown in Table 1. These values were obtained with the measurements of the beam made with the DLP and PμSL methods. The same design was used for both methods. Nevertheless, as the supporting structures cannot be fabricated with the PμSL, the measurements concerning the support are shown in Section 3. Photopolymer materials were used for this study. These materials are frequently utilized in the field of MEMS because of their essential physical and electrical features.

Table 1. Description of the micro beam.

Parameter	Symbol	Value (μm)
Length of the feet	L_f	25
Length of the arm	L_a	50
Height of the arm	w_a	25
Height of the feet	w_f	50
Thickness of the beam	t	30

2.2. Digital Light Processing Method

Digital light processing (DLP) is a rapid additive fabrication technology with superior sensitivity. The processing basis of DLP technology is explained in [51,52] and can be briefly abstracted as follows. A commercial 3D printer, MiiCraft 125 (Rapid City, Canada), was utilized in this paper, as shown in Figure 2b.

As given in Figure 2a, the 3D pattern of the matter is first sliced into layers horizontally (in the x-axis). Thin layers are then transformed into 2D mask images. A light projection device is utilized to harden the photopolymer resin. This device employs a digital masking method to reflect a dynamically described mask image on the resin plane. With respect to [51], a bottom-up projection system has many benefits compared to a top-bottom system. In this system, the mask image is reflected on the bottom of a resin tank with cured resin at the bottom of the tank. This process continues until the desired design is created.

We established a bottom-up DLP system, as shown in Figure 2b. The DLP projector was equipped to provide a 400 nm full HD ultraviolet light source. The contrast ratio of the DLP system projector is 900:1. The XY resolution of the 3D printing device is 65 μm and the maximum building size 125 × 70 × 120 mm. With respect to [52,53], the masking method primarily occurs as three types: liquid crystal display (LCD), digital micro-mirror device (DMD), and liquid crystal on silicon (LcoS). In our improvement, the projector uses the DMD method. An optical reflector is employed to set the direction of the UV light. A position adjusting device is employed to set the position and behavior of the reflector. The adjustable angle range is ± 15°, and the accuracy is 0.003°. A flexible compressing device is used to press the resin vat. This device allows the vat to be lifted to a specific height.

For the DLP method, IP-S resin, which is a photopolymer, was used as the material. This resin was designed to tend the double function of index-matching the dip fluid for final focusing of the object and photo-polymerizable, thus enabling the highest resolution at a given magnification. The elemental composition and fundamental features of the resin are given in Table 2 [54]. The composition was decided upon by using the procedure discussed in [55] and is significant for this study since it decided the x-ray absorption characteristics of the foam [56]. The foam structure extracted from the glass substrate was qualified using optical microscopy and scanning electron microscope (SEM). SEM sample preparation included a sputter coating of 30 nm thick gold to allow electrical charge conductivity while imaging.

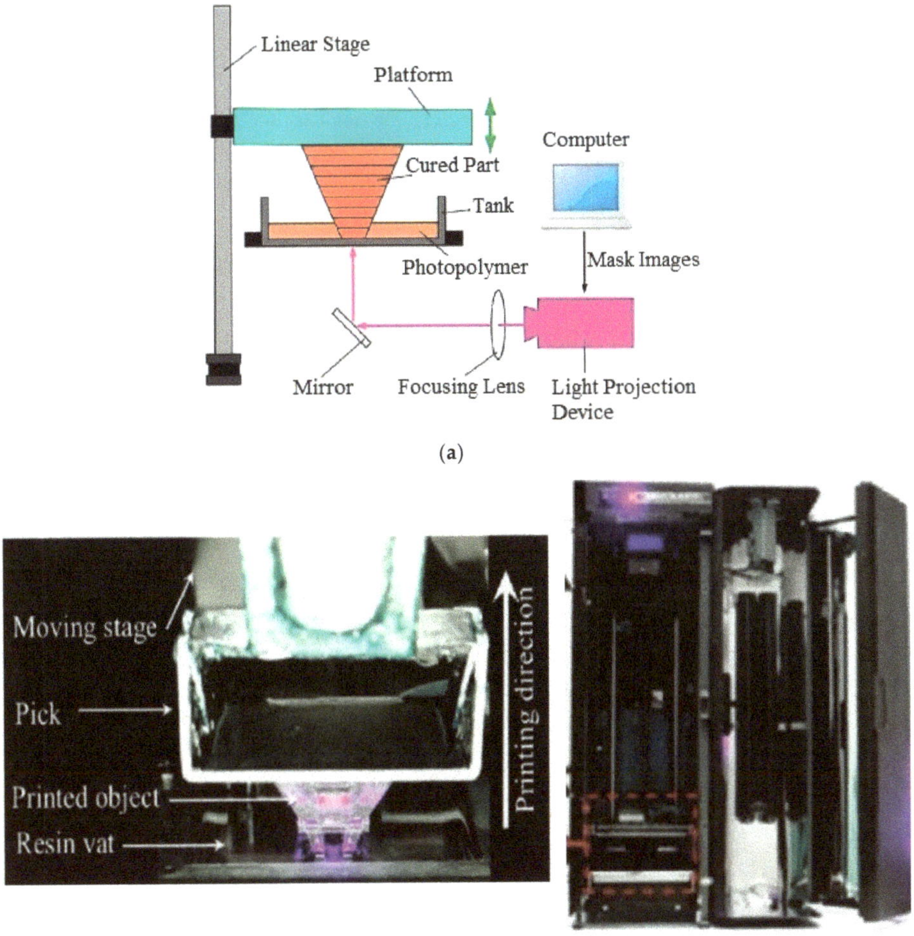

Figure 2. (**a**) Schematic of a digital light processing (DLP) system [53]; (**b**) 3D printing process in progress; (**c**) MiiCraft 3D printer.

Table 2. The elemental composition and fundamental features of the IP-S resin [54].

Chemical Properties of Resin				
Carbon (at%)	**Hydrogen (at%)**	**Nitrogen (at%)**	**Oxygen (at%)**	**Empirical Formula**
31.45	54.07	5.75	11.7	$CH_{1.71}N_{0.085}O_{0.35}$
Physical and Mechanical Features				
Density (liq) g/(cm^3)	**Density (s) g/(cm^3)**	**Young's Modulus (GPa)**	**Hardness (MPa)**	**Refractive Index**
1.3	1.5	3.8	150	1.51

2.3. *PμSL Method*

Projection micro-stereolithography (PμSL) is a sophisticated 3D printing method because of its low cost, precision, velocity, and also the variety of the materials such as ceramics, biomaterials, curable photopolymer, polymer, and nanoparticle composites [54]. This method has demonstrated potential in different implementations for example micro-resonators, micro-grippers, micro-optics, biomedical micro devices, micro-fluidics, and so on [57–59]. Studies on PμSL are ongoing in terms of the quality and accuracy of the construction process, which affects the production of complex 3D microstructures and makes it attractive enough to be considered for commercial applications [60]. This technology begins by creating a 3D construction via a computer-assisted design program and then transforms the construction into a set of digital mask images. The working basis of PμSL is shown in Figure 3 [61].

Using a digital micro mirror device as the dynamic mask eliminates the cost of manufacturing a mask for each layer. Besides, the PμSL method reduces production time since each layer is produced in one exposure, and the time for mask alignment is eliminated. Moreover, this method has the least mechanical moving parts and requires only one accurate z-axis motorized linear stage. Consequently, PμSL decreases the cost of construction and protection [62].

All images symbolize a thin layer of the 3D structure. Along a production period, a single image is demonstrated on the reflective LCD panel. The image from the LCD is then mirrored on the liquid surface. All layers (ranging between 5–40 μm thick) are polymerized. When the layer has been solidified, it is dipped in the resin to allow a new thin layer of liquid to form. Repeating the loop forms a 3D microstructure from a layer stack. For the PμSL method, IP-S resin, which is a photopolymer, was used as the material. The properties of the photopolymer resin, which were specially developed for this 3D printer and utilized in the fabrication of the micro beam, are shown in Table 2 [54].

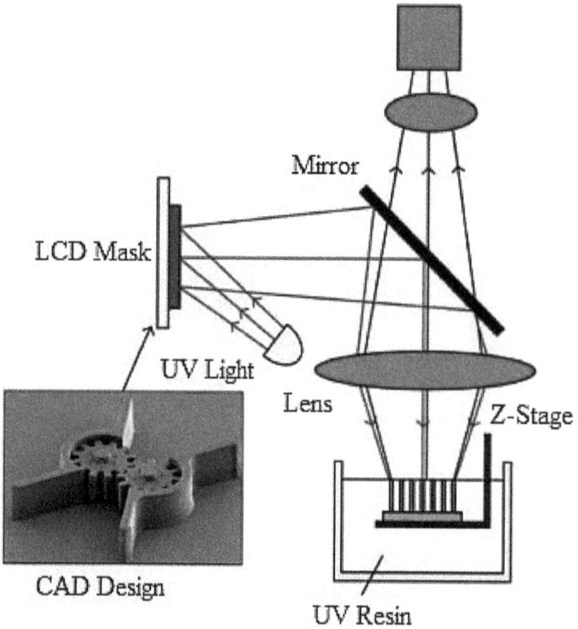

Figure 3. Schematic of the projection micro-stereolithography (PμSL) method [62].

3. Fabrication

3.1. Fabrication with the DLP Method

The micro beam shown in Figure 4 was fabricated by a DLP technology-based MiiCraft 3D printing device. 3D printing technology entails an input CAD model of the parts that may be designed in software or obtained from reverse engineering such as 3D scanners. When the CAD model of the micro beam is completed, it is transformed into standard STL format, which is most commonly used to represent 3D CAD models in 3D printing. In an STL file, the CAD model is symbolized using triangular facets, which are described by the x-, y-, and z-coordinates of the three vertices. The step-by-step schema of the 3D printing operation is displayed in Figure 5. The slicer first divides the object into a stack of flat layers, followed by describing these layers as linear movements of the 3D printer extruder, fixation laser, or equivalent. All these movements, together with some specific printer commands like the ones to control the extruder temperature or bed temperature, are finally written in the g-code file, that can be transferred after to the printer.

During fabrication of the micro beam, the printing parameters for example the layer thickness (LT), the light intensity (LI), and the curing time (CT) of all layer significantly affect the print quality. In this study, each parameter is selected for the printing material and TL = 30 μm, and CT = 3 s are set. When the first layer is printed, LI is set to 50% of the brightness to provide a layer bond to the platform.

There are supports to the arms of this design. The number of supports to the arm is 20, the diameter is 3 μm, and the height is 5 μm. Some unsuccessful experiments were done before this design. Breakages were experienced during the manufacturing when the number of supports to the arm was low. Concerning the experiments, the average distance between the supports should be 10 μm to avoid breakages. When the supports are not printed correctly, they cause a collapse and break off the micro beam. An image of the micro beam manufactured with the DLP is shown in Figure 6. This image was taken with the microscope of the 3D printer device.

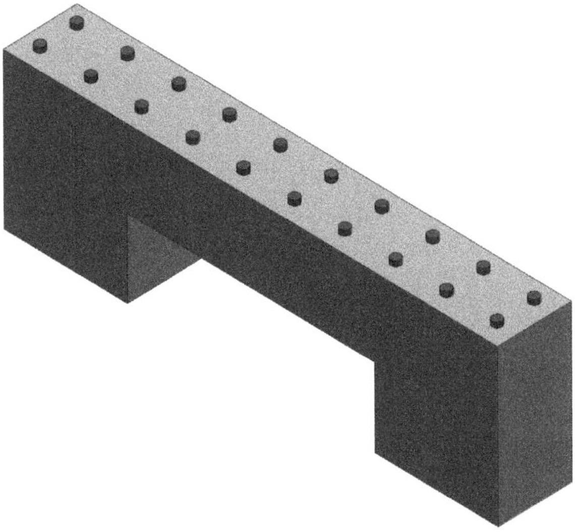

Figure 4. CAD design of the micro beam. The structures at the top of the design were designed as support.

3.2. Fabrication with the PμSL Method

The micro beam shown in Figure 7 was manufactured by a projection micro-stereolithography (PμSL) method based MiiCraft 3D printing device. The manufacturing of the supported design with the PμSL method was not possible for two reasons. First, the supported structures represent the system as a 3D design. However, it is not possible to fabricate 3D structures with devices of MiiCraft 3D printing, based on PμSL technology. Second, these devices have a resolution of 65 μm and can fabricate a minimum thickness of up to 30 μm.

When the support structures are removed, as shown in Figure 7, it is possible to perform the fabrication as the micro beam design can be introduced to the device (a PμSL technology-MiiCraft 3D printer) in two dimensions. An image of the micro beam manufactured with the PμSL is shown in Figure 8. This image was taken with the microscope of the 3D printer device.

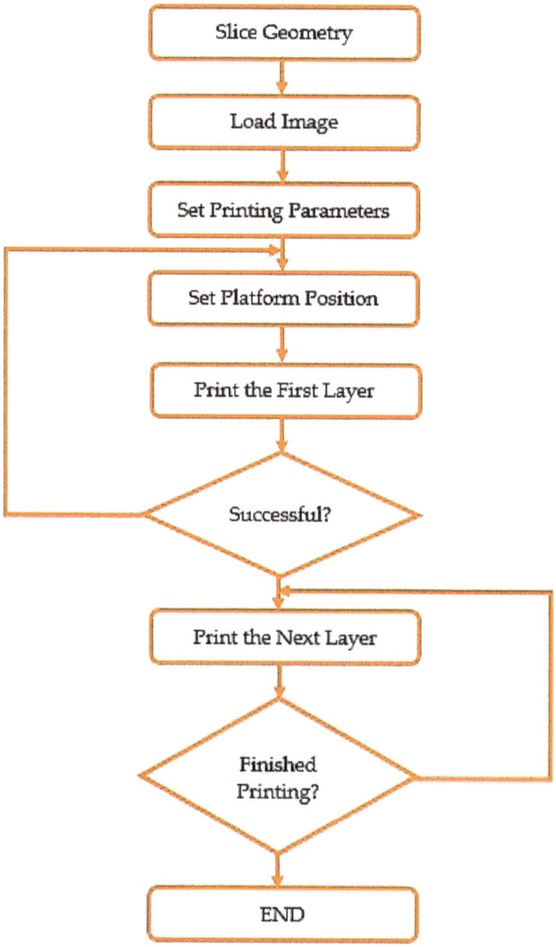

Figure 5. The step-by-step diagram of the 3D printing operation.

Figure 6. Image of the micro beam fabricated with the DLP method.

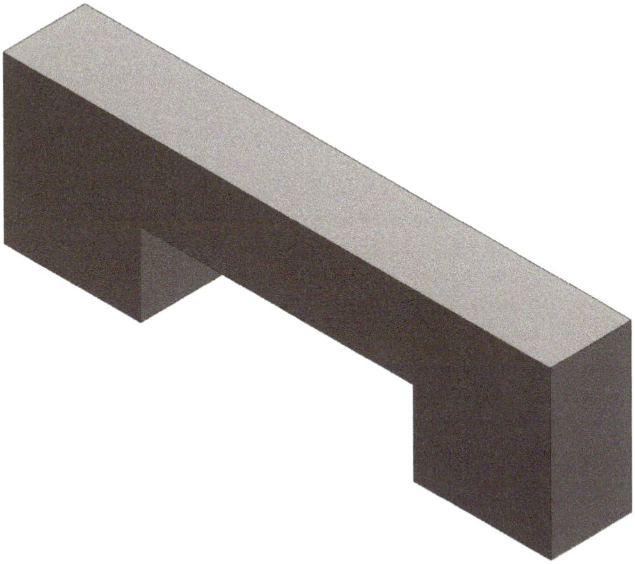

Figure 7. CAD design of the micro beam. The support structures under the micro beam are removed.

Figure 8. Image of the micro beam fabricated with the PμSL method.

4. Conclusions

In this study, a micro beam fabricated with conventional MEMS methods, was manufactured for the first time using the DLP and PμSL methods, which are 3D printing procedures. First, the printable scale of the DLP 3D printing method was evaluated, and it demonstrated that the printer could produce structures with a size of 86.7 μm.

The experimental studies showed that 3 μm diameter supports were fabricated with the DLP method. However, they could not be fabricated with the PμSL method even when the diameters of the supports were 3 μm. After these support structures were removed, the micro beam was fabricated with PμSL. It was determined that PμSL was not suitable for complex structures. The results show the success of the 3D printer and the suitability of manufacturing a micro beam using the DLP printing method with fast and high sensitivity.

As a result of this study, it was found that DLP is more appropriate because it allows the manufacturing of complex 3D structures with smaller dimensions, while PμSL is only suitable for simple 2D microstructures. It is expected that this paper will contribute to the literature in terms of fabrication of a micro device through the use and comparison of different techniques.

Funding: This research received no external funding.

Conflicts of Interest: The authors declare no conflict of interest.

References

1. Baldini, G.; Steri, G.; Dimc, F.; Giuliani, R.; Kamnik, R. Experimental identification of smartphones using fingerprints of built-in micro-electro mechanical systems (MEMS). *Sensors* **2016**, *16*, 818. [CrossRef] [PubMed]
2. Ertugrul, I.; Ulkir, O. MEMS tabanlı mikro rezonatörün tasarımı ve analizi. *Avrupa Bilim Teknoloji Dergisi* **2020**, *18*, 25–29. [CrossRef]
3. Sabato, A.; Niezrecki, C.; Fortino, G. Wireless MEMS-based accelerometer sensor boards for structural vibration monitoring: A review. *IEEE Sens. J.* **2016**, *17*, 226–235. [CrossRef]
4. Liu, H.; Zhang, L.; Li, K.H.H.; Tan, O.K. Microhotplates for metal oxide semiconductor gas sensor applications—Towards the CMOS-MEMS monolithic approach. *Micromachines* **2018**, *9*, 557. [CrossRef] [PubMed]
5. Xu, W.; Yang, J.; Xie, G.; Wang, B.; Qu, M.; Wang, X.; Tang, B. Design and fabrication of a slanted-beam MEMS accelerometer. *Micromachines* **2017**, *8*, 77. [CrossRef]

6. Cao, H.; Liu, Y.; Kou, Z.; Zhang, Y.; Shao, X.; Gao, J.; Liu, J. Design, Fabrication and Experiment of Double U-Beam MEMS Vibration Ring Gyroscope. *Micromachines* **2019**, *10*, 186. [CrossRef]
7. Pu, J.; Mo, Y.; Wan, S.; Wang, L. Fabrication of novel graphene–fullerene hybrid lubricating films based on self-assembly for MEMS applications. *Chem. Commun.* **2014**, *50*, 469–471. [CrossRef]
8. Choudhary, N.; Kaur, D. Shape memory alloy thin films and heterostructures for MEMS applications: A review. *Sens. Actuators A Phys.* **2016**, *242*, 162–181. [CrossRef]
9. Guarnieri, V.; Biazi, L.; Marchiori, R.; Lago, A. Platinum metallization for MEMS application: Focus on coating adhesion for biomedical applications. *Biomatter* **2014**, *4*, 12–24. [CrossRef]
10. Dinh, T.; Phan, H.P.; Qamar, A.; Woodfield, P.; Nguyen, N.T.; Dao, D.V. Thermoresistive effect for advanced thermal sensors: Fundamentals, design considerations, and applications. *J. Microelectromech. Syst.* **2017**, *26*, 966–986. [CrossRef]
11. Chin, T.S. Permanent magnet films for applications in microelectromechanical systems. *J. Magn. Magn. Mater.* **2000**, *209*, 75–79. [CrossRef]
12. Grella, L.; Carroll, A.; Murray, K.; McCord, M.A.; Tong, W.M.; Petric, P. Digital pattern generator: An electron-optical MEMS for massively parallel reflective electron beam lithography. *J. Micro/Nanolithogr. MEMS MOEMS* **2013**, *12*, 10–21. [CrossRef]
13. Qu, H. CMOS MEMS fabrication technologies and devices. *Micromachines* **2016**, *7*, 14. [CrossRef]
14. Lau, G.K.; Shrestha, M. Ink-jet printing of micro-electro-mechanical systems (MEMS). *Micromachines* **2017**, *8*, 194. [CrossRef]
15. Thuau, D.; Ducrot, P.H.; Poulin, P.; Dufour, I.; Ayela, C. Integrated electromechanical transduction schemes for polymer MEMS sensors. *Micromachines* **2018**, *9*, 197. [CrossRef] [PubMed]
16. Phan, H.P.; Nguyen, T.K.; Dinh, T.; Iacopi, A.; Hold, L.; Shiddiky, M.J.; Nguyen, N.T. Robust free-standing nano-thin SiC membranes enable direct photolithography for MEMS sensing applications. *Adv. Eng. Mater.* **2018**, *20*, 1700858. [CrossRef]
17. Hu, Z.X.; Gallacher, B.J.; Burdess, J.S.; Bowles, S.R.; Grigg, H.T.D. A systematic approach for precision electrostatic mode tuning of a MEMS gyroscope. *J. Micromech. Microeng.* **2014**, *24*, 125003. [CrossRef]
18. Yang, S.; Xu, Q. A review on actuation and sensing techniques for MEMS-based microgrippers. *J. Micro Bio Robot.* **2017**, *13*, 1–14. [CrossRef]
19. Mukherjee, P.; Nebuloni, F.; Gao, H.; Zhou, J.; Papautsky, I. Rapid prototyping of soft lithography masters for microfluidic devices using dry film photoresist in a non-cleanroom setting. *Micromachines* **2019**, *10*, 192. [CrossRef]
20. Yoon, H.J.; Kim, D.H.; Seung, W.; Khan, U.; Kim, T.Y.; Kim, T.; Kim, S.W. 3D-printed biomimetic-villus structure with maximized surface area for triboelectric nanogenerator and dust filter. *Nano Energy* **2019**, *63*, 103857. [CrossRef]
21. Zhang, Z.; Corrigan, N.; Bagheri, A.; Jin, J.; Boyer, C. A Versatile 3D and 4D printing system through photocontrolled RAFT polymerization. *Angew. Chem.* **2019**, *131*, 18122–18131. [CrossRef]
22. Zhang, B.; Kowsari, K.; Serjouei, A.; Dunn, M.L.; Ge, Q. Reprocessable thermosets for sustainable three-dimensional printing. *Nat. Commun.* **2018**, *9*, 1–7. [CrossRef] [PubMed]
23. Muth, J.T.; Vogt, D.M.; Truby, R.L.; Mengüç, Y.; Kolesky, D.B.; Wood, R.J.; Lewis, J.A. Embedded 3D printing of strain sensors within highly stretchable elastomers. *Adv. Mater.* **2014**, *26*, 6307–6312. [CrossRef] [PubMed]
24. Wehner, M.; Truby, R.L.; Fitzgerald, D.J.; Mosadegh, B.; Whitesides, G.M.; Lewis, J.A.; Wood, R.J. An integrated design and fabrication strategy for entirely soft, autonomous robots. *Nature* **2016**, *536*, 451–455. [CrossRef] [PubMed]
25. Anderson, K.B.; Lockwood, S.Y.; Martin, R.S.; Spence, D.M. A 3D printed fluidic device that enables integrated features. *Anal. Chem.* **2013**, *5*, 5622–5626. [CrossRef]
26. Vatani, M.; Engeberg, E.D.; Choi, J.W. Conformal direct-print of piezoresistive polymer/nanocomposites for compliant multi-layer tactile sensors. *Addit. Manuf.* **2015**, *7*, 73–82. [CrossRef]
27. Kahr, M.; Hortschitz, W.; Steiner, H.; Stifter, M.; Kainz, A.; Keplinger, F. Novel 3D-Printed MEMS magnetometer with optical detection. *Multidiscip. Digit. Publ. Inst. Proc.* **2018**, *2*, 783. [CrossRef]
28. Lee, J.Y.; Tan, W.S.; An, J.; Chua, C.K.; Tang, C.Y.; Fane, A.G.; Chong, T.H. The potential to enhance membrane module design with 3D printing technology. *J. Membr. Sci.* **2016**, *499*, 480–490. [CrossRef]
29. Schwartz, J.J.; Boydston, A.J. Multimaterial actinic spatial control 3D and 4D printing. *Nat. Commun.* **2019**, *10*, 1–10. [CrossRef]

30. Kitson, P.J.; Glatzel, S.; Chen, W.; Lin, C.G.; Song, Y.F.; Cronin, L. 3D printing of versatile reactionware for chemical synthesis. *Nat. Protoc.* **2016**, *11*, 920. [CrossRef]
31. Bracaglia, L.G.; Smith, B.T.; Watson, E.; Arumugasaamy, N.; Mikos, A.G.; Fisher, J.P. 3D printing for the design and fabrication of polymer-based gradient scaffolds. *Acta Biomater.* **2017**, *56*, 3–13. [CrossRef] [PubMed]
32. Antonello, R.; Oboe, R. Exploring the potential of MEMS gyroscopes: Successfully using sensors in typical industrial motion control applications. *IEEE Ind. Electron. Mag.* **2012**, *6*, 14–24. [CrossRef]
33. Samaali, H.; Najar, F.; Choura, S. Dynamic study of a capacitive mems switch with double clamped-clamped microbeams. *Shock Vib.* **2014**, *2014*, 1–7. [CrossRef]
34. Ulkir, O.; Ertugrul, I. Mikro kiriş uzunluğu değişiminin deformasyona etkisinin araştırılması. *Avrupa Bilim Teknoloji Dergisi* **2020**, *18*, 136–141.
35. Tanter, M.; Fink, M. Ultrafast imaging in biomedical ultrasound. *IEEE Trans. Ultrason. Ferroelectr. Freq. Control* **2014**, *61*, 102–119. [CrossRef]
36. Okoro, C.; Levine, L.E.; Xu, R.; Hummler, K.; Obeng, Y.S. Nondestructive measurement of the residual stresses in copper through-silicon vias using synchrotron-based microbeam X-ray diffraction. *IEEE Trans. Electron Devices* **2014**, *61*, 2473–2479.
37. Tang, M.; Ni, Q.; Wang, L.; Luo, Y.; Wang, Y. Size-dependent vibration analysis of a microbeam in flow based on modified couple stress theory. *Int. J. Eng. Sci.* **2014**, *85*, 20–30. [CrossRef]
38. Botchway, S.W.; Reynolds, P.; Parker, A.W.; O'Neill, P. Laser-induced radiation microbeam technology and simultaneous real-time fluorescence imaging in live cells. *Methods Enzymol.* **2012**, *504*, 3–28.
39. Incerti, S.; Douglass, M.; Penfold, S.; Guatelli, S.; Bezak, E. Review of Geant4-DNA applications for micro and nanoscale simulations. *Phys. Med.* **2016**, *32*, 1187–1200. [CrossRef]
40. Liu, X.; Wang, H.; Li, Y.; Tang, Y.; Liu, Y.; Hu, X.; Jin, C. Preparation of single rice chromosome for construction of a DNA library using a laser microbeam trap. *J. Biotechnol.* **2004**, *109*, 217–226. [CrossRef]
41. Wu, J.; Zhang, Y.; Zhang, N. Anomalous elastic properties of attraction-dominated DNA self-assembled 2D films and the resultant dynamic biodetection signals of micro beam sensors. *Nanomaterials* **2019**, *9*, 543. [CrossRef] [PubMed]
42. Guo, X.; Yi, Y.B.; Pourkamali, S.A. Finite element analysis of thermoelastic damping in vented MEMS beam resonators. *Int. J. Mech. Sci.* **2013**, *74*, 73–82. [CrossRef]
43. Ouakad, H.M.; Younis, M.I. On using the dynamic snap-through motion of MEMS initially curved microbeams for filtering applications. *J. Sound Vib.* **2014**, *333*, 555–568. [CrossRef]
44. Singh, K.; Joyce, R.; Varghese, S.; Akhtar, J. Fabrication of electron beam physical vapor deposited polysilicon piezoresistive MEMS pressure sensor. *Sens. Actuators A Phys.* **2015**, *223*, 151–158. [CrossRef]
45. Ngo, T.D.; Kashani, A.; Imbalzano, G.; Nguyen, K.T.; Hui, D. Additive manufacturing (3D printing): A review of materials, methods, applications and challenges. *Compos. Part B Eng.* **2018**, *143*, 172–196. [CrossRef]
46. Ambrosi, A.; Pumera, M. 3D-printing technologies for electrochemical applications. *Chem. Soc. Rev.* **2016**, *45*, 2740–2755. [CrossRef]
47. Ertugrul, I.; Akkus, N.; Yüce, H. Fabrication of MEMS based electrothermal microactuators with additive manufacturing Technologies. *Mater. Tehnol.* **2019**, *53*, 665–670. [CrossRef]
48. Han, S.; Sato, I.; Okabe, T.; Sato, Y. Fast spectral reflectance recovery using DLP projector. In *Asian Conference on Computer Vision*; Springer: Berlin/Heidelberg, Germany, 2010.
49. Ge, Q.; Sakhaei, A.H.; Lee, H.; Dunn, C.K.; Fang, N.X.; Dunn, M.L. Multimaterial 4D printing with tailorable shape memory polymers. *Sci. Rep.* **2016**, *6*, 31110. [CrossRef]
50. Fantino, E.; Chiappone, A.; Roppolo, I.; Manfredi, D.; Bongiovanni, R.; Pirri, C.F.; Calignano, F. 3D printing of conductive complex structures with in situ generation of silver nanoparticles. *Adv. Mater.* **2016**, *28*, 3712–3717. [CrossRef]
51. Pan, Y.; Zhou, C.; Chen, Y. A fast mask projection stereolithography process for fabricating digital models in minutes. *J. Manuf. Sci. Eng.* **2012**, *134*, 10–24. [CrossRef]
52. Zheng, X.; Deotte, J.; Alonso, M.P.; Farquar, G.R.; Weisgraber, T.H.; Gemberling, S.; Spadaccini, C.M. Design and optimization of a light-emitting diode projection micro-stereolithography three-dimensional manufacturing system. *Rev. Sci. Instrum.* **2012**, *83*, 125001. [CrossRef] [PubMed]
53. Ge, L.; Dong, L.; Wang, D.; Ge, Q.; Gu, G.A. Digital light processing 3D printer for fast and high-precision fabrication of soft pneumatic actuators. *Sens. Actuators A Phys.* **2018**, *273*, 285–292. [CrossRef]

54. Stein, O.; Liu, Y.; Streit, J.; Cahayag, R.; Lu, Y.; Petta, N. Handling and assembling of low-density foam structures fabricated by two-photon polymerization. *Nanoeng. Fabr. Prop. Opt. Devices* **2017**, *10354*, 103540.
55. Jiang, L.J.; Campbell, J.H.; Lu, Y.F.; Bernat, T.; Petta, N. Direct writing target structures by two-photon polymerization. *Fusion Sci. Technol.* **2016**, *70*, 295–309. [CrossRef]
56. Gregori, G.; Glenzer, S.H.; Fournier, K.B.; Campbell, K.M.; Dewald, E.L.; Jones, O.S.; Landen, O.L. X-ray scattering measurements of radiative heating and cooling dynamics. *Phys. Rev. Lett.* **2008**, *101*, 045003. [CrossRef]
57. Mao, M.; He, J.; Li, X.; Zhang, B.; Lei, Q.; Liu, Y.; Li, D. The emerging frontiers and applications of high-resolution 3D printing. *Micromachines* **2017**, *8*, 113. [CrossRef]
58. Lee, H.; Fang, N.X. Micro 3D printing using a digital projector and its application in the study of soft materials mechanics. *J. Vis. Exp.* **2012**, *69*, 4457. [CrossRef]
59. Han, D.; Lu, Z.; Chester, S.A.; Lee, H. Micro 3D printing of a temperature-responsive hydrogel using projection micro-stereolithography. *Sci. Rep.* **2018**, *8*, 1–10. [CrossRef]
60. Han, D.; Yang, C.; Fang, N.X.; Lee, H. Rapid multi-material 3D printing with projection micro-stereolithography using dynamic fluidic control. *Addit. Manuf.* **2019**, *27*, 606–615. [CrossRef]
61. Park, I.B.; Ha, Y.M.; Lee, S.H. Dithering method for improving the surface quality of a microstructure in projection microstereolithography. *Int. J. Adv. Manuf. Technol.* **2011**, *52*, 545–553. [CrossRef]
62. Behroodi, E.; Latifi, H.; Najafi, F. A compact LED-based projection microstereolithography for producing 3D microstructures. *Sci. Rep.* **2019**, *9*, 1–14. [CrossRef] [PubMed]

micromachines

Article

Thin-Film MEMS Resistors with Enhanced Lifetime for Thermal Inkjet

Elkana Bar-Levav, Moshe Witman and Moshe Einat *

Department of Electrical and Electronic Engineering, Ariel University, Ariel 4070000, Israel; elkanabar@gmail.com (E.B.-L.); moshew@ariel.ac.il (M.W.)
* Correspondence: einatm@ariel.ac.il

Received: 26 April 2020; Accepted: 12 May 2020; Published: 14 May 2020

Abstract: In this paper, the failure mechanisms of the thermal inkjet thin-film resistors are recognized. Additionally, designs of resistors to overcome these mechanisms are suggested and tested by simulation and experiment. The resulting resistors are shown to have improved lifetimes, spanning an order of magnitude up to 2×10^9 pulses. The thermal failure mechanisms were defined according to the electric field magnitude in three critical points—the resistor center, the resistor–conductor edge, and the resistor thermal "hot spots". Lowering the thermal gradients between these points will lead to the improved lifetime of the resistors. Using MATLAB PDE simulations, various resistors shapes, with different electric field ratios in the hot spots, were designed and manufactured on an 8″ silicon wafer. A series of lifetime experiments were conducted on the resistors, and a strong relation between the shape and the lifetime of the resistor was found. These results have immediate ramifications regarding the different printing apparatuses which function with thermal inkjet technology, allowing the commercial production of larger thermal printheads with high MTBF rate. Such heads may fit fast and large 3D printers.

Keywords: 2D printhead; 3D printing; thermal inkjet; thin-film resistors

---

1. Introduction

One of the existing technologies for ink/fluid printing onto carrying media is the thermal inkjet printer head [1–3]. This head is composed of a chamber containing the print fluid. On one of the walls of the chamber there is an electrical resistor designed to heat to high temperatures [4]. In addition, there is a nozzle through which the print fluid can be ejected. There was a time where the future of inkjet technology for printing was questionable among printing experts, in light of the laser printing technique. But experience shows that the opposite has happened. Not only is the inkjet concept not vanishing, in fact, to the contrary, it is developing and penetrating new regimes suggesting amazing possibilities such as cell sorting and single-cell lysis [5–7], medical applications [8], fluid micromixer [9], DNA droplets [10], organic transistors [11], silver nanoparticles printing for microelectronics [12] and many more. A company called "Nano-dimension" has developed a printer that prints together conductor and insulator. This technology enables printing an electrical PCB, printed antennas and other applications that may improve electrical design. Similar technology is also used for microvalves and micropumps based on thermal bubble actuated microfluidic chips [13]. A micro-synthetic jet [14] is also based on micro heaters technology. More complicated microheaters are used for atomization of high-viscosity fluids [15]

In a typical thermal inkjet arrangement, when a current is applied to the resistor for a short time (a few microseconds) [16], it heats the surrounding fluid in its immediate vicinity and causes local boiling (micro-boiling, MB). The rapid pressure rise forces liquid at a distance from the resistor into the nozzle and out of the chamber. The rapid temperature rise and resulting temperature gradients on the resistor shorten its lifetime and therefore the lifetime of the entire head. This lifetime is shorter compared to other printer heads, such as the piezoelectric head [17].

As an example of the ramifications of this, when the ink is depleted in a thermal inkjet head the entire head is replaced, as it does not make sense to refill and continue its use. Contrarily, piezoelectric heads are refilled with ink as their lifetime is greater than a single depletion cycle. Despite this and other advantages of the piezo head [18,19], the thermal inkjet head is the more common product in many applications as it is cheaper and simpler to manufacture.

For domestic paper printing the existing lifetime is satisfactory, but for 3D printing (additive manufacturing), where much higher operation cycles are needed for a single print, a higher lifetime is vital.

Since the central disadvantage of the thermal printhead is its shorter lifetime [20–23], the isolation of failure mechanisms and the design of more robust resistors would create a greater incentive for production and manufacturing. Park et al and Lim et al' researches [24,25] present studies on the subject of lifetime enhancement by changing the printhead micro structure and changing the metal composition and thickness of the resistor. McGlone et al' research [26] describes more than 10^7 pulses obtained by the use of amorphous metal thin films. Bendong et al' research [27] presents a concept where the heating is done with an induction element that has no physical connection to the power source.

In this paper, two failure mechanisms of the thermal inkjet resistor are recognized. One is related to the sharp temperature gradient at the conductor–resistor contact, and the second is related to hot spots on the resistor. Both tend to cause a discontinuity in the resistor and, finally, resistor failure. Additionally, designs of resistors to overcome these mechanisms are suggested and tested by simulation and experiment. Various connecting geometries between the conductor and the resistor are tested. The resulting resistors are shown to have an order of magnitude improvement in a lifetime.

2. Thermal Gradients in Resistors

One form of resistor geometry is the rectangular form. It has the advantage of homogeneous current density along its surface, in addition to its geometric simplicity.

Its disadvantage; however, lies in the large temperature gradient between the conductors and itself. These gradients cause mechanical strain and eventually cracking in the connecting media, which ultimately lead to component failure. So, a clear motivation is to find a configuration that reduces the temperature gradient between the conductor and the resistor as it is a failure mechanism.

In order to reduce these gradients, a common solution is a trapezoidal or "ramp-shaped" region connecting the resistor to its conductors. Figure 1 shows the geometry of the ramp leading into the resistor and the resulting lowered temperature gradients from the conductors to the resistor. The gradient smoothing is achieved as a result of the changing resistance along the sloped trapezoidal region. Near the conductor there is lower resistance, whereas near the resistor there is higher resistance. Therefore, the sharp "jump" from the cold conductor to the hot resistor is smoothed, and the weak point of the connection now has a longer lifetime.

However, improving one failure mechanism caused another one to rise. Despite lowering the gradients over the conductor–resistor transition region, Figure 2 shows the resulting formation of "hot spots" at the corners of the ramp–resistor interface (noted with arrows in the figure). This figure describes the results of the electric field simulation that is developed in the resistor. The simulation was carried out using a MATLAB code, written in the PDE tool. The code solved the Laplace equation in the resistor with Dirichlet boundary condition at the conductor–resistor interface, and Neumann boundary conditions everywhere else. The electric field is noted by the red arrows, together with black solid equal-potential lines. The color reflects the magnitude of the electric field. In these "hot spots" there are local maxima of the electric field and, as a result, local maxima of the current and temperature. These hot spots are extreme in both absolute temperature and in local temperature gradient, and are caused by the rise in current density in their vicinity due to the new geometry. Therefore, the hot spots are the first to be destroyed. Once there is a minor destruction and discontinuity at the hot spot point, the current must bend around it and the effect becomes worse. This leads to a rapidly-developing tear in the resistor towards the other hot spot (this is shown later at the experimental part). A full description of this failure mechanism (simulation and experiment) appears with more details in Einat et al' research [28], as it was captured on video in a rare instance.

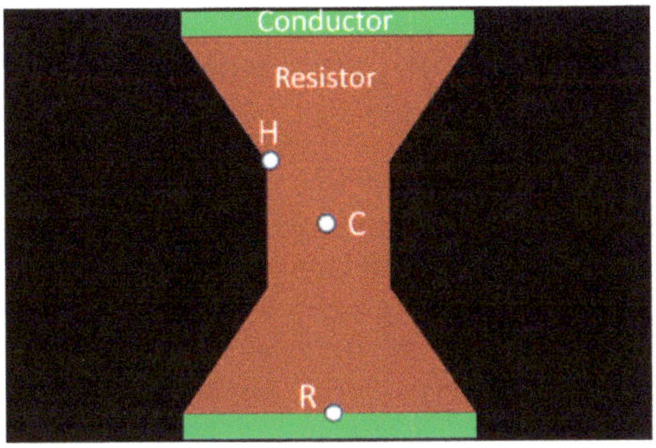

Figure 1. Geometry of the trapezoid resistor, with points H, C and R.

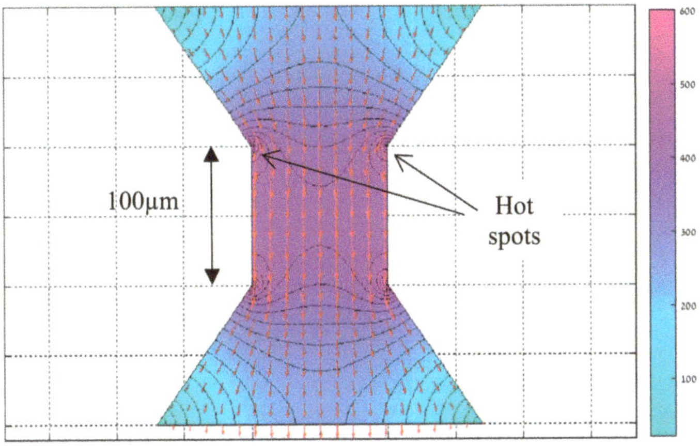

Figure 2. Electric field simulation of Trapezoid resistor.

In order to find the balance between these two failure mechanisms, three points of interest are defined, as seen in Figure 1; (C) at the Center of the resistor body; (*R*) at the center of the conductor–ramp interfaces; and (*H*) at one of the Hot spots. A certain power level and energy are required to obtain the MB effect [28] at the resistor center (C), but it is preferred that the temperature at (*H*), the hot spot, will not rise to much above the temperature at point (C). On the other hand, it is preferred that at point (*R*) the temperature will be as low as possible. Clearly these demands contradict and the goal is to find the optimal compromise that will give the longest lifetime. In order to analyze the points behavior, the electric field magnitude at points (*R*) and (*H*) is normalized to the electric field at point (C) as follows:

$$\mathrm{En}(H) = \frac{E(H)}{\mathrm{E(C)}}, \mathrm{En}(R) = \frac{E(R)}{\mathrm{E(C)}} \tag{1}$$

With E (*H*,*C*,*R*) as the absolute electric field in (V/m) at points *H*, C and *R*, respectively. Since the resistive layer is assumed to be uniform, the local electric field represents the local power and temperature that will develop locally in the resistor layer.

In the following section, simulations are presented which attempt to calculate these ratios, with the ultimate goal of the resulting design being the reduction of these ratios to their minimum. Ideal values are $E_n(H)$ = 100% (meaning that at the hot spot there is the same electric field and temperature as the resistor center) and $E_n(R)$ = 0% (meaning that the conductor is not heated at all). These ideal values actually imply no hotspots and no temperature gradient in the conductor–resistor interface. Approaching these values would improve lifetime characteristics.

As can easily be seen, there is a tradeoff between these two parameters—reducing one increases the other (for example, a square resistor will have no hot spots but will have a maximal gradient between the conductor and resistor border). Therefore, an optimum needs to be found.

3. Resistors Simulation

In an attempt to smooth out the hot spot temperature gradients, a number of configurations were designed, simulated and experimentally tested. Several curved versions replacing the linear connection of the hot spot to the conductor in the trapezoid were tested, as seen in Figure 3. The resistor itself was kept as a 100 × 100 μm rectangular shape in all the versions for a consistent comparison, but the connection to the conductor was done through a different ramp. Figure 3 shows the different configurations, all of which have a filleted transition from ramp to resistor, labeled trapezoid, A1, A2, A3 and A4.

These fillets disperse the current density more evenly around the transition region, thereby considerably reducing the gradient in the electric field and; therefore, the temperature gradient. However, this increases the temperature gradient in the conductor–resistor interface.

Figure 3 shows the simulation results of the electric field of each configuration. The hot spots singularity is clearly seen in the trapezoid. The reduction of the hot spots' singularity of the other shapes can be visually compared.

The simulation results are presented also in Table 1. It can be observed that the minimal $E_n(H)$ is that of A1, whereas the minimal $E_n(R)$ is that of the trapezoidal resistor.

Table 1. Simulation results; the resistor shapes are shown in Figure 3.

Shape	E(C) (v/m)	E(H) (v/m)	E(R) (v/m)	$E_n(H)$	$E_n(R)$
A1	371.3	397	274.2	107%	74%
A2	377.3	443.3	273.2	117%	72%
A3	397.9	587.3	256.6	148%	64%
A4	368.4	404.7	278	110%	75%
Trapezoid	415.2	588.7	247.1	142%	60%

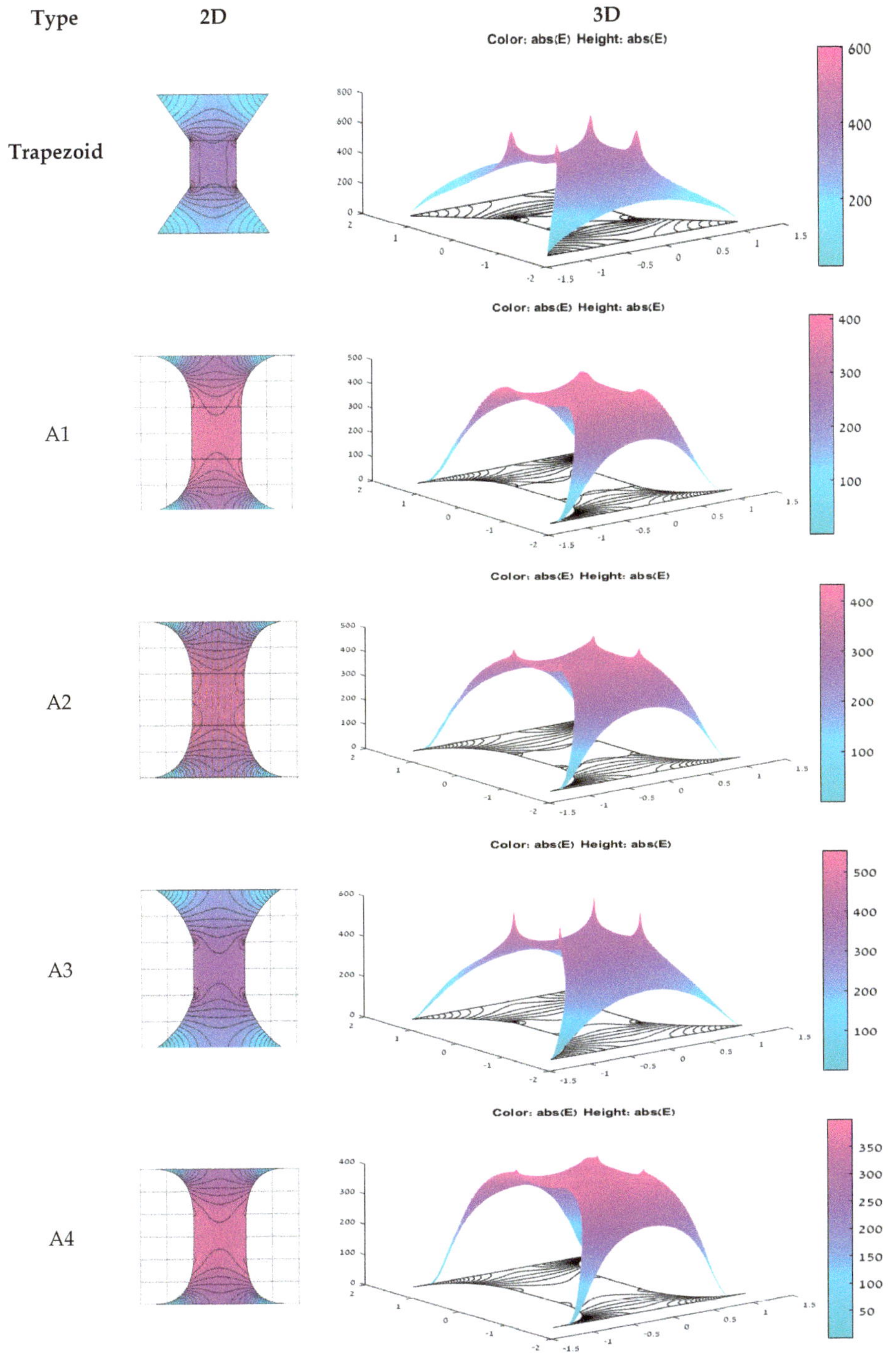

Figure 3. 2D and 3D electric field simulation for: Trapezoid, A1, A2, A3 and A4 resistors.

4. Experiment and Results

The specimens were fabricated as thin-film resistor shapes on an 8″ diameter and 500 μm thickness silicon wafer, with a 1000 μm oxidation layer. On top of the wafer, two consecutive layers were evaporated—the first of a resistive 500 nm tantalum nitride (Ta-N) with a sheet resistance of 30 Ω, and the second of a conductive 100 nm copper layer and a 500 nm gold coating layer. For good adhesion between the resistive layer and the conductive layer, a 10 nm titanium layer was also evaporated. After the evaporation of each layer, photolithography and etching processes were preformed to create the resistor's and conductor's shape. Each resistor body had a 100 × 100 μm rectangular shape, 30 Ω resistance, and the geometry of the different ramps is seen in Figure 3.

The wafer and one of the planned resistors can be seen in Figure 4a,b. For the purpose of experimental testing of the simulation results, a series of experiments were conducted on the different resistors. The wafer was connected to an electric circuit depicted in Figure 5.

(a)

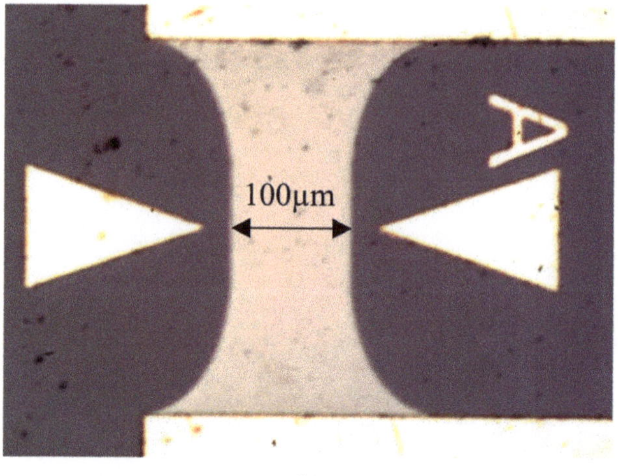

(b)

Figure 4. (a) Test 8″ Wafer, (b) A2 resistor.

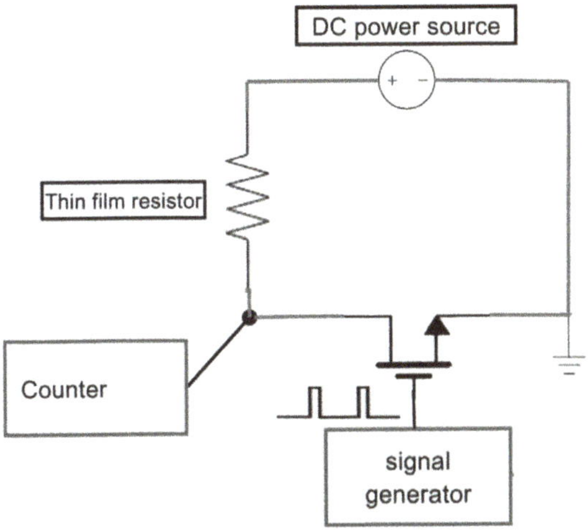

Figure 5. Experiment setup diagram.

The experiment was conducted 5–6 times per resistor, under a constant current of 0.5 A, uniform for all resistor shapes. The circuit supplied a pulse repetition frequency (PRF) of 33 kHz from a signal generator, each pulse having a duration of 5 microseconds. It is important to note that MB had been confirmed [28] with this setup and parameters prior to experimentation, as seen in Figure 6. When a current pulse is given to the resistor, a rapid heating occurs, the fluid above the resistor heats up and a bubble grows above the resistor during the pulse duration (Figure 6). The picture was taken using a stroboscope.

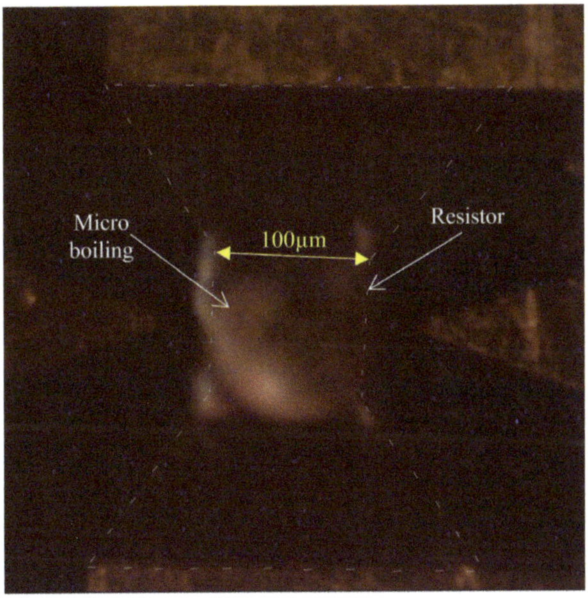

Figure 6. Micro-boiling process.

The number of pulses were counted by a counter mechanism on the circuit, which ceases its operation when the resistor burns out, as can be seen in Figure 7. As explained and seen in Einat et al's research [28], the burnout starts at the hotspot (point H in Figure 1), and continues rapidly by a thermal runaway process until the resistor is disconnected. The experiment results are shown in Table 2, which presents the different resistors and their respective average number of pulses until breakdown. The results in this table are arranged according to the number of pulses (noted as "lifetime") and the simulated results are added again for the convenience of relating the experimental lifetime to the simulated electric fields. The results are also presented graphically in Figure 8.

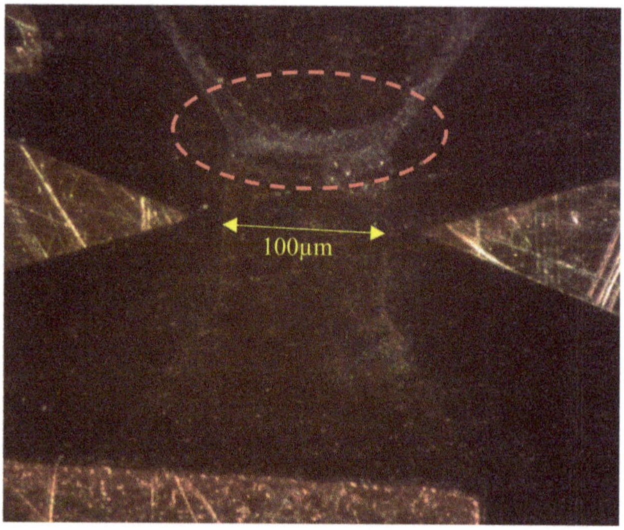

Figure 7. Trapezoid resistor burn out.

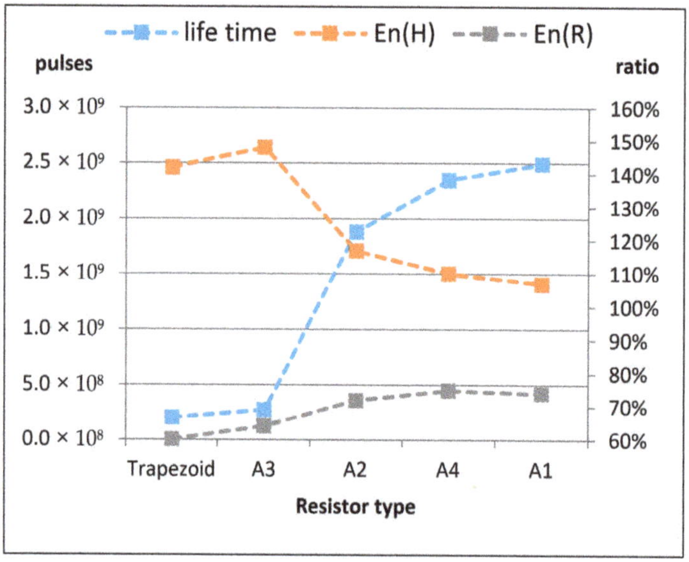

Figure 8. Experimental lifetime results and electric fields simulation results; the resistor shapes are shown in Figure 3.

Table 2. Experiment results sorted by lifetime achieved in comparison to the simulated normalized electric fields.

Resistor Shape	$E_n(R)$	$E_n(H)$	Average Number of Pulses until Breakdown
Trapezoid	60%	142%	2.02×10^{8}
A3	64%	148%	2.75×10^{8}
A2	72%	117%	1.88×10^{9}
A4	75%	110%	2.35×10^{9}
A1	74%	107%	2.49×10^{9}

In Figure 6. The following trends are seen. There is a ~15% degradation in the resistor–conductor interface normalized electric field *En*(*R*), which should optimally be as small as possible. But at the same time, the trend of the normalized electric field *En*(*H*) at the hot spot was improved by ~30%. As seen in the results, with these trends, the number of pulses until breakdown was increased by more than an order of magnitude, from 2 × 108 to 2.5 × 109. This is a major improvement obtained only by choosing the geometry of the resistor ramp properly. It is understood that the hot spot failure is more dominant. Improving the *En*(*H*) factor, even at the expense of the *En*(*R*), leads to an overall major improvement.

5. Discussion and Conclusions

In this research, theoretical and experimental work was done to test the effect of the micro resistor geometry on its lifetime. A clear correlation between the geometry and lifetime of a MB resistor arises from the experimental results, with the differences in lifetime spanning an order of magnitude. These results are supported by a theoretical analysis which identified two failure mechanisms—the extreme field gradients near the ramp–body joining vertex (hot spot, *H*) and near the center of the conductor–ramp interface (*R*).

The experimental results show that the best predicting factor is the value of the normalized field at the hot spot $E_n(H)$. It can be seen that when $E_n(H)$is made smaller, the resistor lifetime improves immensely—even when the normalized field at the conductor–ramp interface $E_n(R)$ is made greater as a result. This tradeoff is apparent in the designs shown in this research, but further modeling and experimentation may yield techniques to improve both normalized field values at once.

The experimental study was done using 30 Ω sheet resistance of the resistive layer, but the same effect is expected regardless of the exact value of the sheet resistance, since it is depended on the geometry.

These results have immediate ramifications regarding the different printing apparatuses which function with thermal inkjet technology. After development and refining, large plates of printer head arrays can be realized with sizes reaching that of any LCD screen. There would be an improvement in both the resolution and speed of the print [29], and the inherently longer MTBF would allow refilling to become practical. In addition, 3D printing apparatuses can greatly benefit from a similar scale-up as entire layers of ink could be printed simultaneously [30], thereby shortening printing time from hours to minutes. The improvement in resistor lifetime can create new opportunities where large-scale, refillable thermal inkjet apparatuses are economically viable.

Author Contributions: Conceptualization, M.E.; methodology, M.E. and E.B.-L.; simulations, E.B.-L.; experimental setup, M.W. and E.B.-L.; running the experiments, M.W. and E.B.-L.; Data curation, M.W. and E.B.-L.; validation, M.E.; formal analysis, E.B.-L. and M.E.; investigation, E.B.-L. and M.W.; resources, M.E.; writing—original draft preparation, M.E. and E.B.-L.; writing—review and editing, M.E. and E.B.-L.; visualization, E.B.-L and M.E.; supervision, M.E.; funding acquisition, M.E.; All authors have read and agreed to the published version of the manuscript.

Funding: This research received no external funding.

Conflicts of Interest: The authors declare no conflicts of interest.

References

1. Hoath, S.D. (Ed.) *Fundamentals of Inkjet Printing: The Science of Inkjet and Droplets*; Wiley-VCH Verlag GmbH & Co. KGaA: Weinheim, Germany, 2016.
2. Zapka, W. (Ed.) *Handbook of Industrial Inkjet Printing: A Full System Approach*; Wiley-VCH Verlag GmbH & Co. KGaA: Weinheim, Germany, 2018.
3. Hutchings, I.M.; Martin, G.D. (Eds.) *Inkjet Technology for Digital Fabrication*; John Wiley & Sons, Ltd.: Chichester, UK, 2013.
4. Le, H.P. Progress and trends in ink-jet printing technology. *J. Imaging Sci. Technol.* **1998**, *42*, 49–62.
5. Hoefemann, H.; Wadle, S.; Bakhtina, N.; Kondrashov, V.; Wangler, N.; Zengerle, R. Sorting and lysis of single cells by BubbleJet technology. *Sens. Actuators B Chem.* **2012**, *168*, 442–445. [CrossRef]
6. De Wijs, K.; Liu, C.; Dusa, A.; Vercruysse, D.; Majeed, B.; Tezcan, D.S.; Blaszkiewicz, K.; Loo, J.; Lagae, L. Micro vapor bubble jet flow for safe and high-rate fluorescence-activated cell sorting. *Lab Chip* **2017**, *17*, 1287–1296. [CrossRef] [PubMed]
7. Majeed, B.; Liu, C.; Van Acker, L.; Daily, R.; Miyazaki, T.; Sabuncuoglu, D.; Lagae, L.; Miyazaki, T. Fabrication of silicon-based microfluidics device for cell sorting application. In Proceedings of the 2014 IEEE 64th Electronic Components and Technology Conference (ECTC), Orlando, FL, USA, 27–30 May 2014; pp. 165–169.
8. Scoutaris, N.; Ross, S.; Douroumis, D. Current Trends on Medical and Pharmaceutical Applications of Inkjet Printing Technology. *Pharm. Res.* **2016**, *33*, 1799–1816. [CrossRef] [PubMed]
9. Tan, H. Numerical study of a bubble driven micromixer based on thermal inkjet technology. *Phys. Fluids* **2019**, *31*, 062006. [CrossRef]
10. Liou, J.-C.; Wu, C.-C. Design and fabrication of microfluidic inkjet chip with high voltage ESD protection system for DNA droplets arrangement and detection. *Microsyst. Technol.* **2015**, *23*, 199–213. [CrossRef]
11. Mattana, G.; Loi, A.; Woytasik, M.; Barbaro, M.; Noël, V.; Piro, B. Inkjet-Printing: A New Fabrication Technology for Organic Transistors. *Adv. Mater. Technol.* **2017**, *2*, 1700063. [CrossRef]
12. Khan, A.; Rahman, K.; Hyun, M.-T.; Kim, N.-S.; Choi, K.-H. Multi-nozzle electrohydrodynamic inkjet printing of silver colloidal solution for the fabrication of electrically functional microstructures. *Appl. Phys. A* **2011**, *104*, 1113–1120. [CrossRef]
13. Huang, C.; Tsou, C. The implementation of a thermal bubble actuated microfluidic chip with microvalve, micropump and micromixer. *Sens. Actuators A Phys.* **2014**, *210*, 147–156. [CrossRef]
14. Sourtiji, E.; Peles, Y. A micro-synthetic jet in a microchannel using bubble growth and collapse. *Appl. Therm. Eng.* **2019**, *160*, 114084. [CrossRef]
15. Law, J.; Kong, K.W.; Chan, C.; Sun, W.; Li, W.J.; Chau, E.B.F.; Chan, G.K.M. Atomization of High-Viscosity Fluids for Aromatherapy Using Micro-heaters for Heterogeneous Bubble Nucleation. *Sci. Rep.* **2017**, *7*, 40289. [CrossRef] [PubMed]
16. Bonfert, D.; Gieser, H.; Bock, K.; Svasta, P.; Ionescu, C. Electrical stress on thin film TaN resistive structures. In Proceedings of the IEEE 17th International Symposium for Design and Technology in Electronic Packaging (SIITME), Timisoara, Romania, 20–23 October 2011; pp. 313–318.
17. Khalate, A.; Bombois, X.; Scorletti, G.; Babuska, R.; Koekebakker, S.; De Zeeuw, W. A Waveform Design Method for a Piezo Inkjet Printhead Based on Robust Feedforward Control. *J. Microelectromech. Syst.* **2012**, *21*, 1365–1374. [CrossRef]
18. Singh, M.; Haverinen, H.M.; Dhagat, P.; Jabbour, G.E. Inkjet Printing-Process and Its Applications. *Adv. Mater.* **2010**, *22*, 673–685. [CrossRef] [PubMed]
19. Khalate, A.; Bomboisa, X.; Babuškaa, R.; Wijshoffb, H.; Waarsingb, R. Performance improvement of a drop-on-demand inkjet printhead using an optimization-based feed forward control method. *Control Eng. Pract.* **2011**, *19*, 771–781. [CrossRef]
20. Baca, A.G.; Overberg, M.E.; Wolfley, S.L.; Fortune, T.R. A thin film TaN resistor reliability evaluation. In Proceedings of the 2017 IEEE International Reliability Physics Symposium (IRPS), Monterey, CA, USA, 2–6 April 2017. [CrossRef]
21. Lee, T.; Watson, K.; Chen, F.; Gill, J.; Harmon, D.; Sullivan, T.; Li, B. Characterization and reliability of TaN thin film resistors. In Proceedings of the 2004 IEEE International Reliability Physics Symposium, Phoenix, AZ, USA, 25–29 April 2004; pp. 502–508. [CrossRef]

22. Dąbrowski, A.; Dziedzic, A. Stability of low ohmic thick-film resistors under pulsed operation. *Microelectron. Reliab.* **2018**, *84*, 95–104. [CrossRef]
23. Liou, J.; Yang, C.; Gong, C. Design and Fabrication of Identification Inkjet Print Head Chip Fuse Sensors. *Sens. Mater.* **2016**, *28*, 493–501.
24. Park, J.-H.; Oh, Y. Investigation to minimize heater burnout in thermal thin film print heads. *Microsyst. Technol.* **2005**, *11*, 16–22. [CrossRef]
25. Lim, J.-H.; Kuk, K.-; Shin, S.-L.; Baek, S.-S.; Kim, Y.-L.; Oh, Y.-S. Investigation of reliability problems in thermal inkjet printhead. In Proceedings of the 2004 IEEE International Reliability Physics Symposium, Phoenix, AZ, USA, 25–29 April 2004; pp. 251–254.
26. McGlone, J.M.; Pommerenck, J.; Graham, M.W.; Wager, J.F. Amorphous Metal Thin Films for Thermal Inkjet Printing. *J. Microelectromech. Syst.* **2018**, *27*, 289–295. [CrossRef]
27. Liu, B.; Hou, Y.P.; Desheng, L.I.; Yang, J.H. A thermal bubble micro-actuator with induction heating. *Sens. Actuators A Phys.* **2015**, *222*, 8–14. [CrossRef]
28. Einat, M.; Grajower, M. Microboiling Measurements of Thermal-Inkjet Heaters. *J. Microelectromech. Syst.* **2010**, *19*, 391–395. [CrossRef]
29. Einat, M.; Einat, N. Two-dimension full array high-speed ink-jet print head. *Appl. Phys. Lett.* **2006**, *89*, 73505. [CrossRef]
30. Einat, M.; Bar-Levav, E. 2D segmented large inkjet printhead for high speed 3D printers. *J. Micromech. Microeng.* **2015**, *25*, 5. [CrossRef]

micromachines

MDPI

Article

On-Substrate Joule Effect Heating by Printed Micro-Heater for the Preparation of ZnO Semiconductor Thin Film

Van-Thai Tran [1], Yuefan Wei [2] and Hejun Du [1,*]

1 School of Mechanical and Aerospace Engineering, Nanyang Technological University, 50 Nanyang Avenue, Singapore 639798, Singapore; vanthai.tran@ntu.edu.sg
2 Advanced Remanufacturing and Technology Centre, 3 Cleantech Loop, Singapore 637143, Singapore; wei_yuefan@artc.a-star.edu.sg
* Correspondence: MHDU@ntu.edu.sg; Tel.: +65-6790-4783

Received: 13 April 2020; Accepted: 9 May 2020; Published: 10 May 2020

Abstract: Fabrication of printed electronic devices along with other parts such as supporting structures is a major problem in modern additive fabrication. Solution-based inkjet printing of metal oxide semiconductor usually requires a heat treatment step to facilitate the formation of target material. The employment of external furnace introduces additional complexity in the fabrication scheme, which is supposed to be simplified by the additive manufacturing process. This work presents the fabrication and utilization of micro-heater on the same thermal resistive substrate with the printed precursor pattern to facilitate the formation of zinc oxide (ZnO) semiconductor. The ultraviolet (UV) photodetector fabricated by the proposed scheme was successfully demonstrated. The performance characterization of the printed devices shows that increasing input heating power can effectively improve the electrical properties owing to a better formation of ZnO. The proposed approach using the on-substrate heating element could be useful for the additive manufacturing of functional material by eliminating the necessity of external heating equipment, and it allows in-situ annealing for the printed semiconductor. Hence, the integration of the printed electronic device with printing processes of other materials could be made possible.

Keywords: inkjet printing; zinc oxide; heat treatment; micro-heater; semiconductor

1. Introduction

Exploration of inkjet printing for the fabrication of electronic devices has recently become a trendy research topic due to the remarkable advantages of the digital-additive fabrication such as saving material and time, high resolution, and compatibility with different materials [1,2]. These advantages of inkjet printing have been widely employed for the fabrication of semiconductor devices [3,4]. Thanks to its outstanding features and abundance, zinc oxide (ZnO) has attracted considerable attention and effort in the additive fabrication of electronics devices such as solar cells [5,6], photodetectors [3,7], and transistors [8,9]. Additive manufacturing of inorganic material from a precursor compound usually requires a heat treatment step in order to convert precursors to the required material [10,11]. However, the annealing process using an external furnace might restrain the progress of 3D printed integrative devices because of the added complexity of the fabrication system.

Low temperature processing of metal oxide has been intensively studied via different approaches, such as ultraviolet (UV) annealing [12,13] and laser sintering [14], in order to facilitate the formation of metal oxide by heating effect of high energy light beam, which requires high complexity setup for the processing system. Furthermore, direct use of nanoparticle ink was considered instead of precursor ink [15]. However, added surfactant to keep nanoparticle ink stable might be an issue for electronic application because it might change the properties of printed material. Alternatively, using local Joule heating to form metal oxide by thermal decomposition of precursor compound is an interesting approach to prepare small-size metal oxide pattern [16].

Joule heating is the phenomenon in which heat is generated from a conductive material when there is an electrical current run through the conductor. The power consumed is proportional to the square of the electrical current and the resistance of the conductor [17]. A micro-heater working on the Joule heating principle possesses advantages such as local heat and optimized energy consumption. Therefore, the resistive heater has been employed in application that requires localized heating and temperature control at small scale, such as activation of gas sensing device [18,19], moisture monitoring [20], and local growth of semiconductor nanostructures [21]. The traditional approach of preparing conductive patterns is using photolithography to deposit and remove certain parts of the conductive film and forming the required shape of the film. As this method has its own drawbacks, such as the high complexity and time-consuming, additive manufacturing could be a promising candidate for fabrication of micro-heaters [22].

In this work, a facile and versatile fabrication process for additive fabrication of semiconductor using inkjet printing and on-substrate heating was proposed. The printed conductive material is further employed for another role, which is the heating element for later processes. Zinc precursor ink is then printed on the same substrate. Eventually, electrical power was applied to the micro-heater to generate heat, which facilitates the decomposition of the zinc salt and the formation of ZnO. Therefore, the necessity for an external bulky furnace is eliminated. The generation of ZnO was examined by the elemental survey of zinc (Zn) and oxygen (O) component in energy-dispersive X-ray spectroscopy (EDS) and the aid of thermogravimetric analysis (TGA). In order to demonstrate the obtained semiconductor film, a UV photodetector application was prepared and characterized. The influence of electrical power during Joule heating to photodetector performance is also evaluated.

2. Materials and Methods

A commercialized Dimatix 2831 inkjet printer (Fujifilm Dimatix, Inc, Santa Clara, CA, USA) was employed in all printing steps described in this work using 10 pL cartridge with 16 nozzles. The silicon/silicon dioxide (Si/SiO_2) substrate (Bonda Technology Pte Ltd, Singapore) was cleaned before the printing of silver ink to construct the electrodes and heater. The substrate was cleaned in acetone and rinsed with isopropanol, then it was dried out by a manual air blower.

A commercial silver nanoparticle ink (silver dispersion 736465, Sigma-Aldrich, St. Louis, MO, USA) was employed for the printing of the silver patterns. Detail of the printing step has been discussed in our previous report [23]. The waveform applied to piezoelectric nozzle to jet ink is shown in Figure A1a (Appendix A). Peak firing voltage was set at 19 V. Drop-spacing was set at 40 μm to ensure the continuity of the printed silver line. Single layer was selected for printing of electrodes and four-layers was selected for printing of micro-heater. Cartridge temperature was set at 35 °C, however, due to the printer platen temperature was set at 60 °C and the close distance of the cartridge and the substrate during printing, the cartridge temperature may rise to about 40 °C during printing. The printed pattern was left on the platen of the printer for 10 min for solvent vaporization.

Printed single line of conductive silver features 100 μm in width, and the thickness is about 200 nm as reported in our previous works [23]. The as-printed single line resistance was measured as 15.5 Ω. Due to the Joule heating process using 4-W power, the resistance was reduced to 5.0 Ω as the effect of sintering.

Figure 1 depicts the fabrication of the device using Joule heating. Printed metal patterns, serving as electrical contacts for the sensor as well as the micro-heater, was printed on the first step of the fabrication process (Figure 1a). The micro-heater composes of two contacting pad and a single line of conductive silver serving as a heating resistor. Then, direct current was applied to the micro-heater, which consumes 4-W power and converts it to heat energy for 5 min (Figure 1b). The generated heat promotes the sintering of silver electrodes and improves the bonding to substrate. The power applied to the micro-heater was manually adjusted by turning the voltage of the direct current (DC) power source when monitoring the current. During the Joule heat treatment, the resistance of the micro-heater changed due to sintering effect which require a careful adjustment of voltage. For example, a 5 W of power was obtained by turning the voltage to 5.0 V while the current reached 1.0 A.

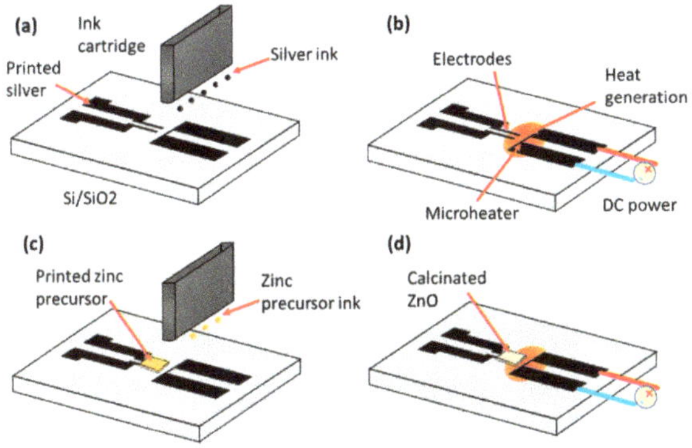

Figure 1. Fabrication step of the sensor using including printed micro-heater. (**a**) Printing of conductive pattern for the metal contacts and heater. (**b**) Applying direct current (DC) power to the micro-heater to generate heat and sinter silver. (**c**) Deposition of zinc salt ink over the printed electrodes. (**d**) Applying DC power again to the micro-heater to calcinate ZnO.

A 50 mM zinc precursor solution was formulated by dissolving of zinc acetate dihydrate ($Zn(CH_3COO)_2{\cdot}2H_2O$) to ethanol. Another cartridge was used to print the zinc precursor after the replacement of the silver ink cartridge by zinc precursor ink cartridge. A rectangular pattern of the zinc salt solution is printed over the metal contacts, which is nearby the micro-heater (Figure 1c). The beneath substrate was kept heated at a temperature of 60 °C during printing to facilitate the evaporation of the solvent and reduce the spreading of solution over the surface.

Ten layers of zinc precursor was printed with a designed pattern of 60 pixels by 50 pixels. Zinc precursor ink printing parameters were optimized to ensure the stability during printing. The waveform was given in Figure A1b (Appendix A) with peak voltage at 19 V. The cartridge was kept at room temperature. Drop spacing, defining by the distance of two nearby droplets, was set at 10 μm, so that it is necessary to calibrate the cartridge holder angle during printing different materials.

After the printing of the precursor, electrical current was applied to the resistive heater again for 5 min (Figure 1d). Two sets of samples which have 4-W and 5-W applied power were studied along with the samples without the treatment. A thermal camera (NEC F30W, AVIO, Turin, Italy) was employed to measure the temperature of the device during the Joule heating process.

Thermogravimetric analysis (TGA) using the equipment TGA Q500 (TA Instruments, New Castle, DE, USA) was utilized for studying the formation of zinc oxide by thermal process. The heating rate we used for TGA is 10 °C/min. Film morphology and elemental study were characterized by field emission–scanning electron microscope (FE–SEM, JOEL 7600F, JEOL Ltd., Tokyo, Japan) and energy dispersive X-ray spectroscopy (EDS) (Oxford Instruments, Abingdon, UK). In order to characterize the performance of the Joule heating processed sensor, an ultraviolet light-emitting diode which emits 365 nm wavelength was used to illuminate the sensor and measure the photocurrent under dark or lighting condition. The photocurrent was recorded using a source metering unit (SMU B2902A, Agilent, Santa Clara, CA, USA).

3. Results

3.1. Sintering of Printed Silver

The structure of the device is depicted in Figure 2, which shows three main components of the devices deposited on Si/SiO_2 substrate. While the common electrodes and the micro-heater were printed on the first layer, the zinc precursor was printed at last. The zinc precursor appears as a white rectangle pattern over the silver electrodes. Thus, the photodetector utilizes the metal-semiconductor contacts with the two-terminals structure. Two low-magnification FE–SEM images of samples before and after heat treatment was presented in Figure A2 (Appendix A) to provide a better understanding of the effect of heat treatment to the film.

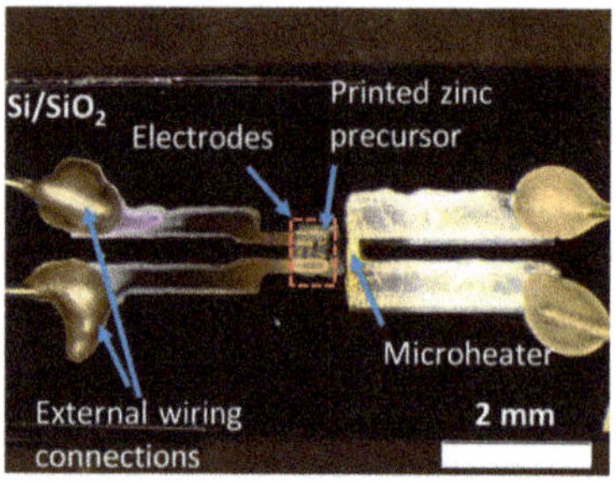

Figure 2. Optical image of the printed Joule heating device showing the three main components of the device, such as electrodes section, printed zinc salt film, and micro-heater. The picture was taken after the heat treatment of the ZnO pattern.

The effect of Joule heating on the nanostructure of printed silver could be observed in Figure 3, which shows the sintering of silver nanoparticles. Printed silver film composes of separated particles after the evaporation of the solvent. The sintering effect is significantly depending on the distance from the heater. At the electrode section (Figure 3b), which is distinct from the radiation source, the particle size of approximately 80 to 100 nm could be observed, which is an increase from the size of about 50 nm of the unsintered particles (Figure 3a).

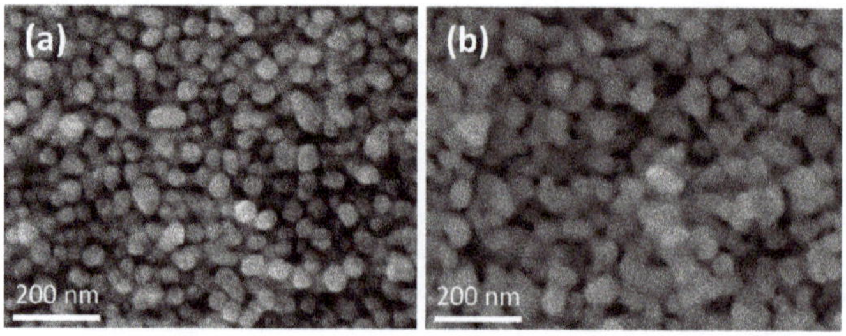

Figure 3. Field emission–scanning electron microscope (FE–SEM) images show the sintering of printed silver nanoparticles ink by the resistive heating. (**a**) The printed silver nanoparticles without annealing. (**b**) The silver nanoparticles at electrode section after annealing.

Heating radiation from resistive effect is the main source that induces the sintering of the printed silver pattern. In Figure 4a, the heater witnesses a notable change of film morphology, such as particle agglomeration up to the size of 200 nm. As the previous discussion has pointed out, the remote pattern exhibits minor sintering effect, while at the center of heat source, major agglomeration could be observed. Later analysis of temperature distribution shows that this is indeed a result of gradually reduce of temperature in the substrate. Although the sintering of silver could improve the electrical properties of the conductive pattern, a severe agglomeration of silver nanoparticles could lead to the interruption of the conductive track and open the circuit. In order to reach a sufficient temperature for the later heat treatment of the sensing material, multi-layers printing of silver was carried on. The heat treatment was conducted after all the layers were printed to reduce dislocation of printed layers due to handling during heat treatment. Figure 4b shows a notable improvement in film morphology when four-layer printing was employed, which does not exhibit severe agglomeration and any gap in film after heat treatment.

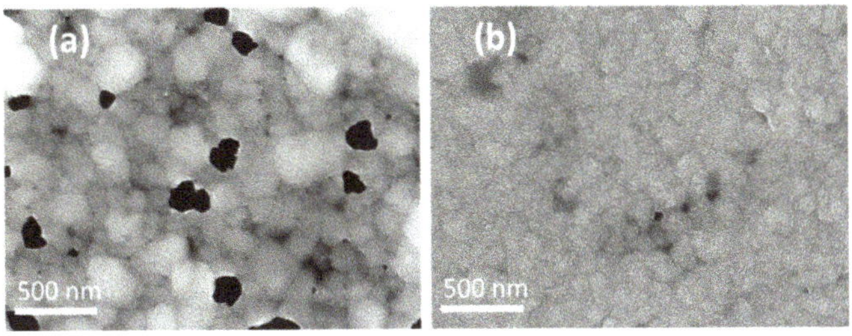

Figure 4. FE–SEM images show the sintering of micro-heater with different numbers of printed silver layers: (**a**) single-layer printing and (**b**) four-layer printing.

3.2. *Temperature Survey of Joule Heating*

The temperature of the device during the Joule heating was investigated using the thermal camera and the results are presented in Figure 5. Thermal photo was taken from the back of the device to avoid the complexity of different emissivity of materials. The temperature was calibrated using the software provided by the thermal camera's manufacturer (NS9205 Viewer, Avio) with applied emissivity of SiO_2 of 0.9 [24].

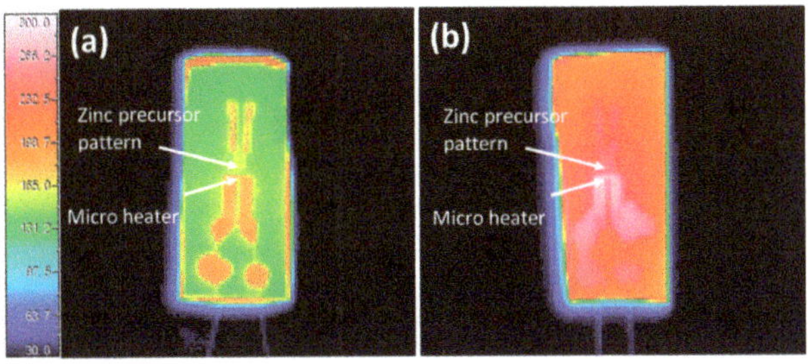

Figure 5. Temperature survey of the device with different heating power by the thermal camera. (**a**) 4-W electrical power. (**b**) 5-W electrical power. The unit of temperature scale bar is degree Celsius.

After calibration using emissivity of 0.9 of SiO_2, an average temperature of 184 °C could be measured at the resistive heater, while that at the zinc precursor film was 171 °C when input electrical power was set at 4 W. There was a significant increase of temperature when raising the input power. When the input power was set at 5 W, the recorded average temperature is 267 °C at the heater and 239 °C at the precursor pattern. The temperature of the resistive heater depends on the input power, and their correlation might possibly be expressed by the relation [25]:

$$P = a(T - T_0) + b(T - T_0)^2 + c(T^4 - T_0{}^4), \quad (1)$$

where P, T, and T_0 are input power, heater temperature and ambient temperature, respectively, and a, b, and c are fitting parameters.

3.3. Generation of ZnO by Joule Heating

Thermogravimetric analysis (TGA) was utilized to investigate the calcination of the precursor and formation of ZnO by high-temperature treatment, which could be cataloged into two stages: the vaporization stage and the decomposition of zinc precursor stage [26]. Figure 6 shows the result of thermal analysis of zinc acetate dihydrate in ambient air. When temperature raising from 60 to 100 °C, water vaporization occurs, which could be correlated to the sharp decline of the salt weight of 15% in TGA result. There is a slight decay from 150 to 200 °C, which denotes the starting of decomposition process. From 200 °C, there is a considerable reduction of weight as the reaction is promoted by temperature. The weight is stable at approximately 23% after 370 °C, indicating that the thermo-decomposition has completed and most of zinc salt has been transformed to ZnO. Although this result recommends that temperature of above 370 °C is necessary to thoroughly decompose the zinc precursor, it also suggests that lower temperature still could partially form ZnO. Starting with zinc acetate dihydrate, the reaction is finished with most of the products of the process are volatile, such as water, acetone ($(CH_3)_2CO$), acetic acid (CH_2COOH), and carbon dioxide (CO_2) [27].

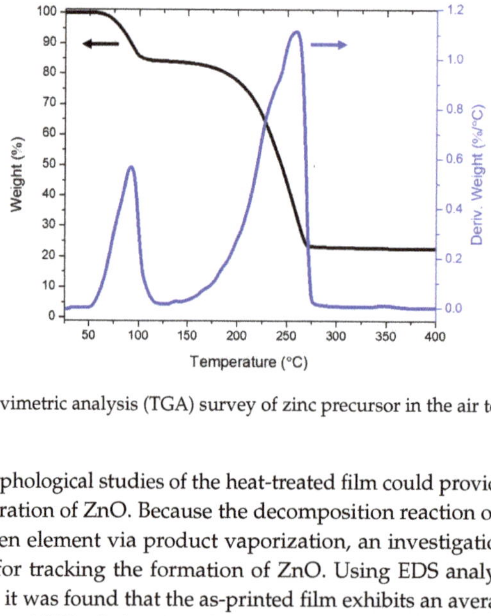

Figure 6. Thermogravimetric analysis (TGA) survey of zinc precursor in the air to study the formation of ZnO.

Elemental and morphological studies of the heat-treated film could provide more evidence for the mechanism of the generation of ZnO. Because the decomposition reaction of zinc acetate dihydrate shows the loss of oxygen element via product vaporization, an investigation of Zn:O atomic ratio would be meaningful for tracking the formation of ZnO. Using EDS analysis, which is shown in Table A1 (Appendix A), it was found that the as-printed film exhibits an average Zn:O atomic ratio of 0.378. In the case of using 4-W Joule heating, the atomic ratio of 0.588 could be observed. This Zn:O atomic ration further increases to 0.605 when 5-W heating power was applied. The rise of Zn:O atomic ratio is an important evidence of thermal decomposition reaction and the formation of ZnO by Joule heating. It could be seen that the Zn:O atomic ratio is not significantly different between 4- and 5-W heating power. Furthermore, there is a noticeable variation in this atomic ratio investigation through the pattern, which could be contributed to the grading of temperature through the surface.

Furthermore, morphological study depicts a remarkable change in film structure after Joule heating process (Figure 7). Figure 7a demonstrates the zinc precursor film before any heat treatment was applied. The film appears to be full of fractures, which could be the result of the solvent evaporation and the condensation of salt. There are remarkable changes in film morphology after the Joule heating treatment, such as the vanishing of notable fractures and the appearance of wrinkles on the surface of the film (Figure 7b,c). However, there is no significant difference between these two treated films. These film structures are commonly observed in the sol-gel derived film as our previous report [3]. These wrinkles could be originated from the internal stress of film during rapid solvent withdrawal, caused by the difference in thermal expansion of the gelation and underlying layer [28]. In addition, the transition from viscous to viscoelastic of the zinc precursor ink also contributes to the formation of these wrinkles [29].

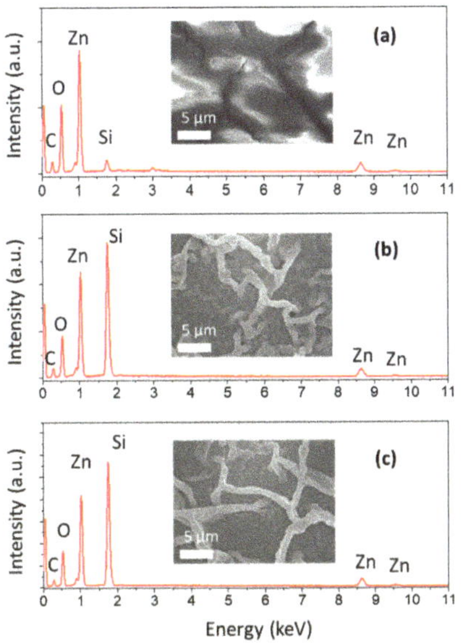

Figure 7. Energy-dispersive X-ray spectroscopy (EDS) analysis of zinc precursor film with different treatment conditions: (**a**) 0 W, (**b**) 4 W, (**c**) 5 W. Insets are FE–SEM images of film morphology according to each condition.

3.4. UV Light Sensing Performance of the Sensor

The sensing of UV light was demonstrated by the fabricated device and the result was presented in Figure 8. A bias voltage of 5 V was applied to the two terminals of device, which also serves as electrode by making metal contacts with the semiconductor film. The electrical current was recorded while the UV light was turned on/off (Figure 8a). It could be noted that the sample without applying heat treatment did not shows a notable response to UV light. On the other hand, sample with the treatment exhibits remarkable response to the short-wavelength light. When UV was turn on, the current raised significantly until it reached the equilibrium. On the other hand, when UV was turned off, the current quickly decay to the initial value. It is also worth noting that there is a remarkable difference of the samples with different heating conditions, such as the sample treated with 5-W power has photocurrent with a maximum value of 1.6×10^{-7} A, which is about ten times higher than that of the sample treated with 4-W power.

The responsivity could be calculated using following expression: [30]

$$R = \frac{J_{ph}}{PS} \quad (2)$$

where *Jph* is the photocurrent, *P* is the light intensity, and *S* is the effective area, which could be determined by the area of the ZnO pattern between two electrodes, such as 0.2 mm × 0.6 mm. Further analysis shows the responsivity values at 5.42 mW/cm^2 light intensity of the 5-W sample and the 4-W sample is 0.029 and 0.0027 A/W, respectively. Although the responsivity is lower than that has been reported in other inkjet-printed ZnO-based UV photodetector [31], it could be due to the obtained temperature is lower than the point where the reaction is totally finished. In Figure 8b, the I–V characteristics of prepared devices were presented, which show a linear current–voltage relationship for both of the devices with treatment. This behavior indeed shows the Ohmic contact nature of the silver–ZnO interface.

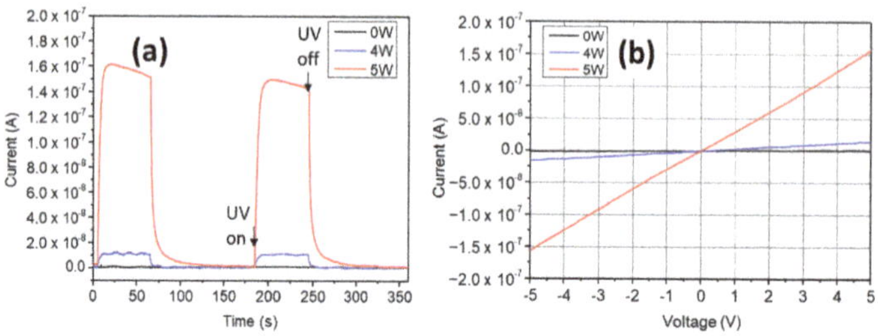

Figure 8. Photo-sensing properties of the fabricated device. (**a**) Sequential illumination of the sensor by UV light and obtained photocurrent. (**b**) *I*–*V* curve of the devices under UV illumination.

4. Discussion

The photosensitivity of the device is strongly influenced by the heat treatment condition. As the discussion in previous section has pointed out, temperature of higher than 150 °C is critical to facilitate the reaction to form ZnO from the zinc salt. It is worth noting that photosensitivity is a common characteristic of a semiconductor. ZnO is a wide bandgap semiconductor (Eg = 3.35 eV at room temperature [32]). Therefore, it only be sensitive with larger photon energy of the incident light, such as the UV light used in this work which has photon energy of 3.40 eV according to 365 nm central wavelength of the LED.

The sample without heat treatment could not form ZnO by the thermal decomposition process. Therefore, the prepared film failed to work as a photodetector. Meanwhile, the sample processed with 4-W power demonstrates a remarkable response to UV illumination. It is because the temperature generated by the Joule heating process has facilitated the formation of ZnO semiconductor. As a wide bandgap semiconductor, the interaction of ZnO crystalline with high energy photons excites the generation of the electron-hole pair which increases the carrier concentration in the lattice [33]. Therefore, film resistance reduces and causes a surge in current running through the device.

Furthermore, the sample processed with 5-W power possesses a better performance in terms of responsivity to UV light. This improvement indicates that the magnitude of input power for Joule heating of semiconductor film could significantly influence the film properties. The root of this enhancement could be originated from the fact that higher temperatures could promote more formation of ZnO as previous discussion, therefore the interaction with photon is improved and more electron-hole pairs could be generated when exposing to the UV illumination. In addition, higher temperatures could also improve the contact between ZnO nanoparticles, therefore reduce the band bending at the interfaces and promote the transportation of electron through those grain boundaries [30].

Current work employed silicon wafer due to its excellent heat resistance, because of the temperature during Joule heating could excess 250 °C, which most of polymer will not be able to withstand. One possible alternative solution is using high temperature polymer such as polyimide. However, due to the low thermal conductivity of polyimide comparing that of silicon wafer [34,35], the structure of device must be changed to reduce the distance from heater to the printed zinc salt in order to obtain a sufficient calcination.

5. Conclusions

In this work, on-substrate heating synthesis of ZnO thin film by the printed silver resistive heater is proposed. Both conductive patterns and the precursor pattern were printed on only one inkjet printer with an exchange of cartridge for each material. Electrical current running through the silver conductive pattern generates heat, which is utilized to facilitate the thermal decomposition of printed zinc precursor film to form ZnO. The magnitude of supplied power for heating process has significant influence on the film formation via the achieved temperature. Therefore, it is found that higher power could produce a better property of printed ZnO semiconductor film. Despite the humble performance of printed photodetector, this work demonstrates a promising approach to additively manufacture electronic devices, which reduces the number of equipment involved and the amount of energy consumed. Hence, integrated 3D printing might also be possible.

Author Contributions: Conceptualization, V.-T.T. and H.D.; methodology, V.-T.T. and Y.W.; formal analysis, V.Y.T.; investigation, V.-Y.T. and Y.W.; writing—original draft preparation, V.-T.T.; writing—review and editing, H.D.; visualization, V.-T.T.; supervision, H.D.; project administration, H.D. All authors have read and agreed to the published version of the manuscript.

Funding: This research was funded by Ministry of Education Academic Research Fund, Singapore.

Acknowledgments: We appreciate the equipment support from Singapore Centre for 3D Printing (SC3DP).

Conflicts of Interest: The authors declare no conflict of interest.

Appendix A

Table A1. EDS Investigation of Zn:O atomic ration at random positions on printed pattern.

	Zn:O Atomic Ratio							
Location	1	2	3	4	5	6	Average	Standard Deviation
0 W	0.430	0.431	0.422	0.390	0.292	0.305	0.378	0.058
4 W	0.536	0.545	0.504	0.705	0.681	0.557	0.588	0.076
5 W	0.553	0.567	0.517	0.691	0.651	0.648	0.605	0.062

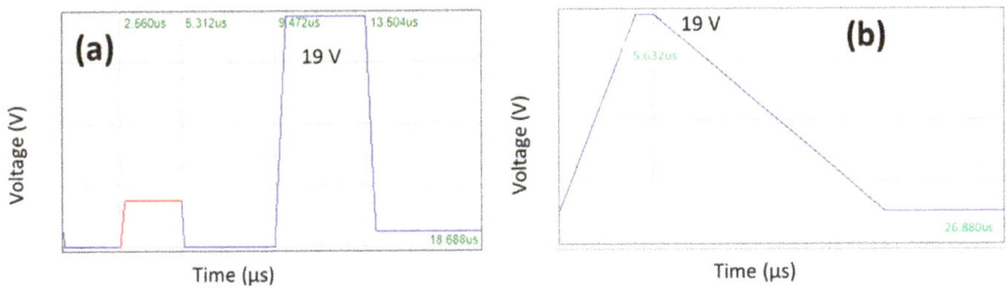

Figure A1. Waveform of signal applying to cartridge nozzles to create jetting of ink: (**a**) silver ink and (**b**) zinc salt ink.

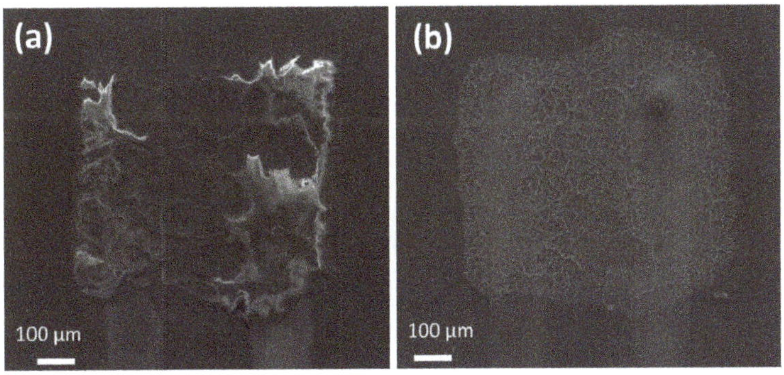

Figure A2. Low-magnification FE–SEM images of two samples before (**a**) and after heat treatment (**b**).

References

1. Zhan, Z.; An, J.; Wei, Y.; Tran, V.T.; Du, H. Inkjet-printed optoelectronics. *Nanoscale* **2017**, *9*, 965–993. [CrossRef]
2. Nayak, L.; Mohanty, S.; Nayak, S.K.; Ramadoss, A. A review on inkjet printing of nanoparticle inks for flexible electronics. *J. Mater. Chem. C* **2019**, *7*, 8771–8795. [CrossRef]
3. Tran, V.-T.; Wei, Y.; Yang, H.; Zhan, Z.; Du, H. All-inkjet-printed flexible ZnO micro photodetector for a wearable UV monitoring device. *Nanotechnology* **2017**, *28*, 095204. [CrossRef]
4. Yus, J.; Gonzalez, Z.; Sanchez-Herencia, A.J.; Sangiorgi, A.; Sangiorgi, N.; Gardini, D.; Sanson, A.; Galassi, C.; Caballero, A.; Morales, J.; et al. Semiconductor water-based inks: Miniaturized NiO pseudocapacitor electrodes by inkjet printing. *J. Eur. Ceram. Soc.* **2019**, *39*, 2908–2914. [CrossRef]
5. Sanchez, J.G.; Balderrama, V.S.; Garduno, S.I.; Osorio, E.; Viterisi, A.; Estrada, M.; Ferre-Borrull, J.; Pallares, J.; Marsal, L.F. Impact of inkjet printed ZnO electron transport layer on the characteristics of polymer solar cells. *RSC Adv.* **2018**, *8*, 13094–13102. [CrossRef]
6. Ganesan, S.; Gollu, S.R.; Alam khan, J.; Kushwaha, A.; Gupta, D. Inkjet printing of zinc oxide and P3HT:ICBA in ambient conditions for inverted bulk heterojunction solar cells. *Opt. Mater.* **2019**, *94*, 430–435. [CrossRef]
7. Hasan, K.u.; Nur, O.; Willander, M. Screen printed ZnO ultraviolet photoconductive sensor on pencil drawn circuitry over paper. *Appl. Phys. Lett.* **2012**, *100*, 211104. [CrossRef]
8. Liang, Y.N.; Lok, B.K.; Wang, L.; Feng, C.; Lu, A.C.W.; Mei, T.; Hu, X. Effects of the morphology of inkjet printed zinc oxide (ZnO) on thin film transistor performance and seeded ZnO nanorod growth. *Thin Solid Films* **2013**, *544*, 509–514. [CrossRef]
9. Jiang, L.; Li, J.; Huang, K.; Li, S.; Wang, Q.; Sun, Z.; Mei, T.; Wang, J.; Zhang, L.; Wang, N.; et al. Low-Temperature and Solution-Processable Zinc Oxide Transistors for Transparent Electronics. *ACS Omega* **2017**, *2*, 8990–8996. [CrossRef]
10. Friedmann, D.; Lee, A.F.; Wilson, K.; Jalili, R.; Caruso, R.A. Printing approaches to inorganic semiconductor photocatalyst fabrication. *J. Mater. Chem. A* **2019**, *7*, 10858–10878. [CrossRef]
11. Gagnon, J.C.; Presley, M.; Le, N.Q.; Montalbano, T.J.; Storck, S. A pathway to compound semiconductor additive manufacturing. *MRS Commun.* **2019**, *9*, 1001–1007. [CrossRef]
12. Chae, D.; Kim, J.; Shin, J.; Lee, W.H.; Ko, S. A low-temperature and short-annealing process for metal oxide thin film transistors using deep ultraviolet light for roll-to-roll processing. *Curr. Appl. Phys.* **2019**, *19*, 954–960. [CrossRef]
13. Jeon, J.B.; Kim, B.J.; Bang, G.J.; Kim, M.-C.; Lee, D.G.; Lee, J.M.; Lee, M.; Han, H.S.; Boschloo, G.; Lee, S.; et al. Photo-annealed amorphous titanium oxide for perovskite solar cells. *Nanoscale* **2019**, *11*, 19488–19496. [CrossRef]
14. Palneedi, H.; Park, J.H.; Maurya, D.; Peddigari, M.; Hwang, G.-T.; Annapureddy, V.; Kim, J.-W.; Choi, J.-J.; Hahn, B.-D.; Priya, S.; et al. Laser Irradiation of Metal Oxide Films and Nanostructures: Applications and Advances. *Adv. Mater.* **2018**, *30*, 1705148. [CrossRef]
15. Sharma, S.; Pande, S.S.; Swaminathan, P. Top-down synthesis of zinc oxide based inks for inkjet printing. *RSC Adv.* **2017**, *7*, 39411–39419. [CrossRef]
16. Rao, A.; Long, H.; Harley-Trochimczyk, A.; Pham, T.; Zettl, A.; Carraro, C.; Maboudian, R. In Situ Localized Growth of Ordered Metal Oxide Hollow Sphere Array on Microheater Platform for Sensitive, Ultra-Fast Gas Sensing. *ACS Appl. Mater. Interfaces* **2017**, *9*, 2634–2641. [CrossRef]
17. Yan, H.; Wu, H. Joule Heating and Chip Materials. In *Encyclopedia of Microfluidics and Nanofluidics*; Li, D., Ed.; Springer: Boston, MA, USA, 2013.
18. Nguyen, H.; Quy, C.T.; Hoa, N.D.; Lam, N.T.; Duy, N.V.; Quang, V.V.; Hieu, N.V. Controllable growth of ZnO nanowires grown on discrete islands of Au catalyst for realization of planar-type micro gas sensors. *Sens. Actuators Chem.* **2014**, *193*, 888–894. [CrossRef]
19. Long, H.; Turner, S.; Yan, A.; Xu, H.; Jang, M.; Carraro, C.; Maboudian, R.; Zettl, A. Plasma assisted formation of 3D highly porous nanostructured metal oxide network on microheater platform for Low power gas sensing. *Sens. Actuators Chem.* **2019**, *301*, 127067. [CrossRef]
20. Dai, C.-L. A capacitive humidity sensor integrated with micro heater and ring oscillator circuit fabricated by CMOS–MEMS technique. *Sens. Actuators Chem.* **2007**, *122*, 375–380. [CrossRef]

21. Nerushev, O.A.; Ek-Weis, J.; Campbell, E.E.B. In situ studies of growth of carbon nanotubes on a local metal microheater. *Nanotechnology* **2015**, *26*, 505601. [CrossRef]
22. Kwon, J.; Hong, S.; Kim, G.; Suh, Y.D.; Lee, H.; Choo, S.-Y.; Lee, D.; Kong, H.; Yeo, J.; Ko, S.H. Digitally patterned resistive micro heater as a platform for zinc oxide nanowire based micro sensor. *Appl. Surf. Sci.* **2018**, *447*, 1–7. [CrossRef]
23. Tran, V.-T.; Wei, Y.; Liau, W.; Yang, H.; Du, H. Preparing of Interdigitated Microelectrode Arrays for AC Electrokinetic Devices Using Inkjet Printing of Silver Nanoparticles Ink. *Micromachines* **2017**, *8*, 106. [CrossRef]
24. Ravindra, N.M.; Abedrabbo, S.; Wei, C.; Tong, F.M.; Nanda, A.K.; Speranza, A.C. Temperature-dependent emissivity of silicon-related materials and structures. *IEEE Tran. Semicond. Manuf.* **1998**, *11*, 30–39. [CrossRef]
25. Lee, S.M.; Dyer, D.C.; Gardner, J.W. Design and optimisation of a high-temperature silicon micro-hotplate for nanoporous palladium pellistors. *Microelectron. J.* **2003**, *34*, 115–126. [CrossRef]
26. Paraguay D, F.; Estrada L, W.; Acosta N, D.R.; Andrade, E.; Miki-Yoshida, M. Growth, structure and optical characterization of high quality ZnO thin films obtained by spray pyrolysis. *Thin Solid Films* **1999**, *350*, 192–202. [CrossRef]
27. Lin, C.-C.; Li, Y.-Y. Synthesis of ZnO nanowires by thermal decomposition of zinc acetate dihydrate. *Mater. Chem. Physi.* **2009**, *113*, 334–337. [CrossRef]
28. Kwon, S.J.; Park, J.-H.; Park, J.-G. Wrinkling of a sol-gel-derived thin film. *Phys. Rev. E* **2005**, *71*, 011604. [CrossRef]
29. Justin Raj, C.; Karthick, S.N.; Hemalatha, K.V.; Kim, S.-K.; Kim, B.C.; Yu, K.-H.; Kim, H.-J. Synthesis of self-light-scattering wrinkle structured ZnO photoanode by sol–gel method for dye-sensitized solar cells. *Appl. Phys. A* **2014**, *116*, 811–816. [CrossRef]
30. Liu, X.; Gu, L.; Zhang, Q.; Wu, J.; Long, Y.; Fan, Z. All-printable band-edge modulated ZnO nanowire photodetectors with ultra-high detectivity. *Nat. Commun.* **2014**, *5*. [CrossRef]
31. Dong, Y.; Zou, Y.; Song, J.; Li, J.; Han, B.; Shan, Q.; Xu, L.; Xue, J.; Zeng, H. An all-inkjet-printed flexible UV photodetector. *Nanoscale* **2017**, *9*, 8580–8585. [CrossRef]
32. Monroy, E.; Omnès, F.; Calle, F. Wide-bandgap semiconductor ultraviolet photodetectors. *Semicond. Sci. Technol.* **2003**, *18*, R33. [CrossRef]
33. Petritz, R.L. Theory of Photoconductivity in Semiconductor Films. *Phys. Rev.* **1956**, *104*, 1508–1516. [CrossRef]
34. Benford, D.J.; Powers, T.J.; Moseley, S.H. Thermal conductivity of Kapton tape. *Cryogenics* **1999**, *39*, 93–95. [CrossRef]
35. Hopkins, P.E.; Reinke, C.M.; Su, M.F.; Olsson, R.H.; Shaner, E.A.; Leseman, Z.C.; Serrano, J.R.; Phinney, L.M.; El-Kady, I. Reduction in the Thermal Conductivity of Single Crystalline Silicon by Phononic Crystal Patterning. *Nano Lett.* **2011**, *11*, 107–112. [CrossRef]

micromachines

Perspective

3D Printed MEMS Technology—Recent Developments and Applications

Tomasz Blachowicz [1] and Andrea Ehrmann [2,*]

[1] Institute of Physics-Center for Science and Education, Silesian University of Technology, 44-100 Gliwice, Poland; tomasz.blachowicz@polsl.pl
[2] Faculty of Engineering Sciences and Mathematics, Bielefeld University of Applied Sciences, 33619 Bielefeld, Germany
* Correspondence: andrea.ehrmann@fh-bielefeld.de

Received: 31 March 2020; Accepted: 18 April 2020; Published: 20 April 2020

Abstract: Microelectromechanical systems (MEMS) are of high interest for recent electronic applications. Their applications range from medicine to measurement technology, from microfluidics to the Internet of Things (IoT). In many cases, MEMS elements serve as sensors or actuators, e.g., in recent mobile phones, but also in future autonomously driving cars. Most MEMS elements are based on silicon, which is not deformed plastically under a load, as opposed to metals. While highly sophisticated solutions were already found for diverse MEMS sensors, actuators, and other elements, MEMS fabrication is less standardized than pure microelectronics, which sometimes blocks new ideas. One of the possibilities to overcome this problem may be the 3D printing approach. While most 3D printing technologies do not offer sufficient resolution for MEMS production, and many of the common 3D printing materials cannot be used for this application, there are still niches in which the 3D printing of MEMS enables producing new structures and thus creating elements for new applications, or the faster and less expensive production of common systems. Here, we give an overview of the most recent developments and applications in 3D printing of MEMS.

Keywords: 3D printing; microelectromechanical systems (MEMS); microelectronics; microfluidics; microsensors; microactuators

1. Introduction

Microelectromechanical systems (MEMS) are miniaturized devices combining electric and mechanical functions. Typical MEMS are, e.g., pressure and gyro sensors, accelerometers, or ink jet heads [1]. MEMS devices are often based on silicon (Si) [2,3]. Nevertheless, in the last decades, polymers were used in MEMS as well, e.g., polydimethylsiloxane (PDMS) for microfluidic devices [4], parylene for valves and sensors [5,6], or epoxy for micromanipulators [7].

Three-dimensional (3D) printing, on the other hand, works typically with polymers or metals and is recently not only applied for rapid prototyping, but also for the rapid production of individual parts or objects that could not be produced with other technologies [8,9]. Typical polymers used in the Fused Deposition Modeling (FDM) technology are acrylonitrile butadiene styrene (ABS), poly(lactic acid) (PLA), polyamide ("nylon"), or polycarbonate [10], while other technologies allow for printing different polymers. The disadvantages of most 3D printing technologies are relatively low mechanical properties, as compared to objects prepared from other technologies [11], which is why sometimes combinations of 3D printed parts with differently prepared objects are suggested [12,13].

As visible in Figure 1, both aforementioned technologies have emerged during the last decades (Figure 1a). MEMS have been reported in the scientific literature since 1980, while it took 10 more years until scientific research in 3D printing started and another decade until the first study on 3D printing of MEMS was reported. In spite of the additional degrees of freedom offered by 3D printing, on average, less than 1% of the studies dealing with 3D printing concentrate on MEMS. This is in contrast to the possible advantages of 3D printing in MEMS production, especially related to avoiding problems with an undesired underetching of 3D structures related to misalignments of the anisotropic etch pattern [14–16]. It can be expected that 3D printing methods allow for tailoring 3D shapes in the desired way, making the structures more reliable when the process parameters are properly adjusted.

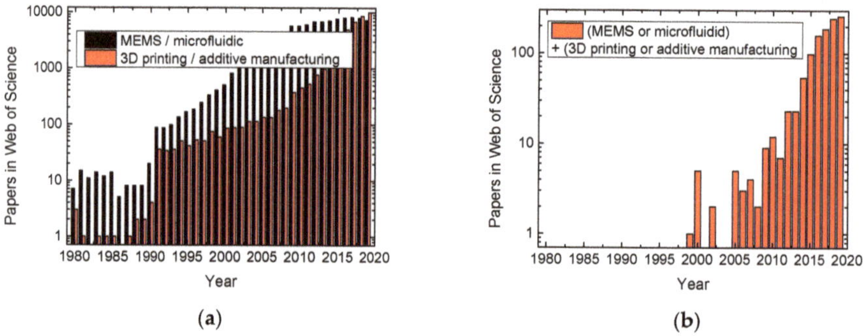

Figure 1. Number of papers found in the Web of Science, dealing with (**a**) microelectromechanical systems (MEMS) or microfluidic systems, and 3D printing or additive manufacturing, respectively; (**b**) the combination of MEMS or microfluidic and 3D printing or additive manufacturing. Data from the Web of Science, accessed on 31/03/2020.

On the other hand, 3D printing processes cause new challenges. The feature sizes that can be reliably reached depend on many parameters, such as the mechanical properties of the printer, the accuracy of stepper motors, but also on the technology used and the material applied for printing. Recently, freestanding polycrystalline copper pillars with diameters between 250 nm and 1.5 µm were reported as result of a 3D electrochemical deposition technique [17], as well as the direct laser writing of 3D submicron pillars [18], while common 3D printing techniques often are limited to the range of 100 µm resolution, combined with roughness or waviness in the order of several 10 µm, which necessitates chemical or heat-based post-treatment [19]. Table 1 gives an overview of typical resolutions reported for different 3D printing technologies. These typical minimum feature widths and printing speeds that can be reached depend on the chosen 3D printing technologies [20].

Table 1. Resolutions of different 3D printing technologies, reported in the literature, sorted from larger to smaller minimum feature sizes.

Technology	Min. Feature Size	Material	Ref.
Selective laser sintering	<400 µm	Div. Polymers	[21,22]
Fused deposition modeling	200 µm	Diverse polymers	[23]
Robot dispensing	200 µm	Hydrogels	[24]
Stereolithography	30–70 µm	Photosensitive polymers	[25]
3D inkjet printing	28 µm	Photoresist	[24]
Resonant direct laser writing	1–4 µm	IP-Dip photoresist	[26]
Multiphoton absorption polymerization	1 µm	SU8 photoresist	[27]
Two-photon polymerization	0.28–1.5 µm	Photoresists	[28]
Direct laser writing	0.085–1.5 µm	Photoresists	[29]

Another problem is the well-known thermal shrinkage occurring in polymer-based 3D printing processes [30,31] as well as after sintering green ceramics or during printing metal objects [32,33].

Here, we give an overview of recent research on 3D-printed MEMS, possible applications of such systems, as well as advantages and challenges connected with this combination of two modern technologies. Before, the most important 3D printing techniques are explained in brief.

2. Typical 3D Printing Techniques

Photolithography can be used by preparing a photomask, a glass plate, or plastic film coated with a non-UV-transparent film. The mask is placed on a photoresist on a silicon wafer, and this photoresist is not exposed to UV light through the open areas of the mask. Depending on the photoresist, either the exposed or the non-exposed areas form the pattern, while the residual photoresist is washed of the wafer, in this way building a master mold [34].

While this technique is well established to prepare three-dimensional objects, it is not necessarily meant when "3D printing" is mentioned. Instead, stereolithography is more often connected with the term "3D printing". This technique is used to print light-sensitive materials from a polymer solution by using a laser to cross-link the polymer at desired positions. It can even be applied for the bioprinting of hydrogels [25].

An even more sophisticated way of using light to cross-link a UV-sensitive material is the two-photon or multi-photon polymerization [27,28]. These techniques need a tightly focused laser beam on a defined volume of the photosensitive polymer, enabling the material to absorb two or more photons simultaneously to reach an excited state. As usual in such second-order processes, its strength is proportional to the squared light intensity. This is why only in the very small focus volume of the laser can the process take place, opposite to the usual single-photon absorption used in stereolithography.

Selective laser sintering again uses a laser to build a 3D object layer by layer, but opposite to the aforementioned stereolithography not to photocure a resin, but rather to fuse small particles of thermoplastic powders, metals, or ceramics [21,22].

Inkjet printing belongs to the techniques that allow for printing biological material such as bioink or hydrogels [24]. Small ink droplets are dispersed on a substrate. Since typical inks have low viscosities; only relatively small 3D structures can be printed, typically in the order of magnitude of some 10 micrometers.

Fused deposition modeling (FDM) belongs to the well-known 3D printing methods since these printers are often available at low cost and are easy to use, but these basic versions are not very accurate. Generally, in this technique, a polymer is molten and pressed through a nozzle to be deposited along defined paths so that 3D structures are formed layer by layer [23].

Besides these often used technologies, there are several others, which are in most cases based on a layer-by-layer setup of a 3D structure. While typical minimum feature sizes are given in Table 1 for several 3D printing techniques, the maximum feature sizes depend on several parameters. Most techniques, besides photolithography, enable in principle printing objects of some centimeters to some 10 cm. The limiting factor is often the time, which depends not only on the technique but also on the desired printing quality or resolution, the chosen printing material, whether a newer or older printer is used, etc. On the other hand, most technologies need a printing bed or have similar size restrictions that cannot be overcome for a specific printer model. Thus, their values depend on too many parameters to be given generally for a certain technique.

3. 3D Printed Microfluidic Devices

The idea of 3D printing different parts of or full microfluidic devices was firstly mentioned in 2013. Leary et al. developed nanoparticles as contrast agents and investigated the possibility of moving them in MEMS structures used as "organs-on-a-chip", including microchannels. In these MEMS channels, human cancer cells and normal cells were grown, and the detectability of small tumors by applying the superparamagnetic nanoparticles was investigated. Here, the authors suggested 3D printing of more complex 3D structures to overcome the limits of the 2D MEMS channels [35].

In 2014, Lifton et al. investigated possibilities to use 3D printing in MEMS technology and concluded that microfluidics and "labs-on-a-chip" would be the most suitable devices to be 3D printed due to the relatively large minimum feature size of 3D printing in the range of 50–500 μm. They compared stereolithography, micro-stereolithography, PolyJet, selective laser sintering, fused deposition modeling, and two less known technologies and suggested concentrating on enhancing the printing resolution toward a range of 1–10 μm to enable the utilization of 3D printing for a broader range of possible MEMS devices [36].

The next sub-sections will give some general remarks on the possible toxicity of the materials used for 3D printing and describe more recent developments in 3D-printed microfluidics, sorted by the respective printing technologies.

3.1. *General Remarks on Toxicity of Materials for Biotechnological Applications*

Concerns regarding the possible toxicity of typical 3D printing polymers used in the FDM technology, but also in stereolithography (SLA) and multi-jet modeling (MJM) were raised in 2015. The authors suggested for biotechnological applications, such as lab-on-a-chip, that polymers should be carefully selected to avoid erroneous results due to not taking into account a possibly reduced biocompatibility of the device [37].

This problem was also mentioned by other authors. Beckwith et al. underlined not only the high resolution, but also the noncytotoxicity of their 3D-printed microfluidic device prepared by SLA, which was used to investigate tumor fragments from biopsy samples [38]. SLA and material-jetting processes were used to prepare molds and afterwards positive replicas by soft lighography and poly(dimethylsiloxane) (PDMS) molding. Using zebrafish for the biotoxicity test, Fuad et al. found no toxicity in replicas from both 3D printing processes [39].

Conductive PLA, including graphite, was found to be biocompatible [40], while Zhu et al. had already shown the same for pure PLA [36]. On the other hand, for typical 3D printing materials, especially commercially available stereolithographic resins, toxic substances were found in diverse photoinitiators, photopolymers, and other compounds used in SLA resins as well as high growth inhibition or mortality in biocompatibility assessments of objects printed with different resins [41].

3.2. *Photolithography*

de Araujo et al. prepared microfluidic devices to create microbubbles, using 3D printing by an OBJET EDEN 250 printer which enables the printing of round channels with a diameter of 0.3 mm. The waxy support used along the channels had to be dissolved and mechanically removed from the channels after printing. In this way, it was possible to prepare microbubbles of very homogenous dimensions, with a standard deviation below 1% [34,42].

For a more often used microfluidic technology, the isothermal titration calorimetry (ITC), Jia et al. used 3D-printed microfluidic devices. Their microdevice combined the 3D-printed microfluidic structure with a polymer MEMS thermoelectric sensor to allow for ITC measurements of biomolecular interactions. Again, the 3D-printed device was found to work leak-free, in addition to the possibility of using non-permeable material and the geometric flexibility [43].

3.3. *Two-Photon and Multi-Photon Polymerization*

To prepare much smaller features, the so-called two-photon polymerization technique can be used, which enables preparing structures of dimensions below the diffraction limit, partly below 100 nm [44–46]. For this technique, usually ultrashort laser pulses are applied.

The polymer material used for this 3D printing process can be, e.g., chemically modified zirconium-based sol–gel composites [44], photoinitiators with highly toxic properties or modified ones with reduced or vanishing toxicity, such as riboflavin and triethanolamine [45], and a broad range of other materials for applications in optics, photonics, electronics, or biotechnology [46]. This technique was also suggested in different patents dealing with microfluidics [47].

3.4. *Inkjet 3D Printing*

An unusual application of microfluidics was reported by Walczak et al. [48]. They used the inkjet 3D printing of a microfluidic device with embedded force sensors to monitor the seed growth and axial growth forces of *Lepidum sativum* (garden cress) seeds, which are used in diverse plant growth investigations [49] and found growth forces around 50 mN for the root and 500 mN for the stalk [48].

3.5. Metal Additive Manufacturing

Huang et al. searched a solution for another problem. Since silicon-based microfluidic devices are often problematic at high temperatures where fluidic interconnects may be leaking, they used binder jet printing, which is a form of metal additive manufacturing. While such metal objects are usually highly porous and thus not suitable for application in microfluidics, the addition of boron nitride supported the sintering process and thus reduced the porosity, resulting in leak-free fitting even at high operating pressures [50].

3.6. Preparing Molds by 3D Printing Methods

Dinh et al. used 3D printing in combination with molding and plasma-assisted bonding to prepare a highly effective microfluidic active heating and cooling module, which was suggested for the large production of SiC power nanoelectronics [51].

A Venturi flowmeter was 3D printed by Adamski et al., combining a commercially available MEMS sensor with a 3D-printed microfluidic structure. Here, microchannel widths of 400–800 μm were reached [52].

Villegas et al. used PDMS molding, similar to Fuad et al. mentioned above [38], but they concentrated on reducing the surface roughness of the PDMS microfluidic channels from 2 to 0.2 μm by coating the mold with a fluorinated-silane and afterwards tethering an omniphobic lubricant to this adhesion layer [53].

3.7. Combining 3D Printing with Other Technologies

Cesewski et al. went one step further and combined a common 3D printing process with robotic handling in the form of an integrated pick-and-place functionality (Figure 2). In this way, they could assemble 3D printed forms with piezoelectric chips, which were used to produce multiple resonant modes to generate bulk acoustic waves, which could again be used to manipulate suspended particles [54].

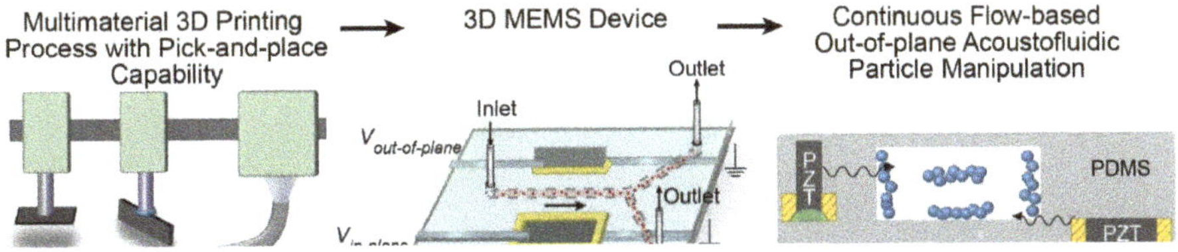

Figure 2. Combination of 3D printing and pick-and-place functionality to produce 3D MEMS devices used for acoustofluidic particle manipulation. Reproduced with permission from [54]. Copyright © The Royal Society of Chemistry 2018.

Another combination of different technologies was suggested by Tamura and Suzuki, using 3D printing and photolithography to prepare the mold of a microfluidic device with micro- and millimeter structures. They showed that the process combination resulted in higher patterning accuracy than pure 3D printing [55].

The common MEMS fabrication technology of a boiler was combined with capillary channels with the 3D printing of a superheater. While the boiler alone could capture 2/3 of the incoming thermal energy, the addition of the superheater increased this value by 10%, in this way increasing thermal energy scavenging [56].

4. 3D-Printed Microelectromechanical Systems (MEMS) Sensors

The idea to use a 3D printing technology, in this case patterned 3D microstructures produced by an electrodepositable photo resist, in MEMS sensors was already mentioned in the scientific literature in 2007 [57]. For the special case of freeform helical microstructures, Farahani et al. investigated in 2014 which 3D printing methods were most suitable and found a high performance of UV-assisted and solvent-cast 3D printing methods, however, with a limited range of usable materials [58].

However, afterwards, it took some time until the first 3D printed MEMS sensors were reported in the literature. Some important chemical and physical sensors are described here.

4.1. Chemical Sensors

3D printing was used to produce a resonant gas cell, which was combined with commercially available fiber optics and a MEMS microphone sensor, in this way creating a miniaturized photoacoustic trace gas sensor [59]. Here, the sensor itself was not 3D printed, but combined with a 3D printed gas cell.

A breath analyzer, measuring CO_2 and oxygen in the exhaled breath of humans, was partly 3D printed. Equipped with a thermopile detector and a MEMS infrared emitter, even low CO_2 concentrations could be detected, and results similar to commercially available breath-by-breath sensors were achieved [60].

Ilke et al., on the other hand, integrated MEMS or electret microphones into 3D printed photo-acoustic gas sensors for the mid-infrared. These sensors could be used to detect methane with detection limits of 90 ppm for the electret microphone and 182 ppm for the MEMS microphone, applying similar integration times around 260 s [61].

4.2. Physical Sensors

Similarly, Valyrakis et al. used 3D printing to prepare a waterproof hollow spherical particle in which a three-axial gyroscope and accelerometer were embedded to measure hydrodynamic forces on a coarse particle during entraining from the riverbed, aiming at understanding sediment transport [62]. Shen et al. also used 3D printing to prepare a fixture for a piezoresistive MEMS flow rate sensor. Both were bonded and integrated on an intravenous tube, allowing for measuring the flow rate in this tube [63]. Similarly, Raoufi et al. developed a MEMS flow sensor embedded in a 3D printed semicircular channel, aiming to enhance current artificial vestibular systems [64].

A 3D-printed housing for an airborne nanoparticle concentration sensor was prepared using a MEMS-based particle growth chip that grows nanoparticles to micro-sized droplets by condensation, in combination with a miniaturized optical particle counter based on a light-scattering method. This system was able to detect nanoparticles of 12.4 nm diameter and showed only small deviations from a reference instrument in high and low concentration environments [65].

A miniaturized transmission electron microscope (TEM) was suggested by Krysztof et al. They prepared a 3D-printed polymer holder for mounting the field emission cathode and electron optics columns and showed effective emission and focusing of the electron beam inside a vacuum chamber as the first step toward a MEMS-TEM [66].

Das et al. used 3D printing in the form of electrohydrodynamic ink jetting to produce PEDOT:PSS strain gauge sensors on kapton and silicone substrates. Comparing COMSOL simulations with the experimental results showed sufficient agreement. The authors suggested using such a system as future piezo-resistive robot skin [67]. However, in a newer work, a novel wet lift-off photolithographic technique was proposed for patterning the poly(3,4-ethylenedioxythiophene) polystyrene sulfonate (PEDOT:PSS) base layer due to the partly problematic electrohydrodynamic jetting technology [68]. Direct ink writing was suggested by Yang et al. instead to pattern 3D conductive circuits on flexible substrates [69].

3D printing was also used to prepare spiral-shaped acoustic resonators, as inspired by the human ear, serving as frequency-selective MEMS microphones. Using these MEMS sensors, they could strongly increase the frequency sensitivity of the investigated resonance frequency of 430 Hz [70].

Tiller et al. used the 3D printing process of digital light processing to prepare a piezoelectric acoustic sensor from piezoelectric and conductive parts with mechanically sensitive membranes of thicknesses as low as 35 μm, allowing for tunable resonant frequencies [71].

Capacitive MEMS vibration sensors were produced by a high-current plasma focused ion beam (FIB) technique and compared with sensors prepared by the common lithography process. While the resonance frequency differed by only 4%, the fabrication time could be reduced by approximately 80% [72].

Park et al. integrated a MEMS pressure sensor based on an inductor–capacitor resonant circuit into metallic and polymer 3D-printed stents for blood vessels. As shown in Figure 3, the sensitivity of the pressure sensor in the polymer stent was much higher than that of the metal stent system, allowing for real-time monitoring [73].

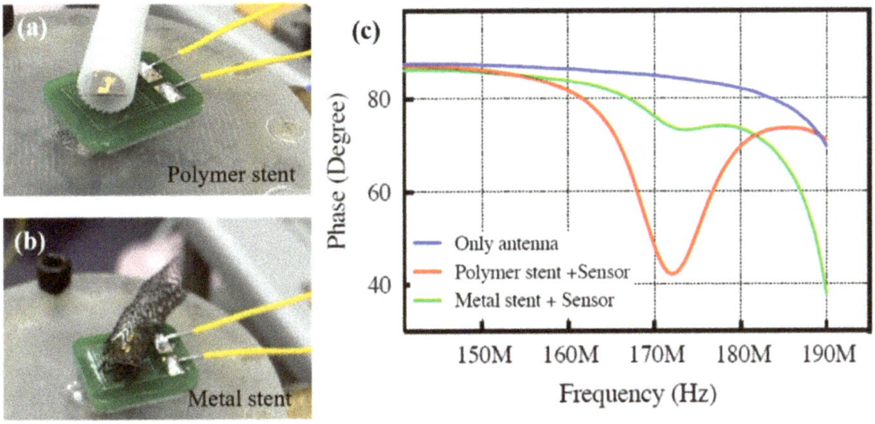

Figure 3. Wireless pressure sensors integrated in (**a**) polymer and (**b**) metal stents; (**c**) sensitivity analysis of these combinations. Reproduced with permission from [73]. Copyright © Elsevier 2019.

Recently, Wang et al. developed a 3D-printed MEMS device, using projection micro-stereolithography, which could be used for in situ tensile tests of micro- or nanowires. This device was used to determine the tensile behavior of SiC nanowires inside a scanning electron microscope (SEM) as well as lead zirconate titanate microwires under an optical microscope, in this way demonstrating micro- and nanomechanical characterization [74].

5. 3D-Printed MEMS Actuators

Identical to sensors, 3D-printed MEMS actuators were firstly suggested in 2007 [57]. Again, the first scientific reports on experimental investigations of such actuators were published several years later. Here, some of the proposed 3D-printed MEMS actuators are depicted, which were sorted according to their applications.

5.1. Switches

A 3D-printed MEMS switch was developed by Lee et al. Using the FDM printing of conductive PLA and poly(vinyl alcohol) (PVA) as water-soluble support, a switch with good electromechanical properties, abrupt switching, and a very high on/off current ratio above 10^6 was realized. The function and printing process are depicted in Figure 4 [75].

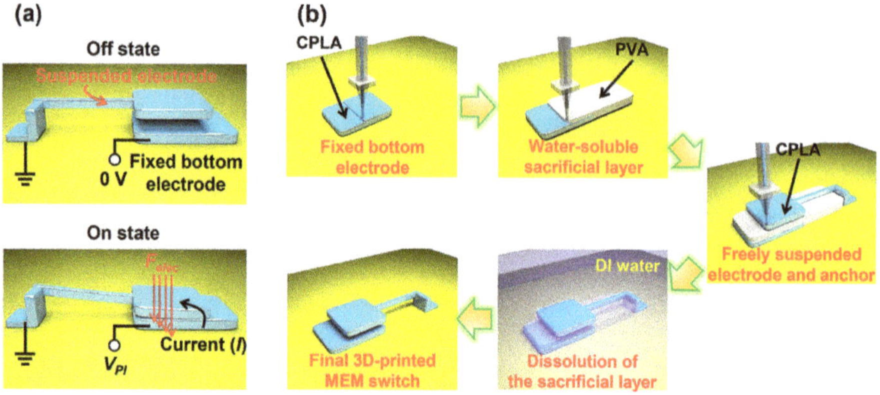

Figure 4. (**a**) Concept of 3D-printed MEMS switch, showing the pull-in phenomenon due to electrostatic forces on the suspended electrode when a voltage is applied to the fixed bottom electrode; (**b**) 3D printing process. Reproduced with permission from [75]. Copyright (2018) American Chemical Society.

5.2. Vibration Actuator

Xie et al. used 3D printing for vibrational tactile actuators for blind people. They combined piezoelectric extensional actuators vibrating in-plane with a scissor mechanism, forming a triangle between two in-plane points and one point above the plane. The oscillations of the two in-plane points in opposite directions led to oscillations of the third point rectangular to the plane. Placing the third point not too high above the plane resulted in conversion of small in-plane actuations into higher-amplitude out-of-plane vibrations, which can be sensed by a finger. These scissor amplifiers were produced by conventional photoresist and 3D printing. However, tests with volunteers showed that the minimum and the comfortable forces needed for the detection of the vibrations was slightly higher for the 3D-printed scissors, which was attributed to sensing being dominated by vibrational amplitudes rather than by forces [76].

5.3. Aeronautics and Astronautics

Combining MEMS dielectric elastomer actuators with a 3D-printed wing skeleton on which a fine Mylar film was glued was suggested to create micro aerial vehicles. While the wings were optimized using computational fluid dynamic simulations, the dielectric elastomer actuator was used due to its high work density, the ability to work at high frequencies, and the easy, low-cost production. Both parts were connected with a four-bar mechanism with a pivot point at the wing base. The design concept is depicted in Figure 5 [77], showing one of the possibilities of combining MEMS with 3D printing to produce miniaturized aerial vehicles.

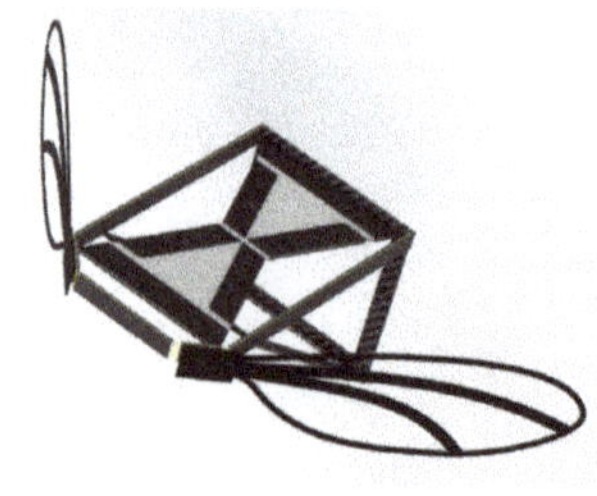

Figure 5. Concept design of a micro aerial vehicle, composed of MEMS and 3D-printed parts. Reproduced with permission from [77].

Khandekar et al. used 3D printing to prepare microthrusters for microsatellites, which can be used for very low thrust in a defined direction for course corrections of nano and microsatellites. Printing ceramic polymer composites, micro-thrusters with sufficient nanomechanical properties to propel nano and microsatellites were developed [78].

5.4. *Nanopositioning*

Fiaz et al. developed cantilevers for MEMS using the electron beam melting of a Ti alloy. The material was chosen due to the good biocompatibility of Ti, its strength, and its corrosion resistance. Comparing these cantilevers with bulk metal ones, the printed cantilevers were softer with a slightly smaller Young's modulus than the bulk material, but they also reached large maximum displacements of nearly 50 μm at resonance frequencies around 1850 Hz and thus relatively fast and accurate positioning of a stage [79].

Using an SLA printer with different commercially available and self-formulated resins in combination with the electrodeposition of different metals, Bernasconi et al. prepared another magnetic actuator. By coating the 3D-printed cantilever with magnetic metals, the cantilever could be deflected by an external magnetic field. The pseudo-linear correlation between the field and deflection was measured and modeled [80].

Magnetic NdFeB microparticles were embedded instead in a nylon 12 matrix to prepare a magnetic actuator, which reached a maximum displacement of 50 μm [81].

Similarly, electrothermal actuators were produced by Fogel et al. They used the 3D printing technology of laser-induced forward transfer (LIFT) to prepare fully metallic microdevices from structural and sacrificial metals, allowing preparing a free-standing structure on the support material. This actuator was deflected by a current, which was applied for a certain pulse duration with a maximum of 1 s [82].

Ertugrul et al. compared the aforementioned two-photon polymerization with projection micro-stereolithography in the case of an electrothermal microactuator. They found that the two-photon polymerization was advantageous since it could produce smaller and more complex, non-symmetric structures [83].

5.5. *Macro-Positioning*

3D printing was used to prepare a hybrid finger composed of hard and soft polymer, equipped with micropumps to enable hydraulic motion control. These MEMS micropumps worked with an electro-conjugate fluid and were found to have sufficient output density to drive the hybrid finger [84].

6. Conclusions

After the first ideas of 3D printing MEMS approximately 20 years ago, much progress was achieved in combining these technologies. Especially microfluidic systems, but also some MEMS sensors and actuators can nowadays be realized by diverse 3D printing technologies. New additive manufacturing techniques, such as the two-photon polymerization technique, allow for preparing smallest features with dimensions below 1 μm. For more established 3D printing techniques, new ideas emerged how to reduce the minimum feature size, in this way making 3D printing more and more suitable for MEMS fabrication.

Table 2 gives a short overview of 3D printing methods and their possible applications in MEMS, indicating the broad bandwidth of technologies and functions that can be reached by them.

Nevertheless, it seems that the research published on the combined technologies has passed its peak. Thus, we hope that this review will stimulate more researchers to investigate new possible applications, enabled especially by newly developed and future 3D printing techniques.

Table 2. Possible applications of typical 3D printing technologies used for MEMS.

Technology	Possible Applications
Fused deposition modeling	Dielectric-conductive systems, switches
Micro-stereolithography	In situ tensile tests of micro- or nanowires, electrothermal microactuator
Stereolithography	Microfluidic devices, conductive parts, molds, cantilevers, magnetic actuators
3D inkjet printing	Microfluidic devices, Venturi microflowmeter, conductive structures, strain gauge sensors
Multiphoton absorption polymerization	Microfluidic devices, photonic crystals, nanophotonic devices
Two-photon polymerization	Microfluidic devices, electrothermal microactuator
Binder jet printing	Microfluidic devices, in-line injection of volatile organic compounds

Author Contributions: Conceptualization, T.B. and A.E.; visualization, T.B. and A.E.; writing—original draft preparation, A.E. and T.B. All authors have read and agreed to the published version of the manuscript.

Funding: This research received no external funding.

Conflicts of Interest: The authors declare no conflict of interest. The funders had no role in the design of the study; in the collection, analyses, or interpretation of data; in the writing of the manuscript, or in the decision to publish the results.

References

1. Tanaka, M. An industrial and applied review of new MEMS devices features. *Microelectron. Eng.* **2007**, *84*, 1341–1344. [CrossRef]
2. Nguyen, C.T.-C. Frequency-selective MEMS for miniaturized low-power communication devices. *IEEE Trans. Microw. Theory Tech.* **1999**, *47*, 1486–1503. [CrossRef]
3. Kahn, H.; Tayebi, N.; Ballarini, R.; Mullen, R.L.; Heuer, A.H. Fracture toughness of polysilicon MEMS devices. *Sens. Actuators A Phys.* **2000**, *82*, 274–280. [CrossRef]
4. Duffy, D.C.; McDonald, J.C.; Schueller, O.J.A.; Whitesides, G.M. Rapid prototyping of microfluidic systems in poly(dimethylsiloxane). *Anal. Chem.* **1998**, *70*, 4974–4984. [CrossRef] [PubMed]
5. Chen, P.-J.; Rodger, D.C.; Meng, E.M.; Humayun, M.S.; Tai, Y.-C. Surface-micromachined Parylene dual valves for on-chip unpowered microflow regulation. *J. Microelectromech. Syst.* **2007**, *16*, 223–231. [CrossRef]
6. Fan, Z.F.; Engel, J.M.; Chen, J.; Liu, C. Parylene surface-micromachined membranes for sensor applications. *J. Microelectromech. Syst.* **2004**, *13*, 484–490. [CrossRef]
7. Chronis, N.; Lee, L.P. Electrothermally activated SU-8 microgripper for single cell manipulation in solution. *J. Microelectromech. Syst.* **2005**, *14*, 857–863. [CrossRef]
8. Ben-Ner, A.; Siemsen, E. Decentralization and localization of production: The organizational and economic consequences of additive manufacturing (3D printing). *Calif. Manag. Rev.* **2017**, *59*, 5–23. [CrossRef]
9. Duarte, L.C.; Chagas, C.; Ribeiro, L.E.B.; Coltro Tomazelli, W.K. 3D printing of microfluidic devices with embedded sensing electrodes for generating and measuring the size of microdroplets based on contactless conductivity detection. *Sens. Actuators B Chem.* **2017**, *251*, 427–432. [CrossRef]
10. Noorani, R. *Rapid Prototyping: Principles and Applications*; John Wiley & Sons: Hoboken, NJ, USA, 2005.
11. Kozior, T.; Blachowicz, T.; Ehrmann, A. Adhesion of 3D printing on textile fabrics—Inspiration from and for other research areas. *J. Eng. Fibers Fabrics* **2020**, submitted. [CrossRef]
12. Korger, M.; Bergschneider, J.; Lutz, M.; Mahltig, B.; Finsterbusch, K.; Rabe, M. Possible applications of 3D printing technology on textile substrates. *IOP Conf. Ser. Mater. Sci. Eng.* **2016**, *141*, 012011. [CrossRef]
13. Unger, L.; Scheideler, M.; Meyer, P.; Harland, J.; Görzen, A.; Wortmann, M.; Dreyer, A.; Ehrmann, A. Increasing Adhesion of 3D Printing on Textile Fabrics by Polymer Coating. *Tekstilec* **2018**, *61*, 265–271. [CrossRef]
14. Jung, H.; Kim, C.J.; Kong, S.H. An optimized MEMS-based electrolytic tilt sensor. *Sens. Actuators A Phys.* **2007**, *139*, 23–30. [CrossRef]

15. Schröder, H.; Obermeier, E.; Horn, A.; Wachutka, G.K.M. Convex corner undercutting of {100} silicon in anisotropic KOH etching: The new step-flow model of 3-D structuring and first simulation results. *J. Microelectromech. Syst.* **2001**, *10*, 88–97. [CrossRef]
16. Pandy, A.; Landsberge, L.M.; Kahrizi, P.M. Mask-Under-Etch Experiments of Si{110} in TMAH. In Proceedings of the 1999 IEEE Canadian Conference on Electrical and Computer Engineering, Edmonton, AB, Canada, 9–12 May 1999; pp. 1621–1626.
17. Eliyahu, D.; Gileadi, E.; Galun, E.; Eliaz, N. Atomic force microscope-based meniscus-confined three-dimensional electrodeposition. *Adv. Mater. Technol.* **2020**, *5*, 1900827. [CrossRef]
18. Nouri-Goushki, M.; Mirzaali, M.J.; Angeloni, L.; Fan, D.; Minneboo, M.; Ghatkesar, M.K.; Staufer, U.; Fratila-Apachitei, L.E.; Zadpoor, A.A. 3D printing of large areaas of highly ordered submicron patterns for modulating cell behavior. *ACS Appl. Mater. Interfaces* **2020**, *12*, 200–208. [CrossRef] [PubMed]
19. Kozior, T.; Mamun, A.; Trabelsi, M.; Sabantina, L.; Ehrmann, A. Quality of the surface texture and mechanical properties of FDM printed samples after thermal and chemical treatment. *Strojniški Vestnik J. Mech. Eng.* **2020**, *66*, 105–113.
20. Moroni, L.; Boland, T.; Burdick, J.A.; de Maria, C.; Derby, B.; Forgacs, G.; Groll, J.; Li, Q.; Malda, J.; Mironov, V.A.; et al. Biofabrication: A guide to technology and terminology. *Trends Biotechnol.* **2018**, *36*, 384–402. [CrossRef]
21. Tan, K.H.; Chua, C.K.; Leong, K.F.; Cheah, C.M.; Gui, W.S.; Tan, W.S.; Wiria, F.E. Selective laser sintering of biocompaticle polymers for applications in tissue engineering. *Bio-Med. Mater. Eng.* **2005**, *15*, 113–124.
22. Zhou, W.Y.; Lee, S.H.; Wang, M.; Cheung, W.L.; Ip, W.Y. Selective laser sintering of porous tissue engineering scaffolds from poly(L-lactide)/carbonated hydroxyapatite nanocomposite microspheres. *J. Mater. Sci. Mater. Med.* **2008**, *19*, 2535–2540. [CrossRef]
23. Moroni, L.; de Wijn, J.R.; van Blitterswijk, C.A. 3D fiber-deposited scaffolds for tissue engineering: Influence of pores geometry and architecture on dynamic mechanical properties. *Biomaterials* **2006**, *27*, 974–985. [CrossRef]
24. Malda, J.; Visser, J.; Melchels, F.P.; Jüngst, T.; Hennink, W.E.; Dhert, W.J.A.; Groll, J.; Hutmacher, D.W. Engineering hydrogels for biofabrication. *Adv. Mater.* **2013**, *25*, 5011–5028. [CrossRef] [PubMed]
25. Wang, Z.J.; Abdulla, R.; Parker, B.; Samanipour, R.; Ghosh, S.; Kim, K.Y. A simple and high-resolution stereolithography-based 3D bioprinting system using visible light crosslinkable bioinks. *Biofabrication* **2015**, *7*, 045009. [CrossRef] [PubMed]
26. Pearre, B.W.; Michas, C.; Tsang, J.-M.; Gardner, T.J.; Otchy, T.M. Fast micron-scale 3D printing with a resonant-scanning two-photon microscope. *Addit. Manuf.* **2019**, *30*, 100887. [CrossRef]
27. Kumi, G.; Yanez, C.; Belfield, K.D.; Fourkas, J.T. High-speed multiphoton absorption polymerization: Fabrication of microfluidic channels with arbitrary cross-sections and high aspect ratios. *Lab Chip* **2010**, *8*, 1057–1060. [CrossRef] [PubMed]
28. Straub, M.; Gu, M. Near-infrared photonic crystals with higher-order bandgaps generated by two-photon photopolymerization. *Opt. Lett.* **2002**, *27*, 1824–1826. [CrossRef]
29. Thiel, M.; Fischer, J.; von Freymann, G.; Wegener, M. Direct laser writing of three-dimensional submicron structures using a continuous-wave laser at 532 nm. *Appl. Phys. Lett.* **2010**, *97*, 221102. [CrossRef]
30. Kampker, A.; Triebs, J.; Kawollek, S.; Ayvaz, P.; Beyer, T. Direct polymer additive tooling—Effect of additive manufactured polymer tools on part material properties for injection moulding. *Rapid Prototyp. J.* **2019**, *25*, 1575–1584. [CrossRef]
31. Zander, N.E.; Park, J.H.; Boelter, Z.R.; Gillan, M.A. Recycled cellulose polypropylene composite feedstocks for material extrusion additive manufacturing. *ACS Omega* **2019**, *4*, 13879–13888. [CrossRef]
32. Maniere, C.; Kerbart, G.; Harnois, C.; Marinel, S. Modeling sintering anisotropy in ceramic stereolithography of silica. *Acta Mater.* **2020**, *182*, 163–171. [CrossRef]
33. Tan, P.F.; Shen, F.; Li, B.; Zhou, K. A thermo-metallurgical-mechanical model for selective laser melting of Ti_6Al_4V. *Mater. Des.* **2019**, *168*, 107642. [CrossRef]
34. de Araujo Filho, W.D.; Morales, R.E.M.; Schneider, F.K.; de Araujo, L.M.P. Microfluidics device manufacturing using the technique of 3D printing. In Proceedings of the ASME 12th International Conference on Nanochannels, Microchannels, and Minichannels 2014, Chicago, IL, USA, 3–7 August 2014; p. V001T15A002.
35. Leary, J.F.; Key, J.; Vidi, P.A.; Cooper, C.L.; Kole, A.; Reece, L.M.; Lelievre, S.A. Human organ-on-a-chip BioMEMS devices for testing new diagnostic and therapeutic strategies. *Proc. SPIE* **2013**, *8615*, 86150A.

36. Lifton, V.A.; Lifton, G.; Simon, S. Options for additive rapid prototyping methods (3D printing) in MEMS technology. *Rapid Prototyp. J.* **2014**, *20*, 403–412. [CrossRef]
37. Zhu, F.; Skommer, J.; Friedrich, T.; Kaslin, J.; Wlodkowic, D. 3D printed polymers toxicity profiling—A caution for biodevice applications. *Proc. SPIE* **2015**, *9668*, 96680Z.
38. Beckwith, A.L.; Borenstein, J.T.; Velasquez-Garcia, L.F. Monolithic, 3D-printed microfluidic platform for recapitulation of dynamic tumor microenvironments. *J. Microelectromech. Syst.* **2018**, *27*, 1009–1022. [CrossRef]
39. Fuad, N.M.; Carve, M.; Kaslin, J.; Wlodkowic, D. Characterization of 3D-Printed Moulds for Soft Lithography of Millifluidic Devices. *Micromachines* **2018**, *9*, 116. [CrossRef]
40. Sun, Z.M.; Velasquez-Garcia, L.F. Monolithic FFF-Printed, Biocompatible, Biodegradable, Dielectric-Conductive Microsystems. *J. Microelectromech. Syst.* **2017**, *26*, 1356–1370. [CrossRef]
41. Carve, M.; Wlodkowic, D. 3D-Printed Chips: Compatibility of Additive Manufacturing Photopolymeric Substrata with Biological Applications. *Micromachines* **2018**, *9*, 91. [CrossRef]
42. Araujo Filho, W.D.; Araujo, L.M.P.; Mauricio, C.R.M. 3D printing techniques in the manufacture of microfluidic devices for generation of microbubbles. *SCIOL Biomed.* **2019**, *3*, 143–151.
43. Jia, Y.; Su, C.; He, M.G.; Liu, K.; Sun, H.; Lin, Q. Isothermal titration calorimetry in a 3D-printed microdevice. *Biomed. Microdev.* **2019**, *21*, 96. [CrossRef]
44. Emons, M.; Obata, K.; Binhammer, T.; Ovsianikov, A.; Chichkov, B.N.; Morgner, U. Two-photon polymerization technique with sub-50 nm resolution by sub-10 fs laser pulses. *Opt. Mater. Express* **2012**, *7*, 942–947. [CrossRef]
45. Nguyen, A.K.; Narayan, R.J. Two-photon polymerization for biological applications. *Mater. Today* **2017**, *6*, 314–322. [CrossRef]
46. Kübler, S.M.; Xia, C.; Sharma, R.; Digaum, J.L.; Martinez, N.P.; Valle, C.L.; Rumpf, R.C. Fabrication of Functional Nanophotonic Devices by Multi-Photon Lithography. *Proc. SPIE* **2019**, *10915*, 1091502.
47. Jui, C.-W.; Trappey, A.J.C.; Fu, C.-C. Discover Patent Landscape of Two-photon Polymerization Technology for the Production of 3D Nano-structure Using Claim-based Approach. *Recent Patents Nanotechnol.* **2018**, *12*, 218–230. [CrossRef] [PubMed]
48. Walczak, R.; Kawa, B.; Adamski, K. Inkjet 3D printed microfluidic device for growing seed root and stalk mechanical characterization. *Sens. Actuators A Phys.* **2019**, *297*, 111557. [CrossRef]
49. Storck, J.-L.; Böttjer, R.; Vahle, D.; Brockhagen, B.; Grothe, T.; Dietz, K.-J.; Rattenholl, A.; Gudermann, F.; Ehrmann, A. Seed germination and seedling growth on knitted fabrics as new substrates for hydroponic systems. *Horticulturae* **2019**, *5*, 73. [CrossRef]
50. Huang, X.L.; Bauder, T.; Do, T.; Suen, H.; Boss, C.; Kwon, P.; Yeom, J. A Binder Jet Printed, Stainless Steel Preconcentrator as an In-Line Injector of Volatile Organic Compounds. *Sensors* **2019**, *19*, 2748. [CrossRef]
51. Dinh, T.; Phan, H.P.; Kashaninejad, N.; Nguyen, T.K.; Dao, D.V.; Nguyen, N.T. An On-Chip SiC MEMS Device with Integrated Heating, Sensing, and Microfluidic Cooling Systems. *Adv. Mater. Interfaces* **2018**, *5*, 1800764. [CrossRef]
52. Adamski, K.; Kawa, B.; Walczak, R. Inkjet 3D printed Venturi Microflowmeter. In Proceedings of the 2018 XV International Scientific Conference on Optoelectronic and Electronic Sensors (COE), Warsaw, Poland, 17–20 June 2018.
53. Villegas, M.; Cetinic, Z.; Shakeri, A.; Didar, T.F. Fabricating smooth PDMS microfluidic channels from low-resolution 3D printed molds using an omniphobic lubricant-infused coating. *Anal. Chim. Acta* **2018**, *1000*, 248–255. [CrossRef]
54. Cesewski, E.; Haring, A.P.; Tong, Y.X.; Singh, M.; Thakur, R.; Laheri, S.; Read, K.A.; Powell, M.D.; Oestreich, K.J.; Johnson, B.N. Additive manufacturing of three-dimensional (3D) microfluidic-based microelectromechanical systems (MEMS) for acoustofluidic applications. *Lab Chip* **2018**, *18*, 2087–2098. [CrossRef]
55. Tamura, T.; Suzuki, T. Seamless fabrication technique for micro to millimeter structures by combining 3D printing and photolithography. *Jpn. J. Appl. Phys.* **2019**, *58*, SDDL10. [CrossRef]
56. Thapa, S.; Borquist, E.; Weiss, L. Thermal energy recovery via integrated small scale boiler and superheater. *Energy* **2018**, *142*, 765–772. [CrossRef]
57. Feldmann, M.; Waldschik, A.; Büttgenbach, S. Technology and application of electro-depositable photo resists to create uniform coatings needed for complex 3D micro actuators and sensors. *Microsyst. Technol. Micro Nanosyst. Inf. Storage Process. Syst.* **2007**, *13*, 557–562. [CrossRef]

58. Farahani, R.D.; Chizari, K.; Therriault, D. Three-dimensional printing of freeform helical microstructures: A review. *Nanoscale* **2014**, *6*, 10470–10485. [CrossRef]
59. Bauer, R.; Stewart, G.; Johnstone, W.; Boyd, E.; Lengden, M. 3D-printed miniature gas cell for photoacoustic spectroscopy of trace gases. *Opt. Lett.* **2014**, *39*, 4796–4799. [CrossRef]
60. Vincent, T.A.; Gardner, J.W. A low cost MEMS based NDIR system for the monitoring of carbon dioxide in breath analysis at ppm levels. *Sens. Actuators B Chem.* **2016**, *236*, 954–964. [CrossRef]
61. Ilke, M.; Bauer, R.; Lengden, M. Performance of a 3D Printed Photoacoustic Sensor for Gas Detection in Mid-Infrared. In Proceedings of the 2017 IEEE Sensors, Glasgow, UK, 29 October–1 November 2017; pp. 1482–1484.
62. Valyrakis, M.; Pavlovskis, E.; Alexakis, A. Designing a "smart pebble" for sediment transport monitoring. In Proceedings of the 36th IAHR World Congress: Deltas of the Future and What Happens Upstream, The Hague, The Netherlands, 28 June–3 July 2015; pp. 2843–2850.
63. Shen, Z.Y.; Kottapalli, A.G.P.; Subramaniam, V.; Miao, J.M.; Asadnia, M.; Triantafyllou, M. Biomimetic Flow Sensors for Biomedical Flow Sensing in Intravenous Tubes. In Proceedings of the 2016 IEEE SENSORS, Orlando, FL, USA, 30 October–3 November 2016; IEEE: Piscataway, NJ, USA, 2016; pp. 1–3.
64. Raoufi, M.A.; Moshizi, S.A.; Razmjou, A.; Wu, S.Y.; Warkiani, M.E.; Asadnia, M. Development of a Biomimetic Semicircular Canal With MEMS Sensors to Restore Balance. *IEEE Sens. J.* **2019**, *19*, 11675–11686. [CrossRef]
65. Kwon, H.B.; Yoo, S.J.; Hong, U.S.; Kim, K.; Han, J.; Kim, M.K.; Kang, D.H.; Hwang, J.; Kim, Y.J. MEMS-based condensation particle growth chip for optically measuring the airborne nanoparticle concentration. *Lab Chip* **2019**, *19*, 1471–1483. [CrossRef]
66. Krysztof, M.; Grzebyk, T.; Gorecka-Drzazga, A.; Adamski, K.; Dziuban, J. Electron optics column for a new MEMS-type transmission electron microscope. *Bull. Pol. Acad. Sci. Tech. Sci.* **2018**, *66*, 133–137.
67. Das, S.K.; Baptist, J.R.; Sahasrabuddhe, R.; Lee, W.H.; Popa, D.O. Package Analysis of 3D-Printed Piezo-Resistive Strain Gauge Sensors. *Proc. SPIE* **2016**, *9859*, 985905.
68. Baptist, J.R.; Zhang, R.S.; Wie, D.M.; Saadatzi, M.N.; Popa, D.O. Fabrication of strain gauge based sensors for tactile skins. *Proc. SPIE* **2017**, *10216*, 102160F.
69. Yang, Z.W.; Yu, H.B.; Zhou, P.L.; Wang, J.Y.; Liu, L.Q. Fabrication of Three-dimensional Conductive Structures Using Direct Ink Writing. In Proceedings of the 2017 IEEE 7th Annual International Conference on Cyber Technology in Automation, Control, and Intelligent Systems (CYBER), Honolulu, HI, USA, 1 July–4 August 2017; pp. 1562–1565.
70. Kusano, Y.; Segovia-Fernandez, J.; Sonmezoglu, S.; Amirtharajah, R.; Horsley, D.A. Frequency selective MEMS microphone based on a bioinspired spiral-shaped acoustic resonator. In Proceedings of the 2017 19th International Conference on Solid-State Sensors, Actuators and Microsystems (Transducers), Kaohsiung, Taiwan, 18–22 June 2017; pp. 71–74.
71. Tiller, B.; Reid, A.; Zhu, B.T.; Guerreiro, J.; Domingo-Roca, R.; Jackson, J.C.; Windmill, J.F.C. Piezoelectric microphone via a digital light processing 3D printing process. *Mater. Des.* **2019**, *165*, 107593. [CrossRef]
72. Watanabe, K.; Kinoshita, M.; Mine, T.; Morishita, M.; Fujisaki, K.; Matsui, R.; Sagawa, M.; Machida, S.; Oba, H.; Sugiyama, Y.; et al. Plasma ion-beam 3D printing: A novel method for rapid fabrication of customized MEMS sensors. In Proceedings of the 2018 IEEE Micro Electro Mechanical Systems (MEMS), Belfast, UK, 21–25 January 2018; pp. 459–462.
73. Park, J.S.; Kim, J.-K.; Kim, D.-S.; Shanmugasundaram, A.; Park, S.A.; Kang, S.; Kim, S.-H.; Jeong, M.H.; Lee, D.-W. Wireless pressure sensor integrated with a 3D printed polymer stent for smart health monitoring. *Sens. Actuators B Chem.* **2019**, *280*, 201–209. [CrossRef]
74. Wang, Y.J.; Gao, L.B.; Fan, S.F.; Zhou, W.Z.; Li, X.; Lu, Y. 3D printed micro-mechanical device (MMD) for in situ tensile testing of micro/nanowires. *Extrem. Mech. Lett.* **2019**, *33*, 100575. [CrossRef]
75. Lee, Y.; Han, J.; Choi, B.; Yoon, J.; Park, J.; Kim, Y.; Lee, J.; Kim, D.H.; Kim, D.M.; Lim, M.; et al. Three-dimensionally printed micro-electromechanical switches. *ACS Appl. Mater. Interfaces* **2018**, *10*, 15841–15846. [CrossRef]
76. Xie, X.; Zaitsev, Y.; Velasquez-Garcia, L.F.; Teller, S.J.; Livermore, C. Scalable, MEMS-enabled, vibrational tactile actuators for high resolution tactile displays. *J. Micromech. Microeng.* **2014**, *24*, 125014. [CrossRef]
77. Joshi, N.; Köhler, E.; Enoksson, P. MEMS based micro aerial vehicles. *J. Phys. Conf. Ser.* **2016**, *757*, 012035. [CrossRef]

78. Khandekar, P.; Biswas, K.; Kothari, D.; Muthurajan, H. Nano Mechanical Properties of Ceramic Polymer Composite Micro Thruster Developed Using 3D Printing Technology. *Adv. Sci. Lett.* **2018**, *24*, 5884–5890. [CrossRef]
79. Fiaz, H.S.; Settle, C.R.; Hoshino, K. Metal additive manufacturing for microelectromechanical systems: Titanium alloy (Ti-6Al-4V)-based nanopositioning flexure fabricated by electron beam melting. *Sens. Actuators A Phys.* **2016**, *249*, 284–293. [CrossRef]
80. Bernasconi, R.; Credi, C.; Tironi, M.; Levi, M.; Magagnin, L. Electroless metallization of stereolithographic photocurable resins for 3D printing of functional microdevices. *J. Electrochem. Soc.* **2017**, *164*, B3059–B3066. [CrossRef]
81. Taylor, A.P.; Cuervo, C.V.; Arnold, D.P.; Velasquez-Garcia, L.F. Fully 3D-Printed, Monolithic, Mini Magnetic Actuators for Low-Cost, Compact Systems. *J. Microelectromech. Syst.* **2019**, *28*, 481–493. [CrossRef]
82. Fogel, O.; Winter, S.; Benjamin, E.; Krylov, S.; Kotler, Z.; Zalevsky, Z. 3D printing of functional metallic microstructures and its implementation in electrothermal actuators. *Addit. Manuf.* **2018**, *21*, 307–311. [CrossRef]
83. Ertugrul, I.; Akkus, N.; Yuce, H. Fabrication of MEMS-based electrothermal microactuators with additive manufacturing technologies. *Mater. Tehnol.* **2019**, *53*, 665–670. [CrossRef]
84. Han, D.; Gu, H.R.; Kim, J.W.; Yokota, S. A bio-inspired 3D-printed hybrid finger with integrated ECF (electro-conjugate fluid) micropumps. *Sens. Actuators A Phys.* **2017**, *257*, 47–57. [CrossRef]

MDPI
St. Alban-Anlage 66
4052 Basel
Switzerland
www.mdpi.com

Micromachines Editorial Office
E-mail: micromachines@mdpi.com
www.mdpi.com/journal/micromachines

www.ingramcontent.com/pod-product-compliance
Lightning Source LLC
LaVergne TN
LVHW070712100526
838202LV00013B/1081
* 9 7 8 3 0 3 6 5 9 7 7 0 6 *